NITRATE, AGRICULTURE AND THE ENVIRONMENT

This book is dedicated to the memory of
Sir John Bennet Lawes

Scientist
Entrepreneur
Man of Principle

His memorial in Harpenden Parish Church reads:

In affectionate memory of Sir John Bennet Lawes Baronet FRS only son of the above John Bennet Lawes born at Rothamsted Dec. 28th 1814 died at Rothamsted Aug. 31 1900. His long life and his great knowledge were to the last devoted to far-reaching investigations into all scientific matters which could affect agriculture. The results obtained were freely given for the benefit of his fellow men both in his own country and in all parts of the world. This tablet is erected by Parishioners of Harpenden and others who deeply feel his loss as an example and a friend.

Sir John Bennet Lawes, LLD, FRS, FCS. (Photograph courtesy of Rothamsted Research.)

Nitrate, Agriculture and the Environment

T.M. Addiscott

Lawes Trust Senior Fellow
Rothamsted Research,
Harpenden, UK

CABI Publishing

CABI Publishing is a division of CAB International

CABI Publishing
CAB International
Wallingford
Oxfordshire OX10 8DE
UK

Tel: +44 (0)1491 832111
Fax: +44 (0)1491 833508
E-mail: cabi@cabi.org
Website: www.cabi-publishing.org

CABI Publishing
875 Massachusetts Avenue
7th Floor
Cambridge, MA 02139
USA

Tel: +1 617 395 4056
Fax: +1 617 354 6875
E-mail: cabi-nao@cabi.org

A catalogue record for this book is available from the British Library, London, UK.

Library of Congress Cataloging-in-Publication Data

Addiscott, T. M. (Tom M.)
Nitrate, agriculture and the environment / by T.M. Addiscott
p. cm.
ISBN 0-85199-913-1 (alk. paper)
1. Nitrogen fertilizers. 2. Nitrogen fertilizers--Environmental aspects. I. Title.

S651.A3199 2004
631.8´4--dc22

2004017344

ISBN 0 85199 913 1

Typeset by MRM Graphics Ltd, Winslow, Bucks
Printed and bound in the UK by Biddles Ltd, King's Lynn

Contents

Contributors

Tom Addiscott had worked at Rothamsted for 36 years when he retired in 2002. He keeps in contact with the institute as a Lawes Trust Senior Fellow associated with the Agriculture and Environment Division and remains involved in the organization of the 'Manor Recitals' – concerts at Rothamsted Manor. Much of his research at Rothamsted concerned the development and use of simple computer models for the leaching of nitrate and phosphate and associated processes such as mineralization, crop growth and nutrient uptake. He also worked on model validation (evaluation) and the problems that arise from the interaction between (statistical) error in parameters and non-linearity in models. One of his retirement projects is to apply Complexity Theory in a soils context. He has been writing on the 'nitrate problem' for nearly 20 years, and a previous book, *Farming, Fertilizers and the Nitrate Problem,* written with Andy Whitmore and David Powlson, was published by CAB International in 1991.

On leaving Oxford in 1964, Tom Addiscott spent a year as a UNA International Service volunteer working on soil phosphate in Tanzania. He then worked for just over a year for the Lawes Chemical Company before moving to Rothamsted in 1966. He retains an interest in Africa and is a member of the FARM-Africa charity, with whom he has travelled in recent years to Kenya, Tanzania and Ethiopia. He was licensed as a Reader in the Church of England in 1978.

Tom Addiscott claims the distinction of being one of the very few, possibly only two, people to have worked at both of the institutions founded by Sir John Bennet Lawes, the Lawes Chemical Company and Rothamsted Experimental Station. He was also presented with the Gold Medal of the Royal Agricultural Society of England in 1991, appointed a Visiting Professor at the University of East London in 1997 and awarded a DSc degree by Oxford University in 1999.

Art Gold is a Professor in the Department of Natural Resource Sciences, College of the Environment and Life Sciences, in the University of Rhode Island, Kingston, Rhode Island, USA.

Candace Oviatt is Professor of Oceanography at the Graduate School of Oceanography in the University of Rhode Island, Narragansett, Rhode Island, USA.

Nigel Benjamin was until recently Head of the William Harvey Research Institute, Queen Mary College, London and has moved to join the Peninsula Medical School.

Ken Giller is Professor of Plant Production Systems in the Department of Plant Sciences at Wageningen Agricultural University, The Netherlands. He was until recently Professor of Soil Science at the University of Zimbabwe, Harare.

Preface

Another book about nitrate? The most important incentive for this new book lies in the realization by Vaclav Smil and David Jenkinson that the world is collectively in a state of dependence on the production of nitrogen by the Haber–Bosch process. We may not like the fact but, because of population growth, we can no more give up nitrogen fertilizer than an addict can give up heroin, less so in fact. The production and use of nitrogen fertilizer inevitably lead to losses of nitrate and nitrous oxide to the environment and we have to learn to deal with them. There can be no return to a state of 'innocence'.

The preface to the earlier book that Andy Whitmore, David Powlson and I wrote about nitrate, *Farming, Fertilizers and the Nitrate Problem* (Addiscott *et al.*, 1991), said that book was in response to fears that nitrate from agriculture had found its way into drinking water, causing stomach cancer in adults and cyanosis in infants, and into rivers and seas, where it was causing the growth of toxic algae and general mayhem in otherwise balanced natural ecosystems. There were also fears of a 'nitrate time bomb' moving inexorably towards our taps. Those fears have not entirely abated and still appear regularly in the press and even scientific papers from time to time. But much has changed in our knowledge about nitrate during the past 13 years, and the new understandings have not received the publicity they deserve. I hope, if you read this book, you will see why it was needed. The most startling change in our understanding is the realization that, rather that being a threat to our health, nitrate plays a central role in our bodies' defence system against bacterial gastroenteritis. This may explain why our bodies go to considerable efforts to maintain their supply of nitrate against the attempts by various official bodies to stop them doing so. The limits on nitrate in drinking water were, in the words of J.-L. L'hirondel, 'a world-scale scientific error that has lasted more than 50 years'. These limits can be seen, on re-examining early papers on well-water methaemoglobinaemia, to be the results of the unfortunate juxtaposition of wells and pit privies. Nigel Benjamin played a leading role in the research on the positive role of nitrate in health, and I am grateful to him for supplying most of the information in Chapter 9 and some of that in Chapter 10.

The understanding of the effects of plant nutrients in the environment has also increased during the past 13 years. It is now clear that phosphate rather than nitrate is the limiting factor for algal blooms in fresh water, but that nitrate is responsible for blooms and other problems in estuarine and marine coastal waters. Art Gold and Candace Oviatt kindly contributed the whole of Chapter 8, which deals with nitrate effects in marine waters, and also some of the material on fresh water in Chapter 7.

Another topic on which understanding has increased is nitrous oxide. We know more about its generation in the soil, particularly in agricultural operations, and its less than desirable effects in the atmosphere, particularly the stratosphere. Assessing emissions of this gas is a considerable challenge because of the scale issue involved. Nitrous oxide is a global pollutant and is discussed at various scales in parts of Chapters 3, 5, 6 and 7.

Losses of nitrate and nitrous oxide from grassland are much more difficult to research than those from arable land. Cows are a formidable complication, but a lot of interesting progress has been made on grassland in the last decade. Losses of both pollutants from both types of systems are discussed in Chapters 5 and 6, respectively.

The book covers, in Chapters 1, 2 and 3, the basic soil chemistry, physics and biology discussed in the previous book, and I am grateful to Andy Whitmore and David Powlson for agreeing to my using material from the previous book as well as new material. The understanding of soil biological processes such as mineralization has probably made the greatest advances, and gross mineralization is an important new topic.

Farmers do not bear the sole responsibility for the nitrate problem. Politicians, including a certain Adolf Hitler, have played an important part in it, as is shown in Chapter 10. Indeed, it arguable that, without politicians, there would have been no nitrate problem. Political correctness about nitrate in the Western world has almost certainly discouraged the necessary use of small amounts of fertilizer in Africa and the application of inappropriate market dogma has probably had the same effect. I discussed these problems with Ken Giller during a visit to Zimbabwe, and he kindly provided material for Chapter 11, which discusses these questions.

Chapter 12 is about risk, which may seem odd in a book about a substance claimed to carry no health risks and only limited environmental risks. The point is more that the risk from nitrate was not well managed and we need to do better with future potential risks.

The last two chapters are entitled 'Coming to Terms with Nitrate' but each has a second title. Chapter 13 is concerned with the increasingly important relationship of scientists with politicians and the public, and the importance of communication. It also examines the influence of the environmental pressure groups. Chapter 14 is about changes in land use needed to restrict losses of nitrate and nitrous oxide. It discusses the likely influence of reform of the Common Agricultural Policy and considers how organic farming can best be integrated into the overall picture.

A number of people have helped to make this book happen. Art Gold, Candace Oviatt, Nigel Benjamin and Ken Giller have already been mentioned as proving material. Several people read chapters and made helpful comments: David Powlson (Chapter 3), Steve Jarvis (Chapter 6), Nigel Benjamin (Chapter 9), Ken Giller (Chapter 11) and Tony Trewavas (Chapter 14). Art Gold and Candace Oviatt were very tolerant of editorial changes I made to Chapter 8. I am grateful to all these people but I must take the blame for any mistakes. I thank the publisher, Tim Hardwick, for his remarkable patience and for suggesting the writing of the book in the first place.

One final point. Some readers might wish me to include a statement of my financial relationships with the fertilizer industry. In 1965–1966 I worked for 13 months at the Lawes Chemical Company and my gross pay was £100 per month – total £1300. Subsequently, I personally have received no other money from the industry, although I have lunched at its expense about half a dozen times.

1 Dependence on Nitrogen

The Most Famous Field in the World

John Bennet Lawes made two significant decisions in 1843. He started an experiment with winter wheat on Broadbalk field, which was part of his estate at Rothamsted (Fig. 1.1). And he appointed Joseph Henry Gilbert, a chemist with a doctorate from the laboratory of the celebrated Baron von Liebig, to supervise the chemical analyses and advise on the design of experiments. The latter proved to be an excellent decision, with Gilbert's meticulous attention to detail providing an ideal foil to Lawes's intuitive flair. The partnership, one of the greatest in British science, lasted the 57 years until Lawes died in 1900. Gilbert died a year later.

Fig. 1.1. Harvesting wheat by hand on Broadbalk, probably in the 1860s. The self-binder was not introduced until 1902. (Photograph courtesy of Rothamsted Research.)

Box 1.1. Description of the soil of Broadbalk field.

The soil of Broadbalk field was classified as Batcombe Series by Avery and Bullock (1969). About 70% of the field is 'typical Batcombe' Series, a flinty clay loam to silty clay loam containing 18–27% clay, overlying mottled clay-with-flints containing about 50% clay within 80 cm of the surface. The rest of the field is either a heavier variant of the Batcombe Series, a flinty silty clay loam with gravelly or chalky material in the subsoil, or Hook Series, a slightly flinty silty clay loam. The pH of the topsoil is 7.5–8.0, and it contains about 1% organic carbon in plots receiving mineral fertilizer and 2.7% in plots that have long received farmyard manure. The site has only a very slight slope but is free-draining.

Because of Gilbert's appointment and the establishment of the Broadbalk Experiment, the history of Rothamsted dates from 1843. When, 125 years later, the then director of Rothamsted Sir Frederick Bawden described Broadbalk as 'the most famous field in the world' (Bawden, 1969), nobody argued. It is certainly the longest-running field experiment. At the time of writing, the 160th year of Broadbalk has just passed and the experiment remains useful for studying not only nitrate leaching (Goulding *et al.*, 2000) but also other environmental issues such as phosphate loss (Heckrath *et al.*, 1995) and the build-up of organic contaminants (Jones *et al.*, 1994), unforeseen in 1843. Its soil is described briefly in Box 1.1.

The Broadbalk Experiment

Lawes and Gilbert's experiment on Broadbalk has not changed greatly since 1843. However, in the first 3 years the plots were arranged to answer specific points, which the experimenters soon realized to be a mistake, and after the fourth year the emphasis was on continuity. By 1852, the present pattern had emerged and remained unchanged until 1967 (Johnston and Garner, 1969). This involved growing winter wheat continuously on plots that received either:

- No fertilizer.
- Minerals (P, K, Na, Mg) only.
- Minerals with four different additions of nitrogen, 43, 86, 129 or 172 lb/acre (48, 96, 144 or 192 kg/ha) as ammonium sulphate. Sodium nitrate was also tested for some applications in some years.
- Farmyard manure.

The main difference in the experiment today is that two extra additions of nitrogen (240 and 288 kg/ha) are now made. These reflect the much increased use of nitrogen by today's farmers, although the original additions were huge by the standards of 1843. Even by 1957, the average nitrogen application to winter wheat by farmers was only 75–90 kg N/ha (Garner, 1957). It could be said that Lawes was looking a long way ahead.

Box 1.2. Sir John Bennet Lawes, 1814–1900.

John Bennet Lawes was very much a Victorian in that he was an exact contemporary of the queen. He was 22 when Victoria came to the throne and he died a few months before her. He was unquestionably an eminent Victorian, though not among those listed by Lytton Strachey. Lawes is known as the founder of Rothamsted Experimental Station and for the contribution he made to agricultural science with his colleague Sir Joseph Henry Gilbert. The Broadbalk experiment and the Drain Gauges are mentioned in this book, together with the first recognition of the roles of mobile and immobile water in solute transport, but there was much more besides, several other field experiments and some notable work on animal nutrition. Lawes was also an entrepreneur, founding with little capital one of the first fertilizer businesses in the world and running it profitably for 30 years, before selling it and using one-third of the profits to set up the Lawes Agricultural Trust to maintain the experiments. He became in his later years a father figure in British agriculture, providing sound guidance in difficult times and looking ahead to future problems. Parish magazines of the era record him as a popular figure, and often benefactor, in Harpenden. Lawes was also a man with a social conscience, a model employer. He set aside land for allotments for workers on the estate, so that they could grow their own vegetables, and started a pig club. He also built a club-house by the allotments where those who toiled could buy a pint of beer more cheaply than elsewhere and were under no pressure to buy more than they wanted. Lawes and the club-house were celebrated by the novelist Charles Dickens in 'The Poor Man and his Beer', an article published in *All the Year Round* magazine in April 1859.

The Broadbalk Experiment was to play a key role in a famous dispute of its day.

Lawes and Liebig

Science thrives on disputes and in the mid-1840s there was a vigorous dispute over the source of the nitrogen used by plants between Lawes (Box 1.2) and Justus von Liebig (Box 1.3). On the face of it, the dispute was somewhat unequal. Liebig was one of the leading scientists of the day and the older of the two, as well as being a baron. Lawes was only 31, had left Oxford without bothering to take a degree and, worse still, was 'in trade'. He had started up a successful enterprise, but *in fertilizers*. By the standards of the day, he was socially greatly inferior to a baron, who doubtless was of independent means.

Lawes's biographer George Dyke records the dispute as starting in 1845 (Dyke, 1993). Liebig had already established the important fact that plants obtained their carbon from the air rather than the soil, as had previously been thought. Perhaps because of this he did calculations of the deposition of ammonia in rainfall, on the basis of which he wrote, in 1845, that 'the supply of ammonia to most of our cultivated plants is unnecessary, if only the soil contain a sufficient supply of mineral food'. By mineral food he meant phosphate, potassium, sodium and magnesium. Dyke (1993) records

Box 1.3. Baron Justus von Liebig, 1803–1873.

Justus von Liebig was the foremost agricultural chemist of his time, famous for the Law of the Minimum, the idea that plant growth is limited by the element least available in the soil. He was notable too for his realization that, although carbon dioxide makes up only 1% of the atmosphere, plants obtain their carbon from the air and not from the soil, as previously thought. Liebig was among the first scientists to achieve some kind of understanding of the nitrogen cycle. His almost correct ideas on crop mineral nutrition and his failure to realize that not enough ammonia was deposited in rain to supply the needs of crops are discussed in the main part of Chapter 1.

Liebig's contribution went far beyond agricultural chemistry. He and his friend Wöhler were among the founders of organic chemistry (Box 2.2) and provided improved methods of organic analysis. Liebig discovered chloroform and chloral and introduced the theory of radicals, the idea that a group of atoms forming part of a molecule could act as a unit and take part in chemical reactions without disintegrating, yet be unable to exist alone. Examples include the methyl radical $-CH_3$ and the carboxyl radical $-COOH$. He was also notable for founding a chemical laboratory at Giessen in Germany which he used to train young chemists (among whom was Joseph Henry Gilbert). Liebig must be given some credit for the strength of chemistry in Germany in the latter part of the 19th century. Indeed, his contribution to the subject must have been part of the intellectual heritage that underpinned the development of the Haber–Bosch process.

that Lawes flatly disagreed and wrote in 1846, 'There cannot be a more erroneous opinion than this, or one more injurious to agriculture'. This remark was hardly designed to placate, but it was in the context of an argument with enormous consequences. Liebig claimed that the fields of England were in a state of progressive exhaustion by removal of phosphates. Lawes agreed about the progressive exhaustion but believed it was caused by ammonia carried off in the grain (Dyke, 1991). The underlying problem was that Liebig seriously overestimated the concentration of ammonia in the atmosphere and this ultimately lost him the argument. However, his ideas about minerals, which were based on the composition of crop ash, were largely correct.

Dyke (1993) commented that, 'Liebig was stronger on matters of theory than on practical tests. Lawes and Gilbert had the mixture of both approaches that is the mark of the best scientists in all subjects'. By the autumn of 1844 they also had the results of the first simple experiment on Broadbalk. If Liebig was right, the plots given just minerals on Broadbalk would have out-yielded those without, and the 'minerals+nitrogen' plots would have done no better than the 'minerals only' plot. Table 1.1, based on a report by Garner and Dyke (1969), gives the results of that first experiment, which was designed to address this point, together with means from the experiments from 1856 to 1863 which followed the pattern started in 1852. Even the first simple experiment suggested strongly that Liebig was wrong and the later results confirmed this. Minerals alone did not give much more yield than no manure at all, while 'minerals+nitrogen' clearly

Table 1.1. Yields in early experiments on Broadbalk by Lawes and Gilbert. (From Garner and Dyke, 1969.)

Treatment	Yields (cwt/acre) Grain	Straw
The first harvest in 1844		
No manure	8.2	10.0
14 t farmyard manure	11.4	13.2
Ash from 14 t farmyard manure	7.9	9.9
Minerals only[a]	9.0	10.3
Minerals + 65 lb/acre ammonium sulphate[b]	11.4	12.7
Mean yields, 1856–1863		
No manure	8.8	8.3
Minerals only[a]	10.6	16.1
400 lb/acre ammonium salts only[b]	13.1	22.6
Minerals + 400 lb/acre ammonium salts[b]	21.5	37.3

[a] Minerals were P, K, Na, Mg.
[b] 65 and 400 lb/acre correspond to 73 and 447 kg/ha respectively.

out-yielded minerals alone. Lawes and Gilbert also tested Liebig's patent wheat manure but found it to have no effect (Johnston and Garner, 1969).

Baron von Liebig was not amused. In a confidential letter to a friend quoted by Smil (2001), Liebig wondered how 'such a set of swindlers' could produce research that is 'all humbug, most impudent humbug'. Had he known of the letter, Lawes could have retorted that one of the swindlers had been trained in Liebig's own laboratory. In fact, according to Dyke (1993), Lawes's attitude was entirely courteous. He took pains to preface all criticism of Liebig with a tribute to his contributions to the subject, but the courtesy was not returned. Indeed, Dyke commented that 'some of [Liebig's] writings were simply spiteful', and this seems to be confirmed by Smil. It is hardly surprising that Lawes could not bring himself to use the 'von' when writing of Liebig.

In public Liebig maintained that the Rothamsted results supported his theory. The ammonium fertilizer, he said, acted as a 'facilitator' for the adsorption of the minerals by increasing their solubility. But other Rothamsted results showed that clover grown without any 'facilitation' from ammonium fertilizer removed more minerals than the grain crops did – and took up more nitrogen.

Lawes emerged from the controversy as a man of principle, and certainly not a swindler. He made his living from the sale of phosphate fertilizers (through the Lawes Chemical Company), but when his experiments showed that additions of nitrogen rather than phosphate caused the greatest increase in yield in wheat, he did not hesitate to publicize the fact.

Smil (2001) mentions a rather delightful irony. Whilst propagating his 'minerals' theory, Liebig had told the British prime minister Sir Robert Peel in 1843 that 'the most indispensable nourishment taken up from the soil is the phosphate of lime'. This was, of course, just what the Lawes Chemical Company was selling, but Lawes would shortly be telling the world that nitrogen was the nutrient that had the greatest impact on grain yields. Dyke (1991) notes that, in 1864, Professor J.T. Way commented in an answer to a Select Committee of the House of Commons, '... but there is one great peculiarity in Mr Lawes's career, which is this, that for a great many years he spent a large sum of money annually in making agricultural experiments; and, on the whole, I believe that those experiments, although he has freely reported them to the public, have not been in favour of his trade and position as a manure merchant, but rather to the contrary'.

This story has a sad postscript. I worked for the Lawes Chemical Company in 1965–1966, a few years before it ceased to be an independent company. One reason why the company was not doing well was that it did not have an independent source of nitrogen for its compound fertilizers, those incorporating nitrogen, phosphorus and potassium. It was still obtaining ammonium sulphate from the nearby Becton gas-works, but this was sold by the specialist 'Straights' division of the company which sold single-nutrient fertilizers. For the much larger quantities of nitrogen the company needed for compound fertilizers, it was buying in urea from ICI, and this put it at a disadvantage in an increasingly competitive market, at a time when the government subsidy on fertilizers was to be removed in 1966. The manufacture of nitrogen fertilizers had changed far more than that of phosphate and potassium fertilizers and the scale of production mattered. The Lawes company was not a large one by comparison with its competitors.

The dispute between Liebig and Lawes was only the beginning of the arguments about nitrogen fertilizer, and, with hindsight, we can see in those results from Broadbalk the seeds of the current debate about nitrogen fertilizers.

Why did the Broadbalk wheat crop respond so much to nitrogen?

Nitrogen has many functions in plants. It is found in the chlorophyll that they need for photosynthesis, the nucleic acids in which the programme for their growth and development is encoded, all plant proteins, the enzymes that catalyse all biochemical processes and, not least, the walls that hold the cell together. Much of the biochemistry of life involves nitrogen-containing chemical compounds. Plants with insufficient nitrogen tend to be stunted, yellowish and generally sickly.

Nitrogen augments yield by increasing the leaf area within which photosynthesis takes place. It does so by increasing both the rate at which cells multiply and the surface area of each cell. It also boosts the number of leaves, but plants vary in the way they achieve this. Cereal plants, such as wheat and barley and the grasses to which they are related, produce extra shoots, known as 'tillers'. These probably evolved as a means of exploiting favourable growing conditions and many of them die off if nutrients are in short supply. Applying nitrogen increases the number of tillers that survive, providing extra leaves, more stems and a greater yield. Plants such as potatoes with a more complex leaf structure produce extra branches and leaves in response to extra nitrogen.

Nitrogen also extends the duration of the leaf cover. As older leaves die they release their nitrogen to benefit younger leaves or the developing grain or tubers. Applying nitrogen delays the dying-off of the leaves and maintains a greener, leafier plant.

What benefit did the 'minerals' give?

Phosphate has a key role in the building of the nucleic acids that are central to the growth and development of the plant. It is also very much involved in the storage and use of energy in cells. The adenosine triphosphate system could be said to be the plant's 'energy currency', and other phosphate systems play a vital role in plant function. Potassium has an important role in the mechanism by which carbohydrates are distributed in the plant from the point of production in the leaves to the point of use at growth points. It helps the plant to use water efficiently and to maintain its osmotic and electrical balance, and it is also an enzyme activator. Magnesium is a key constituent of the chlorophyll molecule and it too is an enzyme activator. Calcium plays a part in the structural tissue of the plant and is also involved in cell division and in maintaining the integrity of membranes. Sulphur was not included among the minerals by Liebig, but it is a constituent of certain key amino acids and proteins. It is important for the production of oils in plants and it has been supplied as a fertilizer for oilseed rape in the UK since about 1993 following the removal of most of the sulphur from the emissions of British power stations. About 30% of the British oilseed rape crop now receives it. Iron was also not included the minerals. It is essential for plant growth but not in the same quantities as the other nutrients mentioned. It falls between the major nutrients and the trace elements mentioned below.

Several other nutrients, the 'trace elements' copper, zinc, manganese and boron, are necessary to the plant but only in small quantities. Other trace elements, such as chromium, selenium, iodine and cobalt, are not essential to plants but are essential to animals grazing on the plants. The trace elements copper and zinc, together with nickel, which is not a trace element, become pollutants when too much is there, often after long-term applications of sewage sludge. The trace element then becomes a 'heavy metal'. This is something of a misnomer, because these 'heavy' metals have smaller atomic weights than many other metallic elements, but lead is a heavy metal in both senses.

Apart from nitrogen, the two main nutrients applied in fertilizers are phosphate and potassium. Cereal crops do not need large applications of these nutrients, but potatoes and other crops that form large storage organs need both nutrients. This is as much a matter of arithmetic as plant physiology. A crop of 60 t/ha of potatoes contains roughly 13 t/ha of dry matter, of which about 2% is potassium, so around 260 kg/ha of potassium will need to be supplied as fertilizer or manure just to replace what has been removed from the soil by the tubers. (Most of the potassium in the rest of the plant will have been translocated into the tubers.) Trace-element fertilizers tend to be needed in particular areas, manganese on highly organic soils in the fens and boron for sugarbeet on sandy soils in East Anglia, for example.

The Build-up in Nitrogen Fertilizer Use

1800–1900

Chapter 4 suggests that gas liquor from coal gas production was the first 'chemical' or non-organic nitrogen fertilizer to be used, and that it probably came into use in about 1810–1820. Ammonium salts were made by acidifying gas liquor by at least 1841 and probably earlier, and by 1900, large amounts of ammonium sulphate were produced in this way (Fig. 1.2). By the 1860s ammonium sulphate was also made from ammonia generated in coking ovens (Chapter 4) as well; this process had become important by 1900 (Fig. 1.2) and continues today.

In the 1830s and 1840s, two further sources became available, guano and Chilean nitrate (Chapter 4). Of these, the Chilean nitrate was much the more important (Fig. 1.2) because it produced far more fertilizer nitrate than guano, and also nitrate for explosives, which guano did not provide. Also,

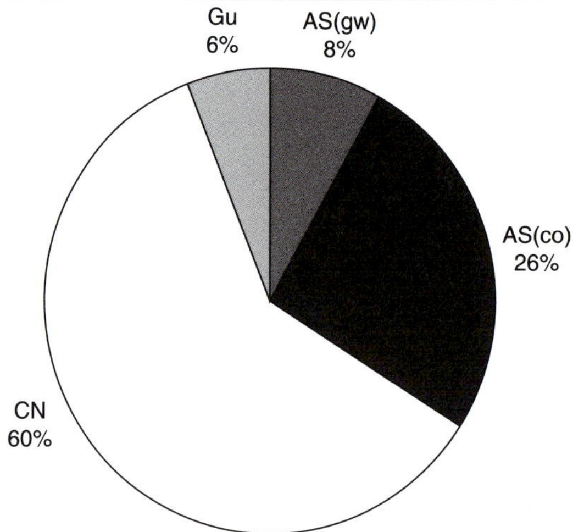

Fig. 1.2. Relative importance of ammonium sulphate from gas-works (AS(gw)), ammonium sulphate from coking ovens (AS(co)), Chilean nitrate (CN) and guano (Gu) as fertilizer sources in 1900. (Based on data from Smil, 2001, pp. 50–51 and p. 240.)

the supply was never exhausted and some production continues to this day. The supply of guano was exploited greedily and contributed in a major way to fertilizer supplies for only about 30 years, from 1842 to 1872 (Fig. 1.2).

Smil (2001, p. 245) calculated that during the second half of the 19th century the world's cumulative production of nitrogen fertilizer was 360 kt of N, of which 220 kt was from Chilean nitrate, 20 kt from guano, and 120 kt of ammonium sulphate from coking. Smil's table did not include gas liquor or ammonium sulphate produced from it, probably because there were no records and such records might have been of limited use because of the variable nature of the product.

1900–2000

Between 1900 and 1980 there was a remarkable increase in the production of nitrogen fertilizers. This is shown in Fig. 1.3, which is based on Appendix L of Smil (2001, p. 245). The term 'exponential increase' is sometimes used rather loosely, but its use here is entirely correct. Production (P) was related very exactly to the number of years after 1900 (t) by the exponential relationship $P = 274 \exp(0.0651t)$.

For the statistically minded, the relation was sufficiently exact to give an r^2 value greater than 0.98 for 13 degrees of freedom. This rate of increase was not sustained, and beyond 1980 production continued to increase but fell away from the exponential relation for the first 80 years of the century. It was in the early 1980s that fertilizer use reached a plateau in the UK (Fig.

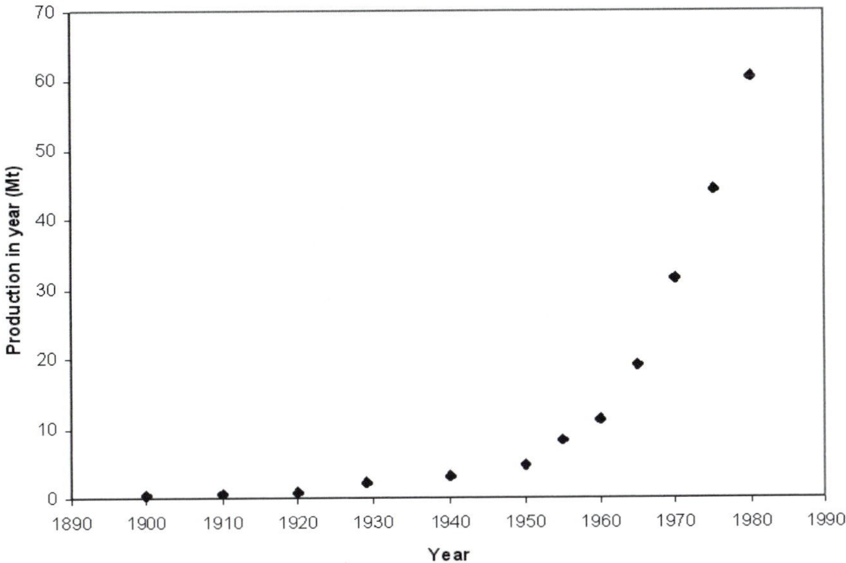

Fig. 1.3. The exponential increase in world fertilizer production between 1900 and 1980. (Based on Appendix L of Smil, 2001, p. 245.)

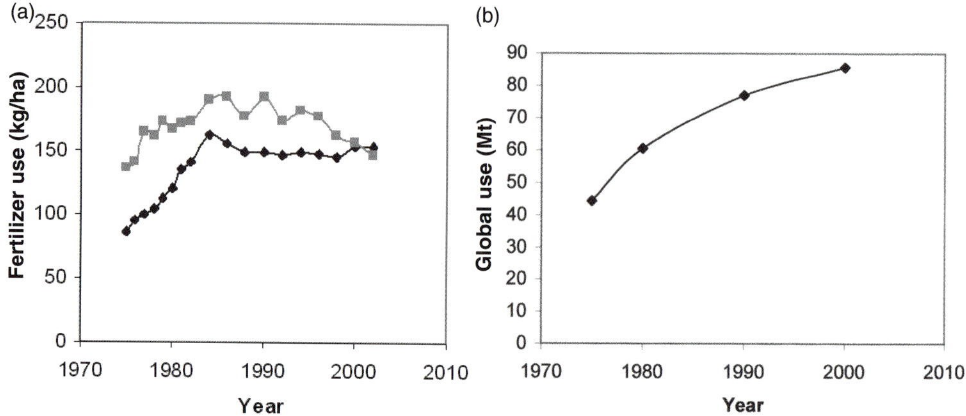

Fig. 1.4. Aspects of the plateau in fertilizer production, 1980–2000. (a) Average applications of fertilizer N in the UK to all tillage (♦) and grass leys of less than 5 years (■). (Data provided by Chris Dawson.) (b) World production of N fertilizer. This is the continuation of the curve in Fig. 1.3. (Data from Appendix F of Smil, 2001, p. 245.)

1.4), and the same was probably broadly true in other Western European countries and North America.

Smil's Appendices F and L show that global production of nitrogen fertilizer rose 238-fold, from 360 kt of fertilizer N in 1900 to 85,700 kt in 2000. Figure 1.5, based on the same Appendices, shows the percentage of N fertilizer production that was achieved by the Haber–Bosch process described in Chapter 4. This increased from zero in 1900 (before the process was invented) to more than 99% in 2000. The latter is a statistic to note. It reflects the increasing dominance of this process in the production of nitrogen fertilizer and, viewed from a different angle, it shows how dependent the world has become on this process.

Dependence on the Haber–Bosch Process?

Smil (2001) and Jenkinson (2001) have drawn to our attention that the population of the world is now collectively dependent on nitrogen fertilizer produced by the Haber–Bosch process. Smil provided two key calculations.

The first concerns staple crops grown intensively on large areas of land, the yields of which could not be maintained without applications of synthetic nitrogen fertilizers. (Synthetic means 'manufactured from its elements' – an exact description for nitrogen from the Haber–Bosch process, as we shall see in Chapter 4.) Smil (2001, p. 155) argued that these fertilizers now provide between 60 and 80% of the nitrogen for the most widely grown staple crops, and that the Haber–Bosch process has recently been supplying about half of the nitrogen input into the world's agriculture. Other things

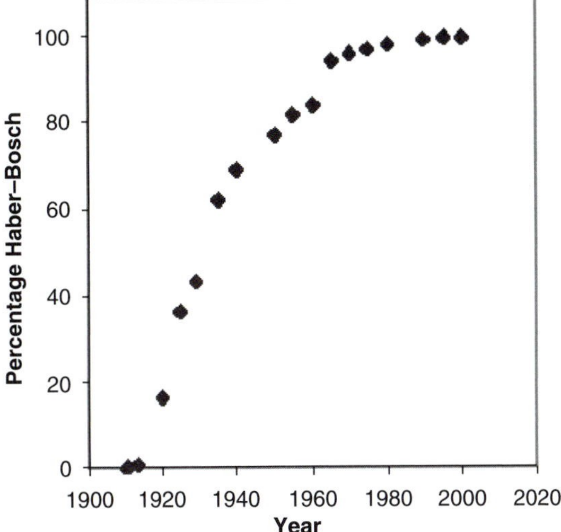

Fig. 1.5. The increase in the percentage of global nitrogen fertilizer produced by the Haber–Bosch process, 1910–2000. (Data from Appendix L of Smil, 2001, p. 245.)

being equal, the world's harvest of staple crops would be halved if nitrogen fertilizer was taken away.

This does not mean that half of all the food produced today would be unavailable or that half the world's population would not be alive without nitrogen fertilizer. Most of us do not eat just staple crops, we eat meat and dairy products as well. The animals that supply these commodities also eat the staple crops, together with grass, which may or may not be fertilized. We also eat fish. It requires, as Smil put it, 'further disentangling of nitrogen's complex pathway in the global agroecosystem'. He has done the 'further disentangling', and his conclusions make stark reading:

> In 1900 the virtually fertilizer-free agriculture (less than 0.5 Mt were applied to crops world wide) was able to sustain 1.625 billion people by a combination of extensive cultivation and organic farming on a total of about 850 Mha. The same combination of agronomic practices extended to today's 1.5 billion hectares would feed about 2.9 billion people, or about 3.2 billion when adding the food derived from grazing and fisheries. This means that only about half the population of the late 1990s could be fed at the generally inadequately per capita level of 1900 diets without nitrogen fertilizer. And if we were to provide the average 1995 per capita food supply with the 1900 level of agricultural productivity, we could feed only 2.4 billion people, or just 40% of today's total.
>
> (Smil, 2001, p. 159)

In short, without synthetic nitrogen fertilizer, agriculture could support only half as many people as are alive today, and then with very basic, predominantly vegetarian diets. Today's average diets could be supplied to only 40% of the population. Collectively we can no more give up nitrogen from the Haber–Bosch process than an addict can give up heroin. The removal of our 'drug' would cause widespread trauma and death.

But what of the future? Smil (2001, p. 175) took the medium version of the UN's long-term population projection which predicts an increase in world population of 3 billion between 2000 and 2050, virtually all in low-income countries. These countries will need to use at least 85% more synthetic nitrogen fertilizer than they do today just to maintain their present average diets. Eliminating malnutrition might need an extra 10% and producing more animal foods a further 10%, with the result that low-income countries may double their 1995 nitrogen fertilizer consumption by 2050. Assuming no change in fertilizer use in the affluent world, which is probably reasonable, Smil calculated that the world might use 140 Mt of N from synthetic fertilizers by 2050. This would mean that the Haber–Bosch process would supply 60% of the nitrogen feeding the world's crops and therefore be essential to the basic nutrition of about 60% of the world's population. Nearly all the protein needed by the 2–4 billion children to be born in the next two generations will have to come from it. Barring some hugely significant technological development or some unimaginably awful human catastrophe, our dependency on synthetic nitrogen fertilizer is set to increase.

Smil (2001) has given us the large-scale picture, painted with a broad brush. There is some detail omitted. Political upheavals also affect the consumption of nitrogen fertilizer. Fertilizer use declined sharply in Eastern Europe after the Berlin Wall fell and the Soviet command economy collapsed. So do fluctuations in supply and demand. There was a surplus of fertilizer produced in the early 1980s in both North and South America and in Western Europe (Jenkinson, 2001) as applications (Fig. 1.4a) levelled out. This led to decreases in prices and a reduction in investment in new plant (Bumb, 1995). These events must have contributed to the falling away of production in the 1980s from the exponential relation for the first 80 years of that century (Fig. 1.3). These details, however, do not greatly change the overall picture of increasing dependence on the Haber–Bosch process or, the other side of the coin, the fundamental importance of this process to humankind.

Dependence on synthetic nitrogen fertilizer causes problems, as do other forms of dependence. Using it inevitably tends to increase losses of nitrate from the soil to natural waters, where it causes concern about the quality of the environment and the health of those who drink the water. It also increases losses of the 'greenhouse gas' nitrous oxide to the atmosphere, which may in the long term prove to be a greater problem than nitrate in water (Jenkinson, 2001).

The environmental problem is complex. Nitrate is rarely lost from the soil in any great quantity directly from nitrogen fertilizer. Much of the added nitrogen is removed in the crop and most of the rest is left in the soil as organic residues. Losses of nitrate usually arise after harvest when soil microbes break down the residues. Substantial amounts of nitrate enter the environment from sources other than fertilizer application. Road traffic and industry both generate nitrogen oxides, which are either deposited as nitrate or are converted to nitrate on deposition. Ammonia can be generated on one part of a farm and deposited on another. To add to the complexity, phosphate is usually responsible for algal blooms in fresh water but nitrate in coastal water. These topics are discussed further in subsequent chapters.

The health issue is also complex, and the conclusion reached by medical researchers (discussed in Chapter 9) that nitrate has several beneficial effects on human health may come as a surprise to some readers.

A state of dependence on synthetic nitrogen fertilizer is not desirable, but it is the state in which we find ourselves as we begin the 21st century. We have to ensure that there is enough food to go round, but we need to make sure that it is produced as efficiently as possible and distributed more effectively than it is at present. Efficiency implies optimizing the ratio of food produced to nitrate released into the environment, rather than just maximizing the yield or minimizing the loss of nitrate per unit area. It means recognizing the different problems and needs of farmers in the richer and poorer parts of the world. It also means determining with certainty in which environmental problems nitrate is the main factor, what role it plays and providing appropriate solutions. Finally, if, as seems increasingly likely, nitrate is beneficial to health rather than a threat to it, we need legislation that recognizes this fact. In short, we need to come to terms with nitrate.

2 The Chemistry and Physics of Nitrate

Chemistry

The chemistry of nitrate is not complicated, and it detains most students of chemistry only briefly.

General chemistry

Nitrate is the most fully oxidized compound of nitrogen and is therefore stable to oxidation but potentially a strong oxidizing agent. Saltpetre (potassium nitrate) has long been the oxidizing constituent of gunpowder and solid ammonium nitrate can explode, with the nitrate moiety oxidizing the ammonium moiety. But this does not mean that nitrate dissolved in water is dangerous. Because of the stability conferred by its structure (see below), nitrate in a very dilute near-neutral solution of its dissociated salts (as found in natural waters) is unreactive chemically and certainly cannot explode. However, it retains the capacity of all oxidizing agents to accept electrons. A brief account of oxidation and its opposite, reduction, is given in Box 2.1.

Box 2.1. Oxidation and reduction.

Oxidation and reduction were originally perceived in terms of the gain or loss of oxygen atoms or conversely the loss or gain of hydrogen atoms. The conversion of ammonia (NH_4^+) to nitrate (NO_3^-) is clearly an oxidation in such terms because hydrogen atoms are lost and oxygen atoms gained. The more fundamental understanding introduced later is that a substance that is oxidized accepts electrons in the process, while the substance that is reduced gives them up. In its conversion to nitrate, ammonia not only loses its hydrogen atoms and gains oxygen atoms; it also gains two electrons, which change it from a positively charged cation to a negatively charged anion.

Nitrate and nitrates

Almost all press comment on nitrate concentrations in natural waters refers to nitrates – in the plural. This is simply wrong. There are admittedly several nitrates – sodium nitrate, potassium nitrate and calcium nitrate for example – but we are concerned with nitrate in water. When potassium nitrate, for example, dissolves in water, it does not remain as potassium nitrate. It undergoes a process known as 'dissociation', in which potassium and nitrate become independent entities in the solution. If potassium nitrate and calcium nitrate are dissolved in the same solution, both dissociate and there is no way of knowing whether a particular nitrate was previously associated with potassium or calcium. A laboratory shelf containing bottles of potassium, sodium, calcium and magnesium nitrates can be said correctly to hold nitrates. If, however, you tip the contents into a large bucket of water and they all dissolve, the water has nitrate in it, not nitrates. There are also a few organic chemicals that are nitrates, but these have totally different properties from nitrate and some are toxic. They are not relevant to this book. (But see Box 2.2.)

Ionic nature

When potassium nitrate (KNO_3) dissociates, the K and the NO_3 both acquire an electric charge. The K becomes K^+ with a positive charge and the NO_3 acquires a negative charge to become NO_3^-. Positive- and negative-charged entities in solution such as these are known as *ions*, positively charged ions being *cations* and negatively charged ions *anions*. The soil solution has to contain the same number of positive and negative charges, but this does not necessarily mean the same number of cations and anions, because some ions carry two or more charges. The calcium ion, for example, is Ca_4^{2+}, while

Box 2.2. Organic and inorganic chemicals.

'Organic' originally referred to substances produced by living organisms, under the influence of some kind of 'vital force'. The 'vital force' idea was shown to be redundant when in 1845 Kolbe synthesized acetic acid from its elements, and again in 1856 when Berthelot did the same for methane. Organic chemistry is now the chemistry of carbon compounds, particularly those in which the carbon forms four covalent bonds. The distinction between organic and inorganic chemistry arguably became redundant with the work of Kolbe and Berthelot but has been retained because the compounds of carbon vastly exceed the known compounds of all the other elements put together, making a separate classification necessary.

The word 'organic' has been now been appropriated by the organic farming movement, but the original use of the word precedes its use by organic farmers by about 200 years. It is difficult to avoid both uses of the word in this book but, where necessary, the text will make it clear which use is intended.

Table 2.1. Solubilities in cold water[a] of the salts of the nitrate ion. (From Weast,1964.)

Cation	Salt	Solubility (g/m^3)	
		Salt	Nitrate[b]
Ca^{2+}	$Ca(NO_3)_2{\cdot}4H_2O$	2.66×10^6	1.40×10^6
Mg^{2+}	$Mg(NO_3)_2{\cdot}6H_2O$	1.25×10^6	0.61×10^6
K^+	KNO_3	0.32×10^6	0.18×10^6
Na^+	$NaNO_3$	0.92×10^6	0.67×10^6
NH_4^+	NH_4NO_3	1.18×10^6	0.91×10^6
Fe^{2+}	$Fe(NO_3)_2{\cdot}6H_2O$	0.84×10^6	0.36×10^6
Fe^{3+}	$Fe(NO_3)_3{\cdot}6H_2O$	1.50×10^6	0.80×10^6
Al^{3+}	$Al(NO_3)_3{\cdot}9H_2O$	0.64×10^6	0.32×10^6

[a]The temperatures at which the solubilities in the table (Weast,1964) had been determined were not all the same and ranged from 0 to 25 K.
[b]The 'nitrate' concentration is that corresponding to the solubility of the salt.

sulphate is SO_4^{2-}. Phosphate can have one, two or three negative charges according to the acidity or alkalinity of the soil. Nitrate, however, always has a single negative charge.

Solubility

The salts formed by nitrate are generally soluble, and calcium nitrate has such a high affinity for water that it is deliquescent, which means that it will pick up moisture from the air and dissolve in it. The main cations in groundwater are likely to be calcium, magnesium, potassium, sodium, iron and aluminium, and the salts they form with nitrate are all highly soluble (Table 2.1). Ammonium nitrate is also highly soluble. Calcium is usually the dominant cation in groundwater, and the nitrate concentration at the limit of solubility for calcium nitrate is 32,000 times greater than the US limit for nitrate concentration in potable water and 28,000 times greater than the European Community limit.

This extreme solubility has two consequences. One is that virtually all the nitrate we encounter in the environment is dissolved in water and the other is that solubility does not limit nitrate concentrations in natural waters.

Structure

The nitrate ion always has the same chemical structure, in which the nitrogen atom and the three oxygen atoms lie in the same plane in a symmetrical trigonal arrangement. For those with an interest in such matters, the nitrogen atom has a formal positive charge, while two negative charges are shared between the three oxygen atoms in a resonance structure comprising three electronic conformations in which each of the oxygen atoms in turn is without charge. The uncharged oxygen atom has two electron pairs and is attached to the nitrogen atom by a π bond, and the charged oxygen atoms have three electron pairs.

This is the only structure that nitrate can have, and it is the same regardless of its origin. Nitrate from a bag of fertilizer and nitrate from a soil on an organic farm both have exactly the same structure and properties, and there is no way in which a chemist, or anyone else, can tell by examining it from which source a particular nitrate ion came.

Sorption by soils

We saw above that nitrate is too soluble for its concentration in water to be limited by solubility. Can it be held by sorption on the soil? By finely divided material, for example? Charcoal is useful for removing impurities from water. The finest particles in soils are the clays. We refer to them in the plural because the clays are a family of materials that have broadly the same general nature but differ in the way in which their constituent elements, aluminium, silicon and oxygen, are put together in the clay structure. Clays can adsorb some substances strongly, but what they adsorb depends on the acidity or alkalinity of the soil as well as the properties of the clay. Clays may bear either a permanent charge (usually negative) or a pH-dependent charge, which arises from the reversible dissociation of hydrogen ions from sites at the edges of the clays, from metal oxide surfaces, or from the phenolic or carboxylic groups of organic matter. Most agricultural soils in the developed world are maintained at pH values of 5.5 to 8.0 by applying lime. This means slightly acid to slightly alkaline, pH 7.0 being neutral. (See Box 2.3.) At these pH values, clays carry an overall negative charge. With electric charges, like repels like; so cations such as potassium and calcium are attracted to the surface of the clay, while anions like nitrate and chloride are not only not adsorbed by the clay but actually repelled by it. This gives rise

Box 2.3. pH.

pH is the measure of acidity and alkalinity and is the negative logarithm (to the base of 10) of the activity of hydrogen ions in a solution. Those unfamiliar with the concept of 'activity' can often replace it by 'concentration' without great error. The pH scale has a range from 0 to 14 in which a pH of 7.0 is neutral. The system becomes more acid as the pH goes below 7.0 and the activity of hydrogen ions increases. It becomes more alkaline as the pH goes above 7.0 and the activity of hydroxyl (OH^-) ions increases. The pH of a soil is measured in a paste made by stirring the soil with distilled water or a dilute electrolyte solution, usually 0.01 M $CaCl_2$. It is important to know in which the pH was measured because the results can differ by up to one pH unit. The pH measured in electrolyte solution will be the greater.

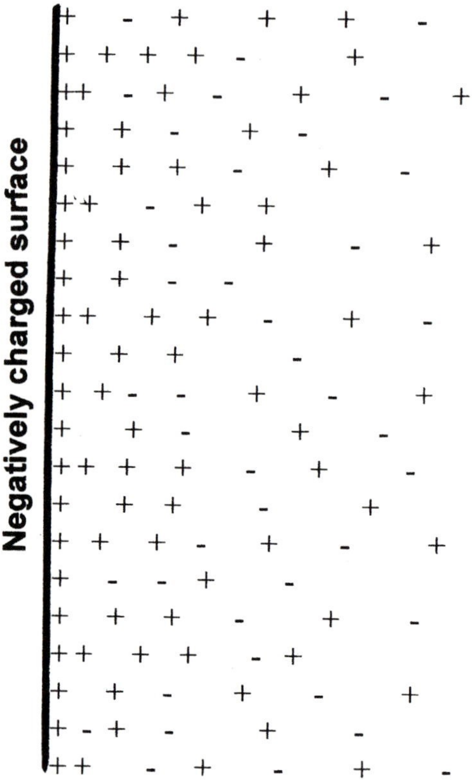

Fig. 2.1. The electrical double layer at a clay surface.

to an electrical double layer (Fig. 2.1) comprising a layer of cations held tightly at the surface (the Stern layer) and a diffuse layer in which cations outnumber anions but to a decreasing extent as we move away from the surface.

Not all soils have an overall negative charge. The humid tropics are hotter and have greater water flows than the temperate regions, so weathering processes proceed faster. Some of the soils, notably those growing tea, are very acid, with pH values around 4.0, and contain oxides of iron and aluminium, which have pH-dependent charge. Soils such as these can carry an overall positive charge. In such soils nitrate is attracted to the clay or oxide surfaces, and Wong *et al.* (1987) and Duwig *et al.* (2003) have shown that this sorption holds it back from being washed out of the soil. The results of Duwig *et al.* suggested two categories of sorption site in a strongly weathered soil, one with strong affinity for nitrate at small concentrations and another with a weak affinity for larger concentrations. In the absence of clear evidence that the soil is positively charged, it will be advisable to assume that sorption, like solubility, does nothing to limit nitrate concentrations in groundwater.

Ammonium

Among soil scientists, ammonium is often lumped together with nitrate as 'mineral nitrogen' because it is so readily converted by soil microbes to nitrate, a process known as nitrification and described in Chapter 3. We therefore need to consider its chemistry in conjunction with that of nitrate, although it is very different from that of nitrate. As its name implies, this ion is a close relative of ammonia, whose pungent smell may be familiar. Like nitrate, ammonium exists as a single entity only in water, and the relation between ammonia, water and ammonium can be summarized as follows, with ammonium hydroxide in the middle:

$$NH_3 + H_2O \leftrightarrow NH_4OH \leftrightarrow NH_4^+ + OH^- \quad (2.1)$$

The formation of the hydroxyl ion OH^- means a solution of ammonia is alkaline (Box 2.3), but the double arrows imply that the change can go in either direction.

Ammonium in the soil

Ammonium is a cation. Its behaviour is therefore in many respects the opposite of that of nitrate. Like potassium and calcium, for example, it is strongly attracted by clays, rather than excluded from them, and is rarely lost in any quantity from non-acid soil unless the soil is very sandy and contains less than 10% clay. We therefore have an incentive to keep it as ammonium by discouraging the process of nitrification. This is discussed further when ammonia is considered as a fertilizer in Chapter 4.

The Root of the Nitrate Issue – Untimely Nitrate

A simple 'equation' summarizes the problem at the root of the nitrate issue:

Availability = Vulnerability

Any nitrogen in the soil that is *available* to crops is usually in one of two forms: ammonium, which is readily converted to nitrate, or nitrate itself. Nitrate is freely soluble and is not adsorbed on to non-acid soils and is therefore *vulnerable* to being washed out of the soil by rain or irrigation water. Ammonium and nitrate together comprise what is commonly known as mineral nitrogen. We need to limit the vulnerability of nitrate.

The surest way to avoid losing money is to carry no more than you need. Much the same is true of nitrate. The less you have in the soil, the less you lose. When crops are growing fast they need a generous supply of nitrate in the soil, but once they have ceased to grow and absorb nitrate we need to ensure that there is as little nitrate there as possible. Any nitrate there is there at the wrong time. So what we have is not so much a nitrate problem as a problem of *untimely nitrate.*

Whether or not untimely nitrate becomes nitrate pollution depends on the fate of rainfall or irrigation water falling on the soil. This is determined largely by the physical behaviour of water in the soil. The remainder of the chapter is therefore devoted to an account of how the physics of the soil influences the downward movement of water and nitrate. But before we turn to soil physics we need to turn briefly to a topic that straddles the realms of chemistry and physics and is important in soil science – diffusion.

Diffusion

Pour some water into a glass and it arranges itself such that the surface is level. The water has distributed itself such that its potential energy, its potential to flow downhill, is the same at all points. Introduce a gas into a closed vessel and something similar happens. The gas distributes itself such that the pressure is the same at all points. Pressure is another form of potential energy. If there is a mixture of gases, each exerts its own 'partial pressure' and distributes itself such that the partial pressure is the same throughout the vessel. If you change the partial pressure in any part of the vessel, there will be a gradient of partial pressure and the gas will flow in such a way as to equalize it. This process is *diffusion* and the flow is *diffusive flow.*

Gaseous diffusion occurs in the soil when bacteria or other organisms in the soil consume oxygen and release carbon dioxide (Currie, 1961; Greenwood and Goodman, 1965). Oxygen diffuses into the soil in response to a gradient in its partial pressure, while carbon dioxide diffuses out of the soil in the same way but in the opposite direction.

A solute in a solution behaves in several ways like a gas in a closed vessel. It has potential energy, its chemical potential, which is related to the partial pressure of a gas. It distributes itself so that its concentration is the same throughout the system, as a gas does with its partial pressure in a closed vessel. And changing the concentration in any part of the solution leads to a gradient in concentration down which solute moves to equalize the concentration. This process is another form of diffusion, and it occurs in the water in the soil where it is important in plant nutrition. A plant root that removes a nutrient from the soil solution creates a concentration gradient down which further supplies of the same nutrient move to replace the nutrient taken by the root. This diffusive supply is usually most important for soil phosphate, because its concentration in the soil is very small. Nitrate is usually present at larger concentrations in the soil solution and is supplied by convective flow.

Diffusion is of particular interest in the context of nitrate losses from the soil as the only process in most soils that can hold nitrate back, albeit temporarily, from being carried away in water passing through the soil. This is discussed later in the chapter.

Physics

In soil science, there is no such thing as the 'physics of nitrate', only the physics of the water in which the nitrate is dissolved – soil water physics.

Dig out a block of soil without disturbing its structure, dry it and weigh it. Then measure the size of the hole by refilling it with a measured amount of water. The bulk density, the ratio of the mass to the volume, will probably be 1.0–1.3 g/cm^3 depending on the type of soil. If you can crush it so effectively that you drive out all the air, which you probably cannot do, the density will be about 2.6 g/m^3. Solid matter therefore only occupies about half the volume of the soil, and the rest is air and water. There seems little to stop water moving through the soil and carrying nitrate with it. In some types of soil this is broadly true, but in others the water moves downward with difficulty or not at all. Much depends on the texture and structure of the soil.

Soil texture

The texture of the soil is the relative proportions of sand, silt and clay in the soil. The structure is related to the texture but is more concerned with how these particles are brought together in aggregates (soil crumbs) or larger soil units.

Soil texture is regarded as sufficiently important for the sizes of the particles have an international classification (Table 2.2). The two sand fractions feel gritty to the touch, as you would expect, but at the other end of the scale the clay particles are so fine that if you wet them thoroughly and rub them between your fingers they feel almost soapy. Most soils contain a mixture of all three constituents, but there are some very sandy soils that contain less than 10% clay, some soils from around the Wash with a large proportion of silt, and some heavy clay soils about which more will be said. It is very common for the texture of the soil to change as you dig deeper, and it may change as you move from one part of the field to another. This 'spatial variability' is so widespread and affects so many soil properties that it has attracted its own discipline of 'geostatistics'.

Soil structure

We met the clays earlier in the chapter in the context of their electrical charge characteristics. They are also remarkable for the way in which their particles adhere to each other. Were it not for this characteristic, there would be no such craft as pottery – or any discernible structure in the soil. Both depend crucially on the adhesion of clays.

Topsoil

In the topsoil, the part that is cultivated, ploughing and harrowing fragment the soil, as do wetting and drying, which make many soils shrink and then swell and crack. The soil may also freeze, which causes the water in it to

Table 2.2. The International Union of Soil Sciences' classification of particle sizes.

Fraction	Particle diameter (μm)
Coarse sand	0.2–2.0
Fine sand	0.02–0.2
Silt	0.002–0.02
Clay	< 0.002

Larger particles are classified, in order of increasing size, as gravel, cobbles, stones and boulders. These are not usually considered to be soil material.

expand and shatter it. There would not be much structure to it were it not for the ability of the clay to cement the particles together. In this, the clay is aided by the organic matter in the soil, the residues of generations of plant remains decomposed by soil microbes as described in Chapter 3 and converted to a form in which it is able to coat and stabilize the clay particles.

The result of all this activity is what we call the soil structure. The main features of the structure in the topsoil are the 'aggregates', or 'soil crumbs', which are the cemented and stabilized fragments whose genesis was described in the previous paragraph. All agricultural soils are aggregated to some extent, but the degree of aggregation depends on the texture of the soil. In general the more clay a soil contains the more aggregated it will be. This is due partly to the cementing effect of the clay and partly to the mutually stabilizing effect of the clay and the organic matter, which means that more clay implies more stabilization by organic matter. Aggregation has less impact on the soil in sandy soils, partly because more sand usually implies less clay, and partly because large sand particles are not easily bound into aggregates and tend to break them apart if incorporated. A farmer wanting a good 'seedbed' that will give the best environment for germination and early growth will aim to produce aggregates of 0.5–2.0 mm.

Subsoil

The main difference between the topsoil and the subsoil in arable land is obviously that the subsoil is not ploughed, unless the farmer has opted for a no-till strategy for the topsoil. And of the natural influences on topsoils, only wetting and drying affect the subsoil and then only in clay subsoils. Topsoil and subsoil usually differ least in sandy soils – the difference may just be the larger amount of organic matter in the topsoil. Silt soils also show little difference where the subsoil is silt, but there are quite large areas where wind-blown silt (loess) is deposited on another material. Where topsoil and subsoil differ most is in a clay soil. The topsoil will comprise aggregates from 0.5 mm upwards, together with larger clods in which aggregates are bound together. But clay subsoil is a much more massive structure whose structural units are measured in centimetres or tens of centimetres rather than millimetres. This is a consequence of the cementing characteristics of clays discussed above.

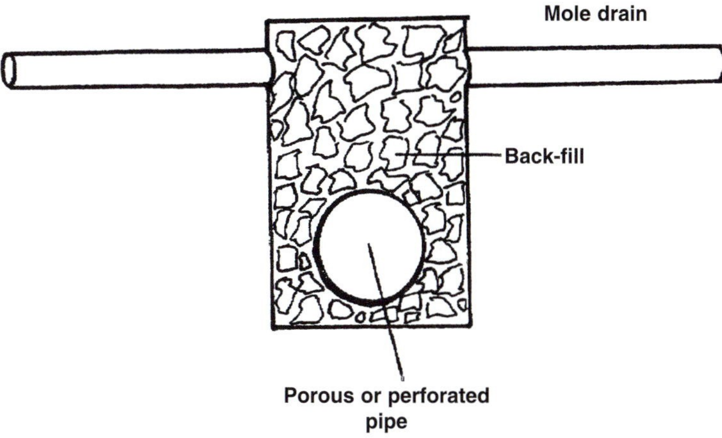

Fig. 2.2. The connection of mole and pipe drains at a trench.

This massive structure can often mean that clay subsoils are virtually impenetrable to water and roots below a certain depth or in a particular layer. In many soils some kind of man-made drainage is needed for the soils to be farmed effectively. This is often of the mole and pipe variety, in which a steel 'mole', shaped like a mole, is drawn through the soil at a depth of approximately 0.6 m. This is attached to a frame on the tractor by a 'leg', a metal connecting bar which leaves behind it a 'leg-slot' and a series of herring-bone cracks, all of which facilitate the entry of the water into the soil. The resulting 'mole drain' interconnects at right angles with a trench filled with gravel (Fig. 2.2) at the base of which is a porous or perforated pipe which carries the water away to a ditch or stream. (See also Fig. 5.6.) This type of drainage has come under question after the flooding in some recent years because it arguably removes the water from the land too rapidly for other parts of the system to dispose of it, but without it some fields simply could not be cultivated. Some subsoils develop a particular layer that is impenetrable to roots and water. This may be tackled by a process known as 'subsoiling' or soil 'busting' in which an implement is torn through the impenetrable layer to make pathways for roots and water. This operation uses a lot of energy and is not undertaken lightly.

Cracking clays

Some clay subsoils have the ability to imbibe water so that they swell when wetted and shrink when dried. If crop roots can gradually draw water out, cracks develop that enable roots and water to penetrate. Such subsoils tend to shrink and crack during summer and then swell and close up again during winter, but in others the cracks become permanent. The Dutch Polders comprise areas of land reclaimed from the sea. During reclamation reeds are grown initially to remove water and salt from the soil, so that cracks form and allow water to drain, a process the Dutch call 'ripening'. I saw one of the polders at an early stage when the reeds were still growing and again 12 years later when the ripening was complete. On the first occasion the soil profile was largely featureless but after the 12 years the cracks were enormous.

Soil water

Water can exist as a solid, a liquid or a gas, and all three phases may be found in soil. The solid (ice) is obviously to be found wherever the soil freezes more than transiently but, as we saw earlier in the chapter, freezing is usually helpful by holding ammonium as ammonium, and ice cannot move or carry nitrate with it. The gas phase (water vapour or steam) is important when soils dry out, but it cannot carry nitrate. We are almost entirely concerned with water in the liquid phase.

We saw earlier in the chapter that solid matter occupies roughly half the volume of the soil, with water and air filling the rest in a network of pores. Having water but no air in the pores means that the soil is *saturated*, but this is unusual. The pores normally contain a substantial amount of air, and the soil is then *unsaturated*. The soil does not need to be saturated for water to flow in it. Indeed, saturation is bad for the soil because air cannot flow in a saturated soil and this restricts plant roots from growing and alters the behaviour of soil microbes.

Water flows in the soil

Water flows in response to a gradient in potential, which may be in any direction but is usually downwards. The rate of flow depends on the transmission of water through the soil pores, but the flow in an individual pore is proportional to the fourth power of its radius. Thus the flow in a pore of 0.1 mm is 10,000 times smaller than the flow in a 1.0 mm pore, other things being equal. We virtually never have measurements of the range of pore sizes in a soil, so the transfer of water in a soil is defined by a parameter, the *hydraulic conductivity*, which lumps all the pore sizes together for a particular soil. It is a highly variable parameter because of the characteristics of the individual pores and can vary by an order of magnitude over a distance of 100 mm. (A difference of an order of magnitude might be, for example, between 10^{-1} and 10^{-2}; that is, 0.1 and 0.01.)

The Richards equation and the convection–dispersion equation

The hydraulic conductivity has been incorporated into solutions of the first of two equations widely used in conjunction with each other – the Richards equation and the convection–dispersion equation. The former describes how water flows through the soil in response to a gradient in the potential of water and the latter how the water carries the solute (convection) and how the solute spreads out within the flow of water (dispersion). The phenomenon of dispersion within a flow is fairly closely related to diffusion. These equations (Box 2.4) have been described by several authors (e.g. Wagenet, 1983). They are fairly easy to use for conditions of steady-state flow, a condition that is attained when an already wet soil experiences an

Box 2.4. The flow equations.

Steady-state flow implies that neither the flow nor the amount of water in the soil changes with time. The theory considers the fluxes of water and solute; that is, the quantities of solute and solute crossing unit area in unit time (Wagenet, 1983). If the fluxes of water and solute are J_s and J_w, J_s is related to J_w as follows:

$$J_s = [\theta D_m(J_w) = D_p(\theta)]\ dc/dz + J_w c \qquad (2.4.1)$$

in which θ is the volumetric water content, c is the concentration of solute and z is the depth from the surface of the soil. D_m is the mechanical dispersion coefficient, and reflects the effects of the differing pore sizes. It depends on the flow and is therefore presented as $D_m(J_w)$. D_p is the diffusion coefficient, and since diffusion depends on the water content of the soil, it is presented as $D_p(\theta)$.

If the amount of solute changes at any depth, z, in the soil, the flux of solute, J_s, is no longer a steady-state flux, and we need to define the rate of change of the amount, s, of solute with time, t:

$$(\delta s/\delta t)_z = -\delta/\delta z\ \{-[\theta D_m(J_w) + D_p(\theta)]\ dc/dz + J_w c\} \qquad (2.4.2)$$

The water flow remains a steady-state flow, so this equation, like the previous one, is amenable to an analytical solution, but it is limited to steady-state water flows, which are rare.

Non-steady-state flows in which both J_w and θ change with time are far more common in nature. They need to be computed at each time interval using equation (2.4.3), in which K is the hydraulic conductivity and H the hydraulic potential (Wagenet, 1983):

$$\delta\theta/\delta t = \delta/\delta z\ [K(\theta)\delta H/\delta z] \qquad (2.4.3)$$

For flows of water in soils K and H are not independent of each other. K depends on θ [hence $K(\theta)$] and θ depends on H. We can only describe non-steady-state flows if the K–θ–H relationship is defined. No analytical solution is possible and a finite element approach is necessary. This has become feasible with modern computers, but the problem does not end there. The parameters that link K, θ and H vary greatly from point to point in a field (e.g. Wagenet and Addiscott, 1987), making the whole enterprise very difficult.

unvarying flow of water, usually downwards. These conditions may exist during irrigation but otherwise they exist rarely, other than in experiments designed to test equations for steady-state water flow (or during camping holidays). In conditions of intermittent flow, such as usually occur in the real world, they become very complicated to use particularly if, as is virtually always the case, the parameters connecting the hydraulic conductivity to the volumetric water content and the hydraulic potential vary in space (e.g. Wagenet and Addiscott, 1987).

The stochastic approach

Some researchers concluded that the soil varies so much that that its variability dominates the leaching of solutes. If that is so, all that is needed is an ultra-simple description of the process such as the travel time, the time taken by the solute to move from the surface to a specified depth, backed up by good statistics. This approach, developed by Jury (1982), considers the probability density function $f_z(w)$ that summarizes the probability P_z that solute applied to the soil surface will arrive at depth z as the water applied increases from w to $(w+dw)$:

$$P_z(w) = \int f_z(w)\, dw \tag{2.2}$$

The underlying concept is of a bundle of twisted capillaries of differing lengths through which solute moves by piston flow (see below). Calibrating the equation for one depth and one quantity of water enables it to estimate the time of arrival of solute at a greater depth when further water is applied. This is subject to the provision that the soil does not change with depth, a condition that may restrict its use to sandy or silty soils.

Piston flow

The simplest concept of solute movement in soil is that water or a solution applied to the surface of the soil simply pushes out the solution already there without mixing with it. This is analogous to a piston pushing out fluid from a cylinder, so it is called 'piston flow'. The depth, z_p, to which the front of the displacing solution penetrates depends on the quantity, Q, of the displacing solution and the volumetric moisture content, θ, the fraction of the soil's volume that can hold water:

$$z_p = Q/\theta \tag{2.3}$$

Obviously z_p and Q have to be in the same units. Rainfall or applications of water are usually expressed in mm (the number of mm that would accumulate on an impermeable surface) and z_p can have the same units. Piston flow never occurs without some spreading of the solute front by dispersion, but the idea can provide a useful rule of thumb for estimating the depth reached by the solute.

Another useful aid to estimating the depth reached by a solute is a leaching equation developed by Burns (1975). This gives the fraction, f, of nitrate or a similar solute that is leached below a depth h (in cm) by the percolation of P cm of rain:

$$f = [P/(P + \theta)]^h \tag{2.4}$$

The author of this equation also developed a leaching model (Burns, 1974) based on the concept of 'temporary over-saturation' of the soil.

Soils with mobile and immobile water

All the approaches to water and solute movement discussed so far presume that the water and the solute move uniformly through the whole soil. This is simply not so. Water does not flow through soil aggregates; it goes round them, while the water inside the aggregate remains largely immobile. The solute moves between the interior of the aggregate and the water on the outside by diffusion (Fig. 2.3). It is the slowness of this diffusion – the time it takes nitrate and other solutes to diffuse out of the aggregate – that helps to protect them from leaching, particularly from larger aggregates and clods when the concentration inside the aggregate is greater than that outside (Addiscott *et al.*, 1983). The slowness of the diffusion is a disadvantage when the concentration is the greater outside the aggregate. This phenomenon is sometimes described as 'hold-back'. Models for soil with mobile and immobile water were developed by Van Genuchten and Wierenga (1976) and

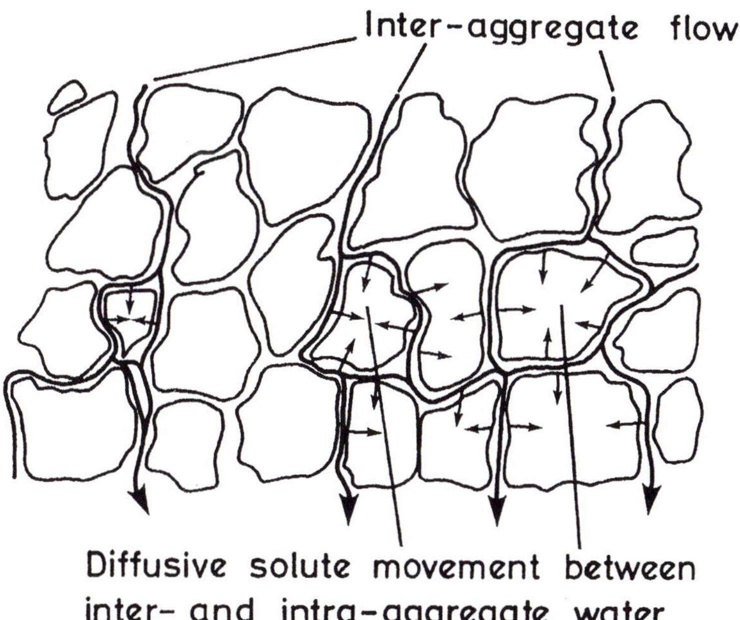

Fig. 2.3. Diffusion of a solute from inside an aggregate, where it is safe from leaching, to water flowing through the soil. (Diagram drawn by Joyce Munden.)

Addiscott (1977), but the principle was established nearly 100 years earlier by Lawes, Gilbert and Warington (1882).

Mobile and immobile water would not be expected in sandy soils with little aggregation, but solute flows studied by Hesketh *et al.* (1998) in such a soil suggested they were there. And when De Smedt *et al.* (1986) did all they could to measure uniform flow by passing tritiated water through a carefully packed sand column, they still found evidence of mobile and immobile water. But these results can be interpreted in terms of the range of pore sizes – and thence velocities of flow in the soil – rather than mobile and immobile water.

By-pass flow

In soils with mobile and immobile water the two categories of water can interact by diffusion. In by-pass flow water moves rapidly through large cracks or channels in the soil, often when more rain falls than can be absorbed by the soil matrix, but without any appreciable interaction with the rest of the soil. This is known as 'channelling' or 'by-pass' flow (Beven, 1981). Such flow may not remove much of the nitrate from within the soil. However, if the nitrate is on the surface rather than in the soil, after fertilizer application for example, it can be carried rapidly through the soil and lost.

This account of the chemistry and physics of nitrate is only the beginning of the story. The effects described interact with many biological processes in the soil, making the behaviour of nitrate remarkably complex for such an essentially simple ion.

3 The Biology of Nitrate

The biology of nitrate is to a large extent the chemistry of nitrate implemented by living organisms. Most of the main processes involved occur only in the soil but some of them also occur in completely different parts of the environment, including our mouths (Chapter 9). All form part of the nitrogen cycle, which is shown at the global scale in Fig. 3.1a and at the scale of a field plot in Fig. 3.1b. Practically all the nitrogen in the soil is organic nitrogen, in the very earliest sense of the word 'organic' – that is, produced by living organisms (Box 2.2). The 1–2% of the nitrogen in the soil that is in inorganic or 'mineral' form, as ammonium- or nitrate-N, is most available to plants but causes most of the environmental problems.

Organic Matter – Dead and Alive

Dead – humus

Humus, or dead soil organic matter, comprises the remains of plants, mostly long dead, that grew in the soil. Plant remains do not remain recognizable for long because they are soon 'processed' by the soil population (see next section). Humus often influences the colour of the soil, and there is usually, but not always, more in the topsoil than in subsoil, making the former darker in colour. The soil is the third largest repository of nitrogen on the planet, coming after the atmosphere and the sea (Table 3.1) and the amounts of organic nitrogen in it are huge. Even a soil with a small amount of organic matter, perhaps because it has been cultivated for many decades, will probably contain 2000–3000 kg/ha of nitrogen, that is 2–3 t. Nearly all of this will be in organic forms and most will be in the plough layer, the top 25 cm of soil. More typical arable soils will contain 3–5 t N/ha and a peat soil will contain much more. These figures are for the UK and other countries in the temperate zone. The hotter the climate, the faster organic matter will be decomposed and the less will remain in the soil.

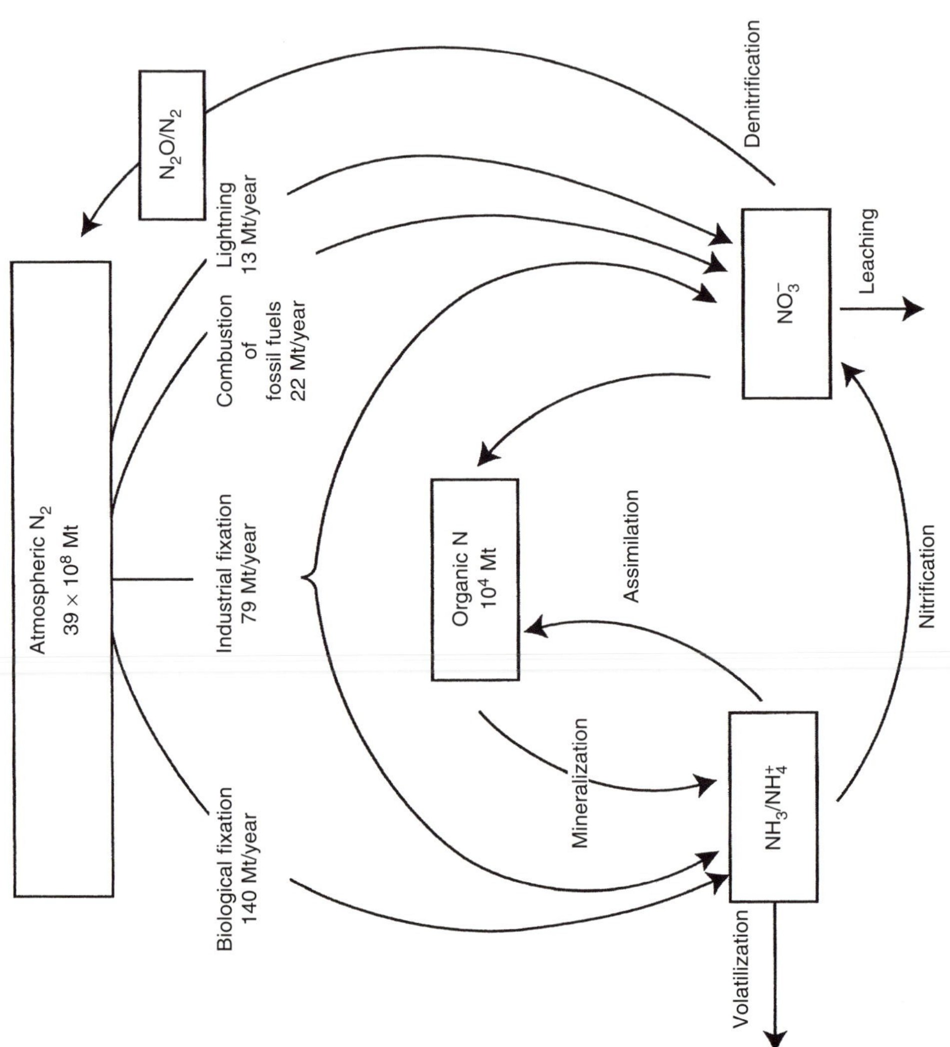

Atmospheric N2
39 × 10^8 Mt
N2O/N2
Lightning
13 Mt/year
Combustion
of
fossil fuels
22 Mt/year
Industrial fixation
79 Mt/year
Biological fixation
140 Mt/year
Denitrification
NO3−
Leaching
Organic N
10^4 Mt
Assimilation
Nitrification
Mineralization
NH3/NH4+
Volatilization
(a)

(b)

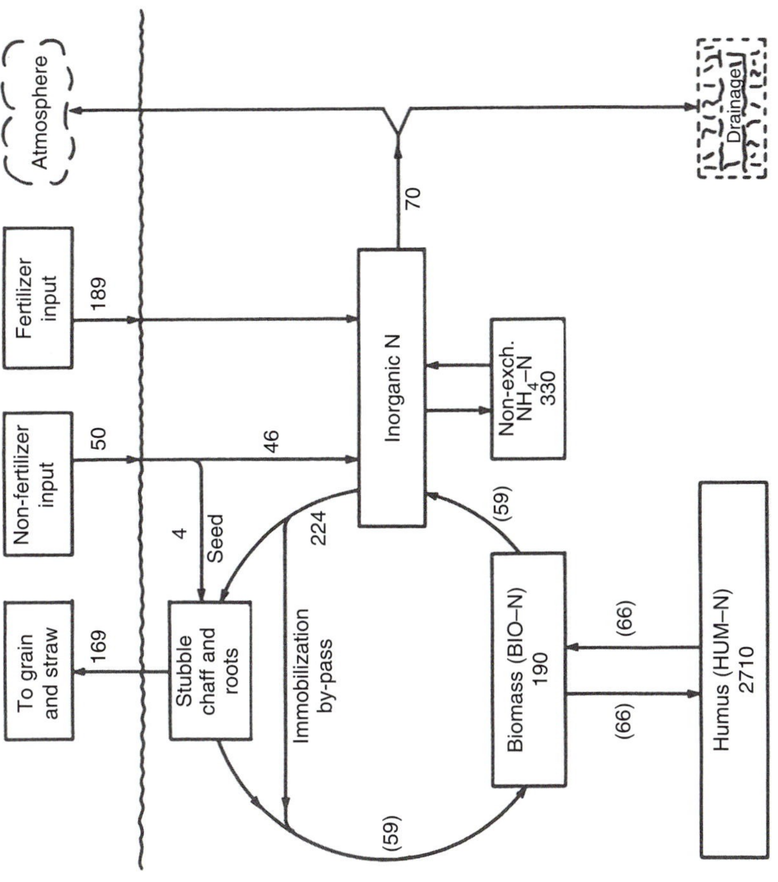

Fig. 3.1. The nitrogen cycle: (a) at the global scale (from L'hirondel and L'hirondel, 2001) and (b) at the scale of a field plot (from Powlson, 1994); the plot was on the Broadbalk Experiment and received 192 kg/ha per year. The amounts of N in (b) are in kg/ha.

Table 3.1. Global distribution of nitrogen. (After Jenkinson, 1990.)

Nitrogen	Amount (t)
The atmosphere	3.9×10^{15}
Soil (non-living)	1.5×10^{11}
Soil microbes	6.0×10^{9}
Plants	1.5×10^{10}
Animals (land)	2.0×10^{8}
People	1.0×10^{7}
Sea (various)	2.4×10^{13}

The quantity of nitrogen in the humus of a soil is 50–100 times greater than the quantity of mineral nitrogen (ammonium-N+nitrate-N), which may be 50–100 kg/ha. It is also much greater than nitrogen in fertilizer applications, which is commonly of the order of 100–200 kg/ha. But these 'small' quantities of nitrogen need to be managed with great care because they are vulnerable to leaching or denitrification, and can have immediate effects on the environment. Humus also needs to be managed, but nothing happens to it rapidly so this is a longer-term project. In recent years there has been great interest in humus as a repository for carbon coming from atmospheric carbon dioxide.

Live – the soil population

The soil contains microbes beyond number. A flock of sheep in a field constitutes a considerable body mass, but there are so many microbes beneath the soil surface that their combined body mass is about the same as that of the sheep above the surface (Jenkinson, 1977). On a global scale, the soil microbes weigh about 30 times as much as the land animals (Table 3.1). The microbes are made up of large numbers of bacteria, fungi, protozoa, algae and – at a slightly larger scale – earthworms and other soil animals. This population beneath the soil surface has a key role in the recycling of plant nutrients on which all other species depend (Edwards and Lofty, 1972; Brookes *et al.*, 1982).

When the plants die, the carbon, nitrogen and sulphur they originally absorbed as carbon dioxide or ions (or possibly other small molecules) are entombed in proteins and other macromolecules, while the phosphate, though retaining its identity, is attached to other macromolecules. The cations are bound to varying extents. If nothing further were to happen in a natural ecosystem, the plants would exhaust the supply of nutrients within a few generations, and in time the ecosystem would collapse. In the longer term, the supply of carbon dioxide for photosynthesis would become depleted too. However, the soil population recycles these resources and makes them available to the plants again, thereby playing an important part in the ability of the ecosystem to organize itself. In doing so, it obtains food and energy.

Recycling the residues

Plant, and occasionally animal, remains on the surface may be attacked first by fungi; you can sometimes see a web of white fungal hyphae on dead leaves or cereal straw on the soil. But the first line of attack often comes from soil animals, millipedes, springtails and earthworms. These break up the residues and bring them into closer contact with the microbes that play the main part in the recycling process. During this process they release carbon dioxide and nutrients, leaving the more resistant material. This resistant material usually undergoes further 'processing', with further releases of carbon dioxide and nutrients before it is left as humus. During all these processes, the organisms gain carbon, nitrogen and other elements essential for their structure and metabolism.

Predation speeds up the recycling. Tennyson's reference to 'nature red in tooth and claw' might make you think of an African game-park, but it is equally applicable to the soil beneath your feet. Some species are the scavengers of the soil, akin to vultures or hyenas, feeding on the dead bodies of other organisms. Other soil species are hunters, corresponding to lions or leopards. Protozoa, for example, prey on bacteria. In each case the predator ingests and uses some of the nitrogen from the prey to build proteins and excretes the rest as ammonium, which is usually nitrified. These effects can readily be demonstrated. Eliminating the protozoa from soil decreases the amount of nitrate in it.

Earthworms play an important part in the recycling by pulling leaves and other plant material down into their burrows, leaving the less decomposable parts such as petioles behind (e.g. Edwards and Lofty, 1972). Darwin (1881), who became very interested in earthworms, found that they would even pull small triangles of paper into their burrows. As leaf material passes through an earthworm, the gut microflora begin to break it down, a process continued by microbes in the surrounding soil. The microbes in the gut and in the soil are indistinguishable from each other (Edwards and Lofty, 1972). The end result is that all the nutrients are released from the plant material, albeit in different ways reflecting the extents to which they were bound. The importance of the combined activities of earthworms and soil microbes is shown by the observation that earthworm casts contain appreciably larger concentrations of nitrate and other nutrients (Lunt and Jacobson, 1944; Nye, 1955). Termites play a somewhat similar role in tropical soils, creating mounds which are richer in nutrients than the surrounding soil.

Although much of the plant material is removed at harvest in a cropped ecosystem, the stubble and roots remain and are recycled by essentially the same processes. But these processes, so essential in a natural ecosystem, can become a problem when the soil is cropped and left bare for part of the year, as we shall see in Chapter 5.

Characteristics of the microbes

The significance of the soil microbes was neatly summarized by Jenkinson (1977), who described them as 'the eye of the needle through which all the organic materials must pass'. Despite this key role in ecosystem function, soil microbes are not as well understood as many microbes studied in other spheres of science such as food contamination. Counting microbes on a film of agar in a Petri dish is satisfactory for medical samples, for example, but with soil samples it can give results that differ greatly from other methods, such as selective staining and counting. This difference occurs mainly because the agar is far richer in microbial nutrients than the soil. Selective staining is the more reliable method, but the counting is tedious and makes the method unsuitable for routine use. Modern methods such as PCR assays now make it easier to study individual organisms (e.g. Mendum *et al.*, 1999), but even these methods have their limitations.

The microbes have been studied collectively as *the soil microbial biomass* by three methods, *fumigation–incubation*, *fumigation–extraction* and *adenosine 5′-triphosphate (ATP)*. The first, based on studies by Jenkinson and Powlson (1976a,b) involves incubating the moist soil with chloroform for 24 h and then incubating it for 10 days and measuring the release of carbon dioxide, from which the biomass carbon is estimated by multiplying the release by a constant (Jenkinson and Powlson, 1976b). Fumigation–extraction methods for carbon (Vance *et al.*, 1987) or nitrogen (Brookes *et al.*, 1985) depend on the observation that the amounts of these elements extractable with 0.5 M K_2SO_4 following fumigation with chloroform provide, with the imposition of suitable constants, a good measure of the amounts of biomass carbon and nitrogen in the soil. The same can be done to measure the phosphorus held in microbes, except that $NaHCO_3$ has to be used for the extraction (Brookes *et al.*, 1985). ATP was described in Chapter 1 as the energy currency of living systems. The amount of ATP in the soil therefore gives a measure of the quantity of microbial biomass. It can be extracted by applying ultrasound to the soil to break open microbial cells in the presence of a strong acid such as trichloroacetic acid to suppress unwanted enzyme activity, and assayed with an enzyme from fireflies (Brookes *et al.*, 1985). Another method involves the measurement of respiration induced by a substrate.

Soil organisms can be classified according to size and whether or not they have a nucleus, but the most helpful classification systems in the present context are those based on nutritional requirements and relationship with oxygen. The nutritional system divides organisms into *heterotrophs* and *autotrophs*. Heterotrophs include many species of bacteria and all the fungi. They need carbon in the form of organic molecules for growth. Autotrophs include the remaining bacteria and most algae, which can synthesize organic molecules from carbon dioxide using energy from the sun. There is a subgroup, the *chemoautotrophs*, which use energy obtained by oxidizing inorganic ions or molecules rather than energy from the sun. The oxygen-based system defines three classes. *Aerobes* need to have O_2 as the terminal acceptor of electrons during respiration. *Facultative aerobes* usually need O_2

but can adapt to anaerobic (more specifically anoxic) conditions by using nitrate and similar inorganic compounds as electron acceptors. *Obligate anaerobes* can grow only in the absence of oxygen because O_2 is toxic to them.

Most soil microbes are hardy species adapted well to life under near-starvation conditions in the soil. Indeed they seem to have developed an interesting mechanism for responding to the possible arrival of substrate (or food in our terms). Most soil microbes exist most of the time in a dormant state with just basal metabolism, but they maintain an adenylate energy charge characteristic of microorganisms undergoing exponential growth. This state of 'metabolic alertness' seems to be a survival strategy that enables the organisms to capture substrate before other organisms. It is costly in terms of energy resources, but De Nobili *et al.* (2001), who identified it, suggested that it is an adaptation that gives an evolutionary advantage over microbes that remain dormant in spores.

Mineralization, Nitrification and Immobilization

The release of carbon dioxide, ammonium, nitrate, sulphate and phosphate from soil organic matter as a result of the activity of the soil microbial biomass and the rest of the soil population is commonly known as *mineralization*. For nitrogen, the process involves two stages, *ammonification* and *nitrification*.

Ammonification is the conversion of the more readily decomposable organic nitrogen compounds such as proteins and nucleic acids into ammonium, and can be effected by a wide variety of bacteria and fungi. It is influenced by temperature (e.g. Addiscott, 1983), moisture and other factors that affect such organisms. A hydroxyl ion is formed, so the process makes the soil slightly more alkaline:

$$\text{Organic N} \rightarrow NH_4^+ + OH^-$$

Nitrification of ammonium to nitrate is an oxidation process that has an acidifying effect because hydrogen ions are formed. It is implemented in two stages by chemoautotrophic bacteria, which derive their energy solely from oxidizing ammonium ions. The first stage converts ammonium to nitrite and can be represented by the equation:

$$2NH_4^+ + 3O_2 \rightarrow 2NO_2^- + 4H^+ + 2H_2O + \text{energy}$$

It was generally believed to be effected by bacteria of the *Nitrosomonas* genera, but recent research using molecular biological techniques (e.g. Mendum *et al.*, 1999) suggests that *Nitrosospira*-type bacteria predominate in arable soils, with *Nitrosomonas* occurring more in ammonium-rich environments such as sewage sludge.

The second stage, for which *Nitrobacter* genera are responsible, is:

$$2NO_2^- + O_2 \rightarrow 2NO_3^- + \text{energy}$$

Ammonium is the final product of mineralization in very acid soils because the bacteria responsible for the first stage of nitrification are sensitive to low pH, hydrogen ions being a product in the first equation. This could be seen as a consequence of the Law of Equilibrium, but it would be an oversimplification. It is also the final product in permanently waterlogged soils because the bacteria in the first stage are aerobic. Nitrite is toxic to most soil organisms, and it is fortunate that it rarely accumulates in soil. It seems most likely to do so in soils that have become over-enriched with nitrate, probably because *Nitrobacter* functions less well in large concentrations of nitrate, which is a product in the second equation. This again could be seen as a result of the Law of Equilibrium. Nitrification is also sensitive to temperature down to about 2.5°C (Tyler *et al.*, 1959; Addiscott, 1983), below which it is inhibited.

Immobilization

Mineralization, as described above, sounds like a one-way street carrying nitrogen from proteins and nucleic acids to ammonium and then to nitrite and nitrate. As in most traffic systems, there is a parallel street running nearby in the opposite direction. This is *immobilization*, in which soil organisms of a wide range of genera take ammonium and nitrate and convert them into organic forms of nitrogen. It is quite normal for some organisms in the soil to be mineralizing nitrogen while nearby organisms are immobilizing it simultaneously. What conventional measurements of mineralization actually measure is *net* mineralization.

Net mineralization was investigated 20–30 years ago by various researchers and found to show first-order kinetics by Stanford and Smith (1972) and zero-order kinetics by Tabatabai and Al-Khafiji (1980) and Addiscott (1983). The difference probably arose because the first authors used dry soil and the later authors used field-moist soils (see Addiscott *et al.*, 1991, p. 87), and neither is necessarily correct. A textbook of physical chemistry warned long ago that simple kinetic relationships are rarely found in nature (Moelwyn-Hughes, 1957, p. 1090). However, Addiscott found that the changes with temperature in the zero-order rate for net mineralization seemed to obey the Arrhenius relationship. The zero-order model with Arrhenius temperature control on rates was used with leaching and crop uptake models by Addiscott and Whitmore (1987), giving satisfactory statistically assessed simulations of changes during winter in soil mineral nitrogen in soils from seven sites for five seasons.

During these investigations of net mineralization, hardly anybody seems to have asked how large the gross rates of mineralization and immobilization were. This was partly because very few people had thought of the question and partly because the techniques for answering it were not in place. Methods have been developed recently for measuring gross mineralization through a pool-dilution approach in which the soil ammonium pool

Box 3.1. A 'pool'.

The term 'pool' is frequently used in the description of flows of organic matter in the soil. It may help to think of an ornamental water garden, in which water enters the highest pool and then flows gently down through a series of pools to the bottom of the system. During this flow some water may be immobilized in a dead-end pool. When the water reaches the bottom of the system, some leaks out, but the rest is re-circulated to the top. If you add dye to one of the pools, you can trace the progress of water from that pool through other pools below it.

For nitrogen in the soil, Jenkinson *et al.* (1985) provided a formal definition of a pool as, 'A compartment containing material that is chemically indistinguishable and equally accessible to plants (or to the soil population)'. This is an exact definition for pools with a clear chemical identity such as ammonium, nitrate or amino acids, but some pools, such as the 'easily decomposable' pool in the SUNDIAL model (Smith *et al.*, 1996a), have a more diffuse identity. The definition can also be applied to nutrients such as sulphur or phosphorus.

The water garden is then an analogy for flows of nitrogen in the soil. Nitrogen enters the system in plant debris and moves through a series of pools with various degrees of chemical identification until it reaches its most oxidized form, nitrate. On the way, some flows into an inert pool of organic matter, where it stays. Some of the nitrogen that becomes nitrate is lost by leaching or denitrification, but the rest is taken up by plants that grow and eventually die. The nitrogen they contain enters the system again in plant debris. If you add ^{15}N-labelled material to one of the pools, you can trace the progress of the material from that pool through subsequent pools. The ^{15}N compound you add must, of course, be chemically indistinguishable from the material in the pool.

is labelled with ^{15}N (e.g. Murphy *et al.*, 1998, 1999). (See Box 3.1 for an explanation of a 'pool' and Box 3.2 for information on ^{15}N-labelling.) The pool initially contains ^{15}N at natural abundance, and adding ^{15}N increases the ^{15}N-enrichment. The enrichment and the size of the ammonium pool are then measured as mineralization releases ammonium at natural abundance and dilutes the isotopic abundance of the pool. Gross mineralization can be inferred from the change in ^{15}N abundance over a few days. These new methods have proved valuable, but need to be treated with care because they involve assumptions, such as perfect mixing between labelled and unlabelled N, that may not be fully met.

Murphy *et al.* (1999) investigated three ways of labelling the pool – adding a mixture of ^{15}N-labelled ammonia and air to the head-space above the soil, injecting the mixture into the soil and adding a solution of ^{15}N-labelled ammonium sulphate to the soil surface. The injection method and the application of the solution gave broadly similar estimates of gross mineralization and showed up the same differences between land uses. The head-space method gave smaller estimates. The rates of gross mineralization were much larger than would be expected from the normal rates of net mineralization. They found gross mineralization rates of 2–3 mg N/kg dry soil per day, roughly 5 kg/ha per day, which must have been countered by comparable rates of immobilization.

Box 3.2. ^{15}N-labelling.

An atom consists of a nucleus surrounded by electrons. The electrons are negatively charged, and their number is the 'atomic number' for the element. The nucleus comprises positively charged protons and uncharged neutrons. The number of protons must be the same as the number of electrons and the number of neutrons is usually, *but not always*, the same as the number of protons. Isotopes of an element are atoms which have the same number of protons but differing numbers of neutrons.

All the isotopes of an element have the same electronic configuration (arrangement of electrons in orbits around the nucleus) and therefore the same chemical properties. Some elements have only one isotope, while others have more. Some isotopes are radioactive. They emit radiation but the strength of the radiation diminishes with time (radioactive decay). This is helpful because the radiation can readily be measured to tell you, if you allow for the lapse of time, how much of the isotope and therefore of the element is present in a sample. But it adds to the problems of the experimenter because radiation is hazardous and there are stringent safety regulations surrounding its use. Other isotopes are not radioactive and are described as stable, but there is no radiation that can be measured and detecting the isotope is much more difficult.

The 'atomic weight' is the sum of the number of protons and the number of neutrons and is often twice the atomic number. Nitrogen is the seventh atom in the periodic table and has an atomic number of 7. Its commonest isotope, ^{14}N, has (as the superscript implies) an atomic weight of 14. Measurements of the atomic weight of nitrogen give a value of 14.008, implying that nitrogen must have a heavier isotope. In fact, nitrogen has both a heavier and a lighter isotope, ^{15}N and ^{13}N, but the former is more common. The ^{13}N isotope is radioactive and ^{15}N stable. Their natural relative abundance is well known, and 0.3663% of all nitrogen atoms are ^{15}N.

The time scale of most agricultural or environmental experiments means that ^{13}N decays far too rapidly to be of use, but it is being used in specialized short-term studies of nitrogen uptake by plants. However, ^{15}N has been a mainstay of nitrogen research for many years. Its stability means that it is not subject to safety legislation and the main limitations on its use are cost of the ^{15}N, which is not usually excessive, and the cost of the equipment and labour needed to measure it, which can be more of a problem. The only thing that differentiates ^{14}N and ^{15}N is atomic weight, and the usual method of discrimination between them is the mass spectrometer, which measures atomic or molecular weight directly. The method is entirely safe, reliable if great care is taken, and tedious. UV spectroscopy has also been used to discriminate between the two isotopes.

Reference was made to this box when the labelling of the soil ammonium pool with ^{15}N was discussed. What this means is that a supply of an ammonium salt has been obtained in which the ammonium ions have been enriched to a specified extent with ^{15}N above the natural abundance of 0.3663 atom%. A fixed amount of this ^{15}N-enriched ammonium has been added to the (chemical) pool of ammonium in the soil, so the soil ammonium pool has been 'labelled' with ^{15}N. The progress of the ^{15}N from this into other pools is followed by sampling those pools and determining the ratio ^{15}N of to ^{14}N. The process is somewhat akin to the tagging of birds and animals. Labelling with ^{15}N has proved invaluable in studies of the fate of fertilizer nitrogen, as we shall see in Chapter 5.

No generalizations can be made about the relation between gross mineralization and net mineralization, except that the latter obviously cannot exceed the former. Net mineralization can be negative, because immobiliza-

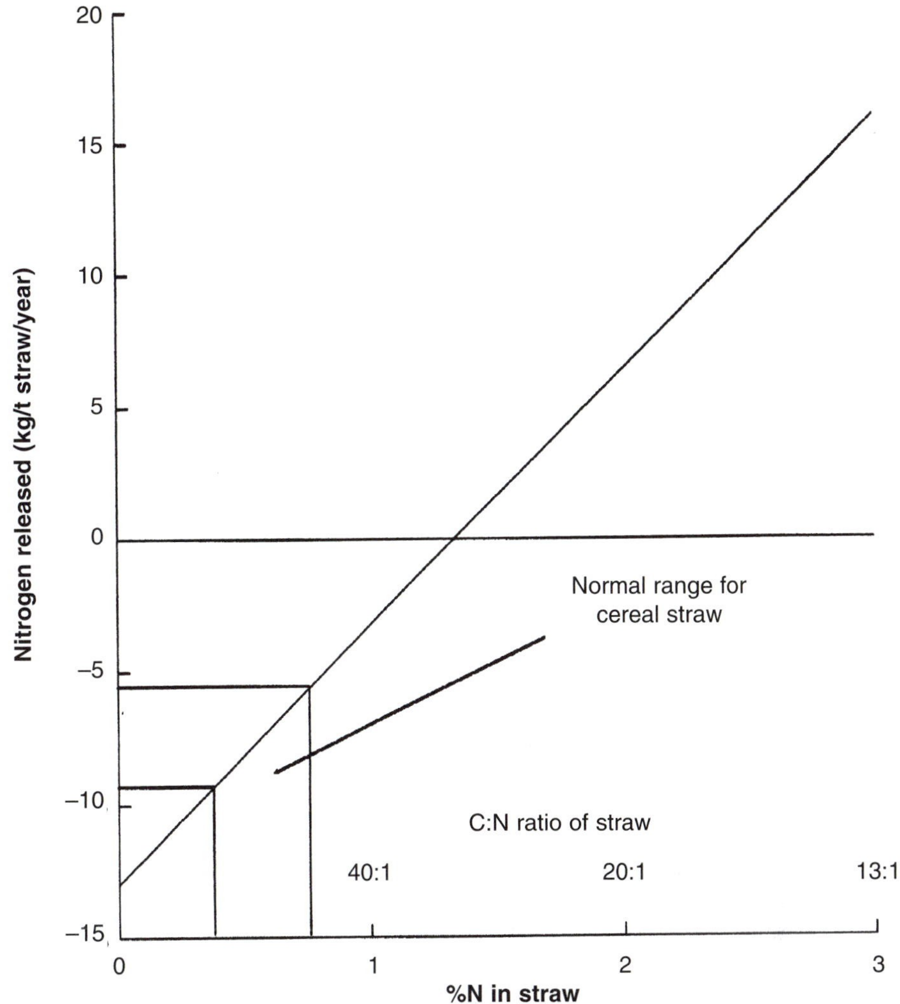

Fig. 3.2. The effect of %N in the straw and thence its C:N ratio on the release of nitrogen from it. A positive release means mineralization and a negative one immobilization. (From Jenkinson *et al.*, 1985.)

tion may exceed mineralization in some circumstances. Immobilization is often caused by the addition to the soil of cereal straw or other material with a high C:N ratio, and this ratio provides a guide as to whether mineralization or immobilization is likely to predominate in a particular soil (Fig. 3.2). Catt *et al.* (1998b) investigated whether straw incorporation was an effective way of immobilizing nitrate and thereby lessening losses by leaching but their results showed no consistent benefit.

Should one measure net or gross immobilization? It depends very much on the purpose of the measurement. If, for advisory purposes, you want to estimate the amount of nitrogen likely to be mineralized in the coming cropping season, a net measurement should be adequate. There is no need for the much greater trouble and expense of a gross measurement. But measuring gross mineralization can tell you far more about the nature of the microbial processes involved. In particular, it can give information on the relation between the nitrogen and carbon cycles in the soil. It is one of the more exciting areas of soil research at the moment.

Denitrification and Related Processes

Nitrogen is the most ephemeral of the major plant nutrients. Nearly 50 years ago, the American soil scientist F.E. Allison published a noted review paper entitled 'The enigma of soil nitrogen balance sheets' (Allison, 1955). The enigma was the failure of so many experiments to balance inputs and outputs of nitrogen to and from the soil. All these experiments showed losses which could not be accounted for by leaching alone. His paper drew attention to denitrification and other processes that release nitrogen gases from the soil. Chemical denitrification is well known in the spontaneous decomposition of ammonium nitrate, in which the nitrate moiety oxidizes the ammonium moiety (Box 4.1). But this would need totally dry soil for it to occur, and the denitrification with which we are concerned happens when the soil is moist and microbes are involved.

We saw above that aerobic microbes need oxygen as a terminal acceptor of electrons. Facultative aerobes can use other molecules, including nitrate, for this purpose. Nitrate is satisfactory as an electron acceptor because it is the most fully oxidized compound of nitrogen, and the reduction of nitrate is the first in a sequence of redox reactions to occur as the soil becomes more anaerobic (Table 3.2). It is carried out mainly by bacteria of the *Pseudomonas* and *Bacillus* species. The equation in Table 3.2 shows the reduction of nitrate proceeding right through to the formation of nitrogen gas, strictly dinitrogen N_2. If this happens, the nitrogen is added to the 78% of nitrogen already in the atmosphere, which is no problem at all. Nitrate has been lost from the pool of plant-available nitrogen, but this is usually a minor problem, except perhaps for paddy rice grown in flooded conditions.

Incomplete denitrification is a major problem. Nitric oxide (NO) and nitrous oxide (N_2O) are formed. The former is a free radical and reacts readily with other free radicals in the atmosphere. Denitrification usually releases only small amounts of nitric oxide from the soil but, when released from larger sources, the gas pollutes the atmosphere by catalysing the formation of photochemical smog and contributing to the formation of ozone in the lower atmosphere (Cotgreave and Forseth, 2002). Nitrous oxide is formed in appreciable amounts and is one of the 'greenhouse gases' as well as facilitating the destruction of the stratospheric ozone layer (Chapter 7).

Denitrification used to be of agronomic concern whether dinitrogen or nitrous oxide was formed, because the nitrate lost would need to be replaced by fertilizer. Today the concern is environmental and it involves

Table 3.2. Redox reactions in the soil. (After White, 1997.)

Redox reactions occur in the following sequence as the soil becomes more anaerobic:

$2NO_3^- + 12H^+ + 10e^- \leftrightarrow N_2(gas) + 6H_2O$

$2MnO_2(solid) + 4H^+ + 2e^- \leftrightarrow 2Mn^{2+} + 2H_2O$

$Fe(OH)_3(solid) + 3H^+ + e^- \leftrightarrow Fe^{2+}$

$SO_4^{2-} + 10H^+ + 8e^- \leftrightarrow H_2S(gas)$

Compounds or ions not marked as being in the gas or solid phase are in the liquid or solution phase.

the formation of nitrous oxide and its loss to the atmosphere. We are concerned in this chapter with the processes themselves. The assessment of nitrous oxide losses from large areas of land is discussed in Chapter 5, and the role of nitrous oxide in the atmosphere is taken up in Chapter 7.

Nitrous oxide can also be formed from nitric oxide produced during nitrification because the nitrite (NO_2^-) formed as an intermediate is not stable (Bremner, 1997). This is a key point because the processes of nitrification and denitrification are not subject to the same controls. Denitrification arises from anoxic conditions, nitrification does not. Nitrification uses ammonium as a substrate, while denitrification obviously needs nitrate. It follows that nitrification is the main source of nitrous oxide when ammonium or ammonium-producing (e.g. urea) fertilizers are used, while denitrification predominates with the application of nitrate fertilizers (Bremner, 1997).

The ratio of nitrous oxide to dinitrogen formed depends on the temperature and pH of the soil. At low soil temperatures and pH < 5, at least as much nitrous oxide as dinitrogen is emitted, but at temperatures of 25°C or above and pH > 6, the emissions are nearly all as dinitrogen. Water-filled pore space influences oxygen supply and thence denitrification, and several authors (e.g. Davidson, 1991) have suggested that nitrification was the dominant source of nitrous oxide when this was less than 60%, while denitrification dominated when it was more than 60%.

Denitrification, oxygen and water

The bacteria that implement denitrification are facultative anaerobes whose activity depends on the temperature and the degree of anoxia in the soil. The movement of oxygen in soil is frequently impeded by water in its pores, because oxygen diffuses about 10,000 times more slowly in water than in air. Thus if, as will usually be the case, direct measurements of soil oxygen are not feasible, the percentage of the pore space filled by water provides a useful surrogate measurement of anoxia. But note that anoxia can be caused by enhanced demand for oxygen as well as an impeded supply.

Can denitrification be estimated from soil nitrate, temperature and percentage water-filled pore space? Smith *et al.* (1998) initially found poor correlations, apparently because different factors were limiting denitrification at different times. However, if they included only the sampling events on which two of the three factors were not limiting, the correlations with the third factor became significant. With this observation in mind, Conen *et al.* (2000) developed a model for nitrous oxide emissions from the soil based on what they described as a 'boundary line' approach. This had a threshold value of 10 mg/kg (dry soil) for nitrate below which no emissions occurred. The other two factors were handled by producing a graph with water-filled porosity on the *y*-axis and temperature on the *x*-axis. On this were plotted the nitrous oxide emissions corresponding to combinations of these factors. The resulting pattern enabled two parallel boundary lines to be superimposed. These separated regions in which the emissions were < 10, 10–100 and > 100 g N_2O-N/ha per day. Figure 3.3 shows the boundary lines in relation to the water-filled pore capacity and the temperature but without the data.

This procedure provided an empirical model that could be used to estimate emissions of nitrous oxide from soils, presumably on the assumption that the nitrous oxide came from incomplete denitrification rather than nitrification. The model was neither mechanistic nor exact and the simulations it gave of experimental data were far from perfect, but this is not surprising given that the measurements on which it was based were not made for the purpose. Its great advantage was that it was simple and its workings illustrate clearly the main factors involved in denitrification. A somewhat more mechanistic, but still relatively simple model is that of Rolston *et al.* (1984), which is described in the context of the problem of estimating emissions of nitrous oxide from large areas of land in Chapter 5.

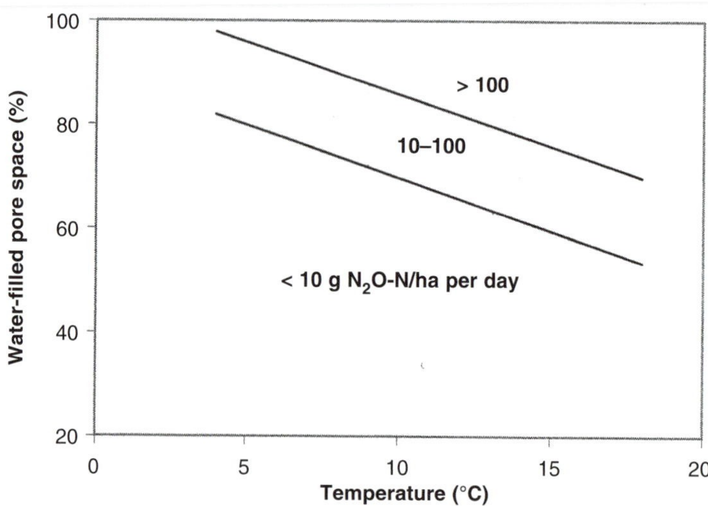

Fig. 3.3. The boundary line approach to emissions of nitrous oxide from soil. For clarity, this diagram shows boundary lines without the data on which they were based. (From Conen *et al.*, 2000.)

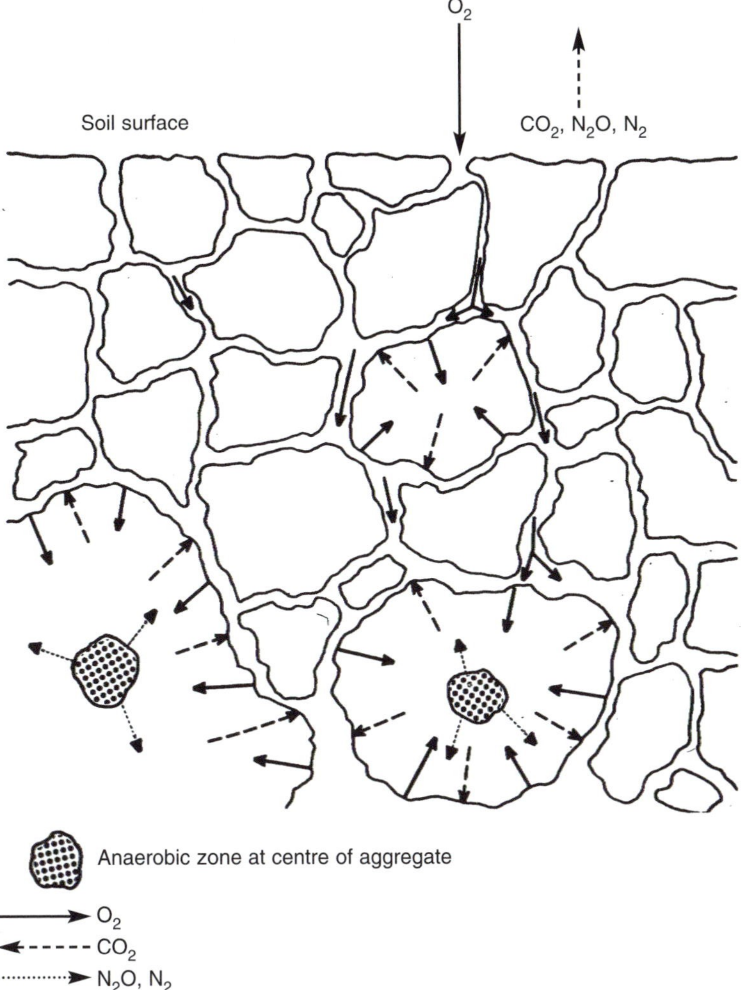

Fig. 3.4. Denitrification in anoxic zones in peds. Oxygen limited by intra-ped diffusion. (From Smith, 1980.)

Even when Conen *et al.* (2000) made allowance for the effects of nitrate, water-filled porosity and temperature, a considerable amount of apparently random nitrous oxide emission remained. This was probably not random but due to localized anoxia caused by the structure of the soil or by 'hot spots' for denitrification (Parkin, 1987). The soil structural effect is likely to arise in heavier, more clayey soils. These tend to have the largest peds (structural units) in which oxygen has to diffuse furthest. As a result, anoxic zones form near the centre of the ped (Fig. 3.4) and denitrification occurs. Smith (1980) described a model for this phenomenon.

Anoxic zones can also occur where there is a strong demand for oxygen that is not met by diffusion. Such a demand could arise from the presence of a substantial fragment of easily decomposed plant material, which could then become the centre of a hot spot for denitrification (Parkin, 1987). Anoxic zones in peds and hot spots both imply great variability in rates of denitrification even in small volumes of soil. As with mineralization and immobilization, nitrification and denitrification can occur simultaneously only a short distance apart in the soil, aerobic and anaerobic processes almost side by side.

The process is not just variable in space; it also varies with time. It happens as a result of a particular set of circumstances and is therefore episodic in nature (e.g. Jarvis *et al.*, 1991). Both forms of variability make it difficult to generalize about the scale of the losses. This is illustrated by some results of Dobbie *et al.* (1999), who reported annual emissions of nitrous oxide ranging from 0.3 to 18.4 kg N_2O-N/ha in a series of experiments run for 3 years with five crops at sites in southern Scotland that had widely varying rainfall and soil types. Total denitrification losses would, of course, have been larger, possibly up to 30–40 kg N/ha. The authors attributed the variations mainly to the number of occasions on which nitrogen fertilizer application was followed by substantial rainfall.

Nitrogen Fixation

Nitrogen gas is chemically inactive. This is just as well, because it makes up 78% of the atmosphere, and life as we know it would not be possible if we were surrounded by a more reactive gas. Imagine living in an atmosphere of chlorine! This inactivity means that nitrogen gas is not directly accessible to plants either, depriving them of a plentiful supply unless something intervenes to help them. That 'something' proved rather elusive when it was sought.

Legumes

Boussingault had shown in 1838 that clover and peas grown in sterilized sand increased the amount of nitrogen in the sand. This was something that crops such as wheat and barley could not do. The legumes must have obtained the nitrogen from the atmosphere, but neither Boussingault nor anyone else could explain how they did so. Lawes and Gilbert showed that the 'Garden Clover' plot at Rothamsted, a small plot of clover which is still maintained in the garden of Rothamsted Manor, obtained huge quantities of nitrogen from somewhere. Around 280 kg/ha were harvested annually in the clover. This was far more than was in the seeds and other inputs, so nitrogen had to be coming from elsewhere. They thought that the nitrogen possibly came from large stores of reactive nitrogen in fertile soil, but if they thought that the clover was using atmospheric nitrogen they did not follow it up, presumably because no one knew how it did so.

Nodules

Curiously enough, neither Boussingault nor Lawes and Gilbert made the connection between this unexplained capacity of leguminous crops to obtain nitrogen and the nodules which could be seen quite clearly on their roots. The issue was not to be resolved until 1886–1888, when two German scientists, Hellriegel and Wilfarth, demonstrated in a series of experiments that the capacity to fix nitrogen involved microbes in the nodules operating in a symbiotic relationship with the plant. A symbiotic relationship is a biological arrangement in which both parties benefit. The plant received nitrogen fixed from the atmosphere, and the microbes in the nodules received a share of the carbohydrates produced by photosynthesis.

But nitrogen gas is dinitrogen, N_2, in which the two nitrogen atoms are held together by a triple bond, N≡N. This is a very powerful bond that needs a great deal of energy to break it in the industrial Haber–Bosch process (Chapter 4), appreciably more than is available in most biological processes. How do the *Rhizobium* microbes in the nodules manage it? They manufacture the *nitrogenase* enzyme, which is one of the very few enzymes able to catalyse the cleavage of a triple bond with the amount of energy available in soils in the field. This enzyme can also catalyse the cleavage of other triple bonds such as those in acetylene (H–C≡C–H) and cyanide (–C≡N).

Nitrogenase is readily deactivated by oxygen, so the microbe has to maintain anaerobic conditions around the enzyme for it to fix nitrogen from the air. This must be one reason why the higher plants failed to evolve the capacity to fix nitrogen. The root nodule provides an environment in which the microbes can control the oxygen concentration around the enzyme while receiving the energy from the plant to do so. Its position also enables it to feed the fixed nitrogen into the same stream as nutrients taken up by the root.

Could the benefits be spread?

The bacteria, the nodules and their plant hosts still attract interest from researchers. The legumes are currently the only plants capable of forming the associations through which nitrogen is fixed and there are obvious benefits to extending this capacity to other plants. However, the specificity of symbiotic relationships means that, even with genetic modification as a tool, we are still a long way from being able to develop a cereal crop able to form a relationship with a bacterium through which it could obtain fixed nitrogen. A cereal crop with this property would be invaluable in developing countries where nitrogen fertilizer is often too expensive for smallholder farmers (Chapter 11), at least partly because of transport costs. It would also save immense amounts of energy in the developed world at a time when we are becoming increasingly aware of the limits to oil and gas reserves. Such a crop would have some limitations, but they are trivial compared with the potential benefits. The bacteria in the nodules use carbohydrate that the crop has photosynthesized and the crop would therefore not yield as much grain as a crop grown with fertilizer. Also the fact that the crop was grown without fertilizer would not mean that no nitrate was lost to the environment!

Our existing nitrogen-fixing crops make a major contribution to nitrogen supply in the temperate zones, particularly in grassland farming. The Canterbury Plains in the South Island of New Zealand depend almost entirely on nitrogen fixed by clover and are highly productive. And in hilly or other areas where the terrain is difficult, clover is often a more practical proposition than nitrogen fertilizer. In arable systems, peas and beans are important crops in their own right and also provide 'break crops' for limiting the transfer of disease from one cereal crop to another. Also nitrogen residues from legumes become available through mineralization to the following crop or to crops grown with the legumes in inter-cropping systems (systems in which the legume and non-legume crop are grown side-by-side in alternate rows). Organic farming systems depend for their nitrogen supply on legumes and atmospheric deposition. But where the legumes really come into their own is in the developing world. For large areas of sub-Saharan Africa, and in many other places, they are unquestionably the main source of nitrogen. Beans, soya and groundnuts have a vital culinary and economic role in the lives of hundreds of millions of people.

Is Nitrate the Preferred Form of Nitrogen for Plant Uptake?

It has long been assumed that nitrate is the main form of nitrogen taken up by plants but that most plants can also take up ammonium. This was so widely accepted that the present section would not have been added were it not for the topic recently becoming a matter of controversy. Perakis and Hedin (2002) reported that nitrogen was lost as dissolved organic compounds in stream waters flowing from pristine South American forests but mainly as inorganic nitrate in streams flowing from North American forests that suffer deposition of nitrogen oxides. This led van Breeman (2002) to claim that the uptake of dissolved organic nitrogen by plants was the norm in unpolluted ecosystems and that nitrate uptake was an adaptation to pollution. He suggested that the standard thinking about how nature deals with nitrogen needed to be re-evaluated. This in turn led *New Scientist* to publish an article entitled 'Botched Botany' (Pearce, 2002) which berated scientists for getting the nitrate story wrong all along.

Addiscott and Brookes (2002), however, pointed out that the presence of dissolved organic nitrogen in the pristine streams did not prove that nitrogen of this kind was that taken up by plants, particularly as 'dissolved' was defined as passing though a filter of 1 μm pore size and would have included molecules up to 1000 larger than those likely to have been taken up by plants, together with colloidal organic matter and bacteria. They suggested that a forest ecosystem with no nitrogen inputs would evolve to recycle all usable nitrogen, inorganic or organic, and minimize its loss in streams. And the dissolved organic nitrogen was in the streams not because it was what the plants took up, but because it was what they were unable to use. No re-evaluation of our current thinking about nitrogen uptake seems necessary at the moment. (And was 'Botched Botany' perhaps a case of botched journalism?)

How nature deals with nitrogen depends greatly on temperature. Ryegrass plants given equal concentrations of ammonium-N and nitrate-N took up an increasing proportion of ammonium as the temperature became cooler (Clarkson and Warner, 1979). Recent research in plant physiology, reviewed by Williams and Miller (2001), has shown that plants are equipped with transport mechanisms for a variety of nitrogen-containing organic solutes. And field research in northern forests with cold temperatures (Persson and Nasholm, 2001) has shown that plants can absorb small organic molecules such as amino acids. This flexibility may have evolved simply because cold unfertilized soils do not contain much nitrate. Nitrifying bacteria become less active as the temperature declines and are inhibited at 3–5°C, leaving ammonium and small organic molecules as the main source of nitrogen. The ammonifying bacteria are probably inhibited at around 0°C, depending on the genera involved.

Plants in the South American forests studied by Perakis and Hedin (2001) may well have taken up some small organic nitrogen molecules, because the temperatures were quite low (4–11°C). But this does not require a re-evaluation of the standard thinking about how nature deals with nitrogen. My guess, based on the evidence above, is that plants generally use nitrate but as the temperature declines they increasingly take up first ammonium and then amino acids and other small nitrogen-containing molecules.

4 Nitrogen Fertilizer

Early Sources of Nitrogen Fertilizer

Gas liquor and ammonium sulphate

The year 1843 saw not only the foundation of Rothamsted Experimental Station but also the start of Lawes's commercial operations. His first advertisement was placed in the *Gardener's Chronicle* of 1 July 1843. It can be read in full in Box 4.1, but note here the sentence which reads, 'The Super Phosphate of Lime alone is recommended for fixing the Ammonia of Dung-heaps, Cesspools, *Gas liquor*, &c.' (italics added).

Gas liquor was a by-product when coal was heated in the absence of air to produce coal-gas. Because coal contains 1–2% N, mainly from protein, the raw gas contained ammonia, and gas liquor was obtained essentially by scrubbing this gas with water. It contained a variety of other chemicals as well as ammonia (Table 4.1), some of them disinfectants but some known or possible carcinogens. According to the *Shorter Oxford English Dictionary*, the first recorded use of the term 'coal-gas' was in 1794. This suggests that gas liquor was the earliest source of *chemical* nitrogen fertilizer (as opposed to *organic* manures) and was available before 1800, but there is no evidence as to how soon it was used as a fertilizer. The smell of gas liquor should have indicated that it contained ammonia, and my guess is that it was beginning

Box 4.1. Lawes's first advertisement.

J.B. LAWES'S PATENT MANURES, composed of Super Phosphate of Lime, Phosphate of Ammonia, Silicate of Potass. &c., are now for sale at his factory, Deptford Creek, London, price 4s. 6d. per bushel. These substances can be had separately; the Super Phosphate of Lime alone is recommended for fixing the Ammonia of Dung-heaps, Cesspools, Gas Liquor, &c. Price 4s. 6d. per bushel.

to be used as a fertilizer by 1810–1820, initially by those living fairly close to a gas-works. In one of his very earliest experiments, reported in 1842 (Dyke, 1991, Addenda, p. 1), Lawes tested the effects of gas liquor neutralized with phosphate of lime or phosphoric, sulphuric, muriatic (hydrochloric), nitric or carbonic acid. This experiment must have been made in 1841 or probably rather earlier. Lawes's advertisement, cited above, suggests that gas liquor would have been familiar to readers of the *Gardener's Chronicle* in 1843.

A considerable amount of the ammonia in gas liquor was probably wasted. As Lawes pointed out (Dyke, 1991), most gas liquor was made in winter, when gas-fires and domestic and street gas-lights were used most, but the ammonia in the gas liquor could not be applied safely until after the winter was over and the leaching risk abated. He suggested various ways in which it could be used, including conserving it as ammonium salts. Not only could these be used when needed, they were also easier to transport than gas liquor. Another advantage was that when the ammonium salt was crystallized from the gas liquor, the less desirable constituents of the liquor would have been left behind. (Phenol and related useful compounds might have been obtained from gas liquor (Table 4.1), but most seem to have been obtained from coal-tar, another by-product of coal-gas production.) The ammonium sulphate used in the first Broadbalk experiment in 1843 and the ammonium salts used from 1852 onwards would have been produced from gas liquor.

Gas liquor was still in use in the 1950s. I remember it being spread from a tanker on to the grass field behind my home. The passage of 50 years has not erased the memory of the appalling stench, which was probably caused as much by the pyridine and other chemicals it contained as by the ammonia (Table 4.1). The ammonium sulphate sold by the Lawes Chemical Company from its Barking plant when I worked there in 1965–1966 was still produced by the nearby Becton gas-works and sold as a straight nitrogen fertilizer (see p. 6).

From the 1860s onwards ammonium sulphate was also produced when coal was heated in coking ovens to produce coke for industrial purposes.

Table 4.1. Some likely constituents of gas liquor (other than water).

Constituent	Formula	Notes
Ammonia	NH_3	Main constituent
Phenol	C_6H_5OH	Disinfectant
Cresol	$CH_3C_6H_4OH$	Disinfectant
Pyridine	$C_5H_5\,N$	Horrible smell
Thiophen	C_4H_4S	
The following are not water-soluble but could have been in the liquor in suspension:		
Benzene	C_6H_6	
Toluene	$CH_3C_6H_5$	
Xylene	$(CH_3)_2C_6H_4$	
Cumene	$(CH_3)_2CHC_6H_5$	

Table 4.2. Comparing the long-term worldwide significance of nitrogen fertilizer from various sources. Approximate total N production, 1800–2000. (Based on Appendices F and L of Smil, 2001.)

Origin	Production (Mt N)	% of total
Gas liquor and ammonium sulphate produced from it	0.2	0.02
Ammonium sulphate from coking	56	6.2
Guano	2.4	0.27
Sodium nitrate	25	2.8
Cyanamide	19	2.1
Electric arc process	0.3	0.03
Haber–Bosch process	795	88.5
Total	897.9	100

(Coke had replaced charcoal for smelting pig-iron about 100 years before but the ammonia released in its production had been allowed to escape into the atmosphere.) Ultimately coking ovens became a much more important source of fertilizer nitrogen than gas-works (Table 4.2), but in 1900 when the UK was the world's largest producer, half the total production still came from gas-works and only one-third from coking ovens (Smil, 2001). Coking became an increasingly sophisticated process and was eventually a far more important source of ammonium sulphate than coal-gas production (Table 4.2), which ceased in the UK when natural gas came on stream in the 1970s. Production of ammonium sulphate in coking ovens continues to this day but it seems to be in decline, presumably because of the general decline in heavy industry. According to Smil (2001), world production reached a peak of 970,000 t N in 1980, but had fallen to 370,000 t N in 2000. Nevertheless, the contribution made by ammonium sulphate from coking to nitrogen fertilizer supplies has been second only to that of the Haber–Bosch process (Table 4.2).

Ammonium sulphate has the advantage that it is the by-product of the manufacture of coal-gas and/or coke, and therefore reasonably cheap. But it has two disadvantages as a nitrogen fertilizer. It contains only 21% N, which makes it expensive to transport by comparison with other sources of nitrogen (Table 4.3), and it tends to acidify the soil. However, the sulphate it contains has recently become of interest in some parts of the UK because of sulphur deficiency resulting from the clean-up of emissions from power stations.

Peruvian guano

The Chincha Islands off the coast of Peru harboured vast numbers of nesting sea-birds who fed off an abundant supply of fish and excreted copiously on to the rocks on which they perched. There was virtually no rain to wash the excrement off, so it accumulated over thousands of years. The Incas used

Table 4.3. The main fertilizers used, their chemical formulae and the %N they contain.

Fertilizer	Formula	%N
Solids		
Ammonium nitrate	NH_4NO_3	34
Ammonium sulphate	$(NH_4)_2SO_4$	21
Calcium nitrate	$Ca(NO_3)_2$	16
Sodium nitrate	$NaNO_3$	16
Urea	$CO(NH_2)_2$	46
Mono-ammonium phosphate	$NH_4H_2PO_4$	12
Di-ammonium phosphate	$(NH_4)_2HPO_4$	20
Liquids/solutions		
Liquified ammonia	NH_3	82
Ammonia in water	NH_3/NH_4OH	25
Other solutions used include:	Ammonia/ammonium nitrate	
	Ammonia/ammonium phosphates	
	Urea/ammonium nitrate	

Fertilizers may contain slightly less nitrogen than is suggested by their formulae. This is because they contain stabilizers, anti-caking agents or similar substances.

it to fertilize their crops, but the Spanish who conquered Peru did not take any back to Spain. It may have crossed their minds that their sponsors would be less than happy if presented with bird excrement rather than gold. However, the explorer von Humboldt took some back to Europe for analysis in 1804, and the analysis attracted immediate interest. Because of its origin, it was a fairly variable commodity, but it contained about 15% N (not a great deal less than ammonium sulphate) together with some phosphate and potash. Farmers in Europe and the USA soon realized its value, and demand reached 200,000 t of guano per annum in England alone in the early 1850s. Smil (2001) noted that Liebig's promotion of mineral fertilizer contributed greatly to the demand for this predominantly nitrogen fertilizer, which was somewhat ironic given his argument with Lawes (Chapter 1), but guano did contain some phosphate and potash. The Peruvian government also realized its value and nationalized the supply in 1842, just as this demand took off. Between 1840 and 1870 Peruvian guano was the world's most important commercial nitrogen fertilizer, with annual exports peaking at about 600,000 t towards the end of the 1860s. But the supply was limited, and in 1872 the Peruvian government stopped further exports and limited extractions to those for domestic use.

The best non-Peruvian source of guano was Ichaboe Island off southwest Africa, and Lawes wrote an article for the 1845 *Agricultural Gazette* in which he compared the merits of Ichaboe and Peruvian guano (Dyke, 1991). But the 300,000 t deposits were soon exhausted in the intense demand for guano and were finished 3 years after they were discovered. Other sources of guano were discovered but never again in the same quantities or of the same quality as the Peruvian guano. On a long-term basis, guano played a relatively minor role in nitrogen fertilizer production (Table 4.2), but the fever for guano was a portent of the demand to come for nitrogen fertilizer. (Much of the detailed information about guano came from Smil, 2001.)

Chilean nitrate

Like guano, Chilean nitrate was found in a very dry part of South America. Very large deposits of sodium nitrate (which had nothing to do with seabirds) were found intermittently dispersed along the arid plateau that runs for much of the length of Chile roughly 100 km inland from the Pacific Ocean. These deposits might seem surprising given the great solubility of all salts of nitrate (Chapter 2), but rainfall usually occurs there at intervals of 3 or more years and the evaporation (if there is any moisture) is considerable. Perhaps not surprisingly, very few plants or even microorganisms are found. The nitrate is also covered by two protective layers, 10–30 cm of sand, silt and rocks at the surface, followed by 0.5–2 m of cemented brittle material. The sodium nitrate mineral *caliche* is at 1–3 m. Some areas contain veins of almost pure sodium nitrate, but the quality is variable, with nitrate accounting for less than 10% of what is extracted in some areas and up to 70% in others. The protective layers had to be dug or blasted away to expose the *caliche*, before the latter could be taken to refineries. There it was crushed and mixed with hot water to dissolve out the sodium nitrate and the latter was concentrated and dried.

Exports of sodium nitrate from Chile began shortly before those of guano, in 1830, and sodium nitrate was used on the Broadbalk Experiment in 1843. These exports were, however, much larger than those of guano (Table 4.2) and lasted a great deal longer, reaching a peak between the First and Second World Wars. There was one key difference between the two materials. Guano was just a fertilizer but sodium nitrate could be used for making explosives. As we saw in Chapter 2, the nitrate ion is a powerful oxidizing agent. This meant that exports were used for both purposes, and Smil (2001) noted that by 1900 the USA was using nearly half its imports of sodium nitrate for making explosives. This dual role made sodium nitrate sufficiently important for a small war to be fought over it from 1879 to 1883 between Chile on the one side and Bolivia and Peru on the other. This was provoked when the Bolivians seized the operations of the Chilean Nitrate Company (carried on under agreement) in a province of Bolivia, and it eventually resulted in a considerable expansion of Chile's boundaries at the expense of both Bolivia and Peru.

In the end, Chilean nitrate was never exhausted as guano was. It was simply superseded by the Haber–Bosch process for ammonia synthesis. (As with guano, much of the detailed information about Chilean nitrate came from Smil, 2001.)

Other early sources

Two other sources of nitrogen fertilizer deserve a brief mention, cyanamide and the electric arc process. The final form of the cyanamide production process reacted calcium carbide with nitrogen at about 1000°C:

$$CaC_2 + N_2 \rightarrow CaCN_2 + 2C \quad (4.1)$$

The product could be used directly as a fertilizer or reacted with superheated steam to produce ammonia. Germany built the most plants for its production, many of them during the First World War. The contribution of cyanamide to long-term nitrogen fertilizer production was comparable with that of Chilean nitrate (Table 4.2).

The electric arc process builds on Priestley's observation that some form of nitrogen reaction occurs when electric sparks pass through air. The reaction is:

$$N_2 + O_2 \leftrightarrow 2NO \quad (4.2)$$

This is a reversible reaction with a small yield and it needed a lot of energy, usually hydroelectricity. It never achieved more than a few per cent of the market at the time, and its long-term contribution was almost negligible (Table 4.2).

The Haber–Bosch Process for Ammonia Synthesis

During a debate in the British House of Commons in 1917, Sir William Pierce described the Haber–Bosch process for the synthesis of ammonia as 'one of the greatest achievements of the German intellect during the [First World] war' (Smil, 2001). Eighty-five years on, the synthesis remains one of the greatest achievements of chemistry anywhere at any time. As the world population expands, so does its importance. Indeed, Smil argued that as a technical invention the process rates higher in importance to the human race collectively than aeroplanes, nuclear energy, space flight, television or computers, and I support this view. Yet, as Smil noted, just a few generations later surprisingly few people know of the discovery, and even fewer are aware of its fundamental and steadily increasing importance for modern civilization. I learnt of the Haber process at school, but this may have been because I had an exceptional chemistry teacher (Box 4.2), and I am certain that Smil's comment is mainly correct. It would be interesting to compare the number of people in the UK who have heard of Haber with the number who have heard of Liebig, both Germans. I suspect that, even though Liebig was two generations further back in time than Haber, his name would be better known than Haber's.

The process is more correctly described as the Haber–Bosch process, because, while the initial discovery was Haber's, the process would not

Box 4.2. Chemical denitrification and explosions.

Chemical denitrification is almost certainly negligible by comparison with its biological counterpart in the soils of the moist temperate zones, but it might be more important in soils subject to strong wetting and drying. Its main impact, all too literally, has been in explosions. This box is essentially a tribute to the best teacher I had at school or university, S.B.C. Williams of Berkhamsted School, who taught me, among many other things, about gas-works and the Haber process. What made his chemistry lessons particularly memorable was that he had worked in the explosives industry during the Second World War. This gave his lessons an additional dimension, both in the experiments he showed us and the stories he told us. I hope that, after about 45 years, I have remembered the following anecdote clearly. I am not certain, but I think he was directly involved.

'SBC' and his colleagues were aware of the explosive power of the chemical denitrification that occurred when the nitrate moiety of ammonium nitrate oxidized the ammonium moiety, and they knew that the reaction was catalysed by dust. This suggested to them that metal powders might be suitable catalysts, and they tested them in an old-fashioned cannon. They put fixed charge of ammonium nitrate into the cannon with each of the metal powders in turn, fired the cannon and measured the distance travelled by the cannonball. Aluminium powder was determined to be the most effective because, when they tried it, they lost the cannonball. To the best of my knowledge, this slightly eccentric experiment was responsible for Ammonal, one of the most effective explosives used in the war.

The explosiveness of ammonium nitrate was, sadly, demonstrated all too powerfully in the Oklahoma bombing, the disasters at fertilizer plants at Oppau in 1992 and near Toulouse in 2002, and in the worst industrial accident the USA has ever had. A French ship just loaded with 2300 tons of ammonium nitrate blew up at Texas City on Galveston Bay in 1947, probably as a result of a carelessly thrown cigarette end. The blast caused explosions at nearby oil refineries, damaged or destroyed thousands of buildings at Texas City, rattled others hundreds of miles away, and knocked aircraft from the sky. The ship's anchor, weighing 3000 lb (1.36 t) was blown a distance of 2 miles (3.2 km).

Needless to say, when ammonium nitrate is sold as fertilizer, a stabilizer is added, mainly to forestall its use for terrorist purposes.

have become an industrial reality without the organizational and managerial skills of Bosch. I shall give a brief account of the process below, but for a full account of the discovery I can unreservedly recommend the book *Enriching the Earth: Fritz Haber, Carl Bosch and the Transformation of World Food Production* by Vaclav Smil, mentioned in the Foreword. Plenty of information on the process can also be found on the World-wide Web. Just enter 'Haber process' and search.

Synthesis of ammonia

The principle of ammonia synthesis is simple. You just attach three hydrogen atoms to a nitrogen atom. The practice, which involves finding the best combination of temperature and catalyst for the reaction, is immensely more difficult. Haber began work on the problem in 1904 and by 1908 he had progressed sufficiently far for the German chemical company BASF to guarantee Haber 6000 marks per annum to use as he saw fit in his nitrogen fixation research (Smil, 2001). It is interesting to note in passing that BASF was far more eager to support the electric arc process for combining nitrogen and oxygen, and they gave Haber and his assistant König 10,000 marks annually for this purpose. Haber commented in his Nobel lecture that the company made every effort to facilitate his work on the electric arc process, but approved his work on the high-pressure synthesis of ammonia 'only with hesitation'.

Haber's two initial tasks were to select a suitable catalyst and to devise a laboratory apparatus for synthesizing ammonia at suitably high temperatures and pressures. He settled on osmium as the catalyst and patented its use for the process in 1909, after which he set up a laboratory-scale process that became operational in the same year. A definitive patent, obtained in 1911, was for 'Process for production of ammonia from its elements under pressure and elevated temperature, whose characteristic is that the synthesis takes place under very high pressure of about 100 atm, but to be useful at 150–200 atm or higher'. BASF was very nervous of the high pressures involved but Bosch, who was in charge of the company's nitrogen fixation research, convinced them that steel suitable for the high-pressure reactor would be developed, and the revolution in the manufacture of nitrogen fertilizer was underway, Bosch leading the industrial application of Haber's discovery.

It is an unfortunate reflection on human nature that the first consequences of the discovery and industrial development of the process were not agricultural but military. It contributed greatly to Germany's war effort in the First World War from May 1915 onwards and also throughout the Second World War, in which it delayed the collapse of the Third Reich. This was particularly ironic given the Nazis' treatment of Haber (Box 4.3). Its impact on agriculture did not come until much later (Smil, 2001). It had little influence on farmers' use of nitrogen fertilizer during the first half of the 20th century, and fertilizer consumption worldwide in 1950 was only about 5% of what it was to be in 2000.

The work of Haber and Bosch was so well founded that the fundamentals of the process have changed little since their time. What has changed, however, is its energy efficiency. Today's methods of production are the result of a long series of improvements from 1915 onwards aimed at maximizing the recycling of heat energy and minimizing losses. Efficiency has improved a great deal, from nearly 100 GJ/t of ammonia initially to 30–40 GJ/t in 2000. A large proportion of ammonia plants now use natural gas as the source of hydrogen, and therefore tend to be situated near primary supplies of gas (Lowrison, 1989; Smil, 2001).

There are a number of variants of the process and details seem to differ according to the author consulted. This account of the process is based upon that of Smil (2001). The natural gas (methane) is filtered to remove dust or liquid particles and sulphur from dihydrogen sulphide is removed by

Box 4.3. Haber and Bosch after the breakthrough.

It was unlikely that Haber and Bosch would ever again to achieve anything as exciting and important as the synthesis of ammonia and the establishment of the process for manufacturing it, but what happened to them was far worse than anticlimax.

The least of their problems was that the fruits of their work formed part of the Versailles Treaty which Germany was forced to sign in 1919. Their lives' work was essentially handed over to the victors. Under the treaty, BASF had to license the construction of ammonia synthesis plants in the UK, France and the USA (Smil, 2001). The UK plant was at Billingham and was eventually owned by ICI. The French plant was at Toulouse (Smil, 2001), and it was at this plant that the disastrous ammonium nitrate explosion occurred in 2002 (Box 4.2). The US plant was at Sheffield, Alabama (Smil, 2001) and must have been the source of the ammonium nitrate in the ship that blew up at Texas City in 1947 (Box 4.2).

Smil (2001) added a postscript to his book which described the lives of Haber and Bosch after their great achievement. Haber's experiences could almost be made into an opera. The great determination he had shown in developing the synthesis was stimulated by intense German patriotism. Once his input into the synthesis was complete, the conscientious Haber proceeded to war work of a more controversial nature. This involved the development of gas weapons including chlorine, which by the end of the war caused about 1.3 million casualties.

Haber was awarded the Nobel Prize in 1918 for his work on ammonia synthesis, but the prize aroused much controversy because of his work on gas. Also perhaps because of the work on gas, his wife shot herself with his army revolver in 1915. He remarried but suffered greatly from depression, which eventually led to a divorce in 1927. Worse was to come when Hitler came to power in 1933 because Haber was from a Jewish family. His war work for his beloved Germany and his Nobel Prize counted for nothing and he had to resign his institute directorship that year. Perhaps mercifully, he died in 1934.

Bosch fared better, playing an important and honoured role in the redevelopment of the German chemical industry in the 1920s and receiving his Nobel Prize in 1932. He did his best to protect Jewish scientists from the Nazis, but he too suffered increasingly from depression, in his case about the fate of Germany under Hitler, and died in 1940.

If, as I trust, there is an afterlife, I hope that Herr Hitler is being made eternally aware that the war effort of his 'Glorious Reich' depended in no small measure on Haber, who was one of the Jews he so much despised and whose lives he destroyed.

passing it over zinc oxide. It is then subjected to a process described as steam reforming in which the 'reforming' is essentially a reorganization of the molecules involved. The methane from the natural gas and superheated steam react over a nickel catalyst at 750–850°C to give carbon monoxide and hydrogen:

$$CH_4 + H_2O \rightarrow CO + 3H_2 \qquad (4.3)$$

This is followed by a secondary reforming process in which any unconverted methane is oxidized in an exothermic reaction at 1000°C:

$$CH_4 + 2O_2 \rightarrow CO_2 + 2H_2O \qquad (4.4)$$

At this point too the nitrogen N_2 is adjusted to give the appropriate 3:1 N_2:H_2 ratio. The mixture is then passed through a heat exchanger to cool it and the carbon monoxide and then the carbon dioxide removed over catalysts. It is then raised to the required pressure, stripped of any moisture using molecular sieves (to protect the catalyst) and then reacted at 400–450°C over a metallic iron catalyst based on magnetite (Fe_3O_4) with catalyst promoters of aluminium oxide, potassium chloride and calcium:

$$N_2 + 3H_2 \rightarrow 2NH_3 \qquad (4.5)$$

Other sources of carbon and hydrogen, such as naphtha, oil and coal, can be used but give poorer energy efficiency than natural gas. What is used to provide energy for the process depends very much on what is available. Table 4.2 shows the extent to which the Haber–Bosch process has dominated the fertilizer industry.

Use of ammonia as a fertilizer

Ammonia can be used directly as a fertilizer. It is liquidized under pressure and then injected directly into the soil using equipment drawn by a tractor. A blade at the front cuts a slit in the soil, into which the ammonia gas is injected through a nozzle. This is followed by a metallic 'sole' which is drawn over the slit to close it up to prevent the ammonia escaping. Once in the soil, the ammonia tends to reach equilibrium with the soil water and most of it becomes ammonium hydroxide. The ammonium is sorbed and the hydroxide makes the soil more alkaline. The practice is technically quite difficult, but has the great advantage that ammonia is 82% N so that very little unnecessary weight has to be transported. It is common in the USA but no longer in use in the UK. Ammonia is also injected into the soil in solution, often with ammonium nitrate or phosphate. The solution can be made simply by partially neutralizing a solution of ammonia with nitric or phosphoric acid.

Nitrifying bacteria normally convert ammonium rapidly to nitrate (Chapter 3), but because nitrate is at risk of leaching it is often an advantage to keep it as ammonium, which is usually strongly sorbed by soil clays. This can be done with nitrification inhibitors, which discourage bacterial activity. One inhibitor is ammonia itself, which affects bacteria either directly or through the alkalinity it causes. Carbon disulphide is another small molecule with inhibitory effects, but it is not easy to use because of its toxicity and foul smell. It is better to use ammonium or sodium trithiocarbamate which release carbon disulphide as they decompose in the soil (e.g. Penny *et al.*, 1984). The chemical companies have developed other nitrification inhibitors, the best known of which are N-Serve and dicyandiamide (DCD).

The effectiveness of these nitrification inhibitors depends on the length of time during which the soil remains frozen in winter. In the northern USA and Canada it remains frozen for much of the winter, and ammonia or ammonium can be applied in autumn with inhibitors that keep it in that form until the soil freezes. The cold then inhibits the nitrifying bacteria and the fertilizer remains safe until spring. Large amounts of water may pass through the soil when the thaw comes, but so long as the fertilizer has remained as ammonium, little should be lost. In much of Western Europe the soil freezes only rarely and remains warm enough for nitrification to occur, albeit in a limited way, for much of the winter. There would usually be little point in applying ammonium in autumn, even with inhibitors, to provide for a crop in spring.

Urea

Liquid ammonia needs sophisticated equipment when it is applied as a fertilizer, and little is applied directly outside the USA. However, ammonia can be converted without too much difficulty into urea, which now accounts for nearly half of the nitrogen fertilizer used worldwide. Urea is found in urine and is noteworthy historically as the first organic compound (Box 2.2) to be synthesized. This was achieved in 1828 by Wöhler, a friend and associate of Liebig (Box 1.3). Wöhler's method involved evaporating a solution containing potassium isocyanate and ammonium sulphate. The resulting ammonium isocyanate undergoes a reversible molecular rearrangement to urea:

$$NH_4NCO \leftrightarrow CO(NH_2)_2 \tag{4.6}$$

Modern methods of urea production are more direct. Liquid ammonia is reacted with carbon dioxide to form ammonium carbamate:

$$2NH_3 + CO_2 \leftrightarrow NH_2CO_2NH_4 \tag{4.7}$$

Dehydration of the carbamate – that is, removing H_2O from its formula – gives urea:

$$NH_2CO_2NH_4 \rightarrow CO(NH_2)_2 + H_2O \tag{4.8}$$

The first reaction is exothermic (it gives out heat) but the second, the removal of water, is endothermic (needs an input of heat). The overall process needs a net input of energy, making urea about 35% more expensive in energy terms than ammonia.

Urea is highly soluble. (This is essential for its role in the body as the main degradation product of protein excreted in liquid form.) It is often sold as a fertilizer material in the form of prills – small approximately spherical agglomerations of solid urea. These are made in a prilling tower, in which a hot, highly concentrated solution of urea is sprayed into the air at the top of the tower. The solution forms spherical globules as it is released and the height of the tower is such that the water has evaporated from them and they have become solid prills by the time they reach the bottom. These prills have several advantages as a fertilizer material. They contain 46% N, more than any other source of nitrogen except liquid ammonia. This is important where the fertilizer has to be transported long distances. The prills are also clean and easy to handle and not prone to 'caking'. (This means sticking together in the fertilizer bag to form a rigid mass. This used to happen particularly with some compound fertilizers, in the bags near the bottom of a stack of bags.) And, unlike ammonium nitrate, they cannot explode.

Like ammonia, urea can be applied as a fertilizer in solution. It is often applied in this way with ammonium nitrate. The two compounds enhance each other's solubility.

Plants cannot use urea directly. It has first to be hydrolysed by the enzyme *urease*, which is very widespread in soils. The hydrolysis is effectively the reverse of the industrial synthesis. Urea acquires a molecule of water to become ammonium carbamate and this is converted to ammonia, which mainly becomes ammonium hydroxide when it meets the soil water. The ammonium is usually nitrified fairly rapidly to nitrate (Chapter 3).

Despite its advantages in nitrogen content and handling, urea did not take on rapidly in Europe. There were suspicions that it was less efficient than other fertilizer sources and that it was toxic to plants. Any inefficiencies were probably caused by slow hydrolysis in cold soils, and ammonia loss to the atmosphere in warmer conditions, particularly in soils that were sandy or had a high pH. Those that fell into both categories were particularly vulnerable. Sandy, as opposed to clay soils, would have contained fewer cation-exchange sites to hold ammonium. And the high concentration of hydroxyl (OH^-) ions in high pH soils would have shifted the equilibrium in equation (2.1) to the left, encouraging ammonia to remain as NH_3 rather than NH_4^+ and therefore vulnerable to gaseous loss. Ammonia losses are worst under drying conditions – sun and wind.

Urea itself is not toxic to plants, although plants grown in hydroponic solution with urea as the sole source of nitrogen do not grow at all normally (D. Cox, Rothamsted, 1973, personal communication). The toxicity problem that delayed the take-up of urea was not caused by urea itself but by an impurity, *biuret* $NH_2CONHCONH_2$.

The dominance of urea in today's fertilizer market has not been the result of acceptance by farmers in Europe and the USA, both of whom still tend to use ammonium nitrate or solutions, but of the rapid increase in the number of rice-growers in Asia who use nitrogen fertilizer. China, India and Indonesia are now the world's largest producers of urea. The high proportion of nitrogen in urea is an advantage in countries that lack good transport systems and urea is a good fertilizer for rice, which does not need nitrate and can take up ammonium. Nitrate fertilizers would be of little use in paddy rice because of denitrification.

Production of nitric acid for nitrate fertilizers

Ammonia is oxidized to nitric acid over a platinum catalyst, which is alloyed with rhodium for strength (Lowrison, 1989). The nitric acid is reacted either with ammonia to give ammonium nitrate or with the appropriate oxide, hydroxide or carbonate to provide the nitrate fertilizer required, often calcium nitrate. Calcium ammonium nitrate is also used. Nitrochalk sounds similar to calcium nitrate but is a granulated mixture of ammonium nitrate and chalk. The ammonia may also be reacted with phosphoric acid to give mono- or di-ammonium phosphate, both of which are used in compound fertilizers (see p. 6). Ammonium nitrate is the most widely used nitrogen fertilizer in Europe, and is commonly used in the form of prills (see above). Calcium nitrate, calcium ammonium nitrate and Nitrochalk are all used too. All tend to raise the pH of the soil slightly. Calcium nitrate, in particular, is hygroscopic – it picks up moisture from the air and will collect enough to dissolve it if left long enough. Granulation or prilling helps to avoid this problem, but some farmers accept the inevitable and use them in solution.

Slow-release nitrogen fertilizers

The slow-release nitrogen fertilizer became something of a 'philosopher's stone' for fertilizer manufacturers in the middle years of the last century. The ideal was a material that released nitrate at the same time and rate as the plant took it up, but this was not feasible and the best that could be achieved was delayed release.

One approach, pioneered in the USA, was to coat urea with sulphur, so that it became available when the soil microbes ate their way through the sulphur. This idea was taken up by ICI in the UK, who produced a sulphur-coated urea marketed as Gold-N. This product was investigated at the company's request at Rothamsted. The problem with the material lay in the sulphur coating. If the coating was completely intact, the urea remained inside and took a very long time to come out, but if there was even a pinhole in the coating, the urea escaped rapidly through it (D. Cox, Rothamsted, 1974, personal communication). This behaviour was largely confirmed in field experiments (Addiscott and Cox, 1976; Cox and Addiscott, 1976).

The other approach tried was to develop nitrogen compounds that dissolved only very slowly. One such was isobutylidene di-urea (IBDU), which was developed in Japan. This behaved reasonably satisfactorily as a slow-release fertilizer (e.g. Penny *et al.*, 1984), but never became widely used in the UK because of the price differential between it and conventional nitrogen fertilizers. But a representative of ICI told me the company made money from it when they patented its use as a cattle feed supplement! At Rothamsted, we investigated another condensation product of urea, glycoluril, as a slow release source and found it was a very good one, more effective than IBDU (Addiscott and Thomas, 1979). But once again the problem would have been price. We tried to find a source of used glyoxal, the compound with which urea was condensed, that had been rendered impure by its first use but from which glycoluril could be precipitated. We found, however, that glyoxal was sufficiently valuable to be purified for re-use by the original user.

Slow-release fertilizers are common in horticulture, particularly when added to high value items such as pot-plants, but in agriculture the most successful one is still injected ammonia, with or without inhibitors. Farmers who do not use injected ammonia seem to find it best to mimic slow release by splitting fertilizer applications.

5 Losses of Nitrogen from Arable Land

Thirty or forty years ago, loss of nitrogen simply implied the loss from the soil of a resource that the farmer would need to replace. Today, we worry not so much about 'loss from' as 'loss to'. Where has the lost nitrogen gone, and in what form? What part of the environment has it reached, and what mischief is it making? The latter questions are addressed in Chapters 7 and 8, and in this chapter we simply ask how it has been lost. Nitrogen can be lost from arable land as ammonium or nitrate, but in this book we are concerned mainly with losses of nitrate. Losses of ammonium from the soil occur by volatilization of ammonia as described in Chapter 4. Ammonia can also be lost from crops during senescence.

Nitrogen Loss – the Underlying Causes

Natural systems of vegetation are parsimonious with their resources. They have evolved to fill a particular ecological niche in the most effective way and there are competitor species well able to replace those that do not use their resources efficiently. Permanent grassland and forest usually lose very little nitrate or other usable forms of nitrogen to the rest of the environment. The annual cycle of growth, senescence and decay includes the production of nitrate by soil microbes from dead leaves and roots (Chapter 3), but this nitrate is captured by the roots of the grass or the trees before it can escape from the system. Nitrate concentrations in drainage from mature grassland or forest are usually small.

Arable farming is not a natural system, but it is a very old system. One of Christ's most memorable parables concerns an arable farmer who went out to sow but achieved very variable yields. Evidence of arable farming in the Middle East can be found far earlier than his time, in the Old Testament and in archaeological records. The effects of arable farming in the UK also go back a long way. Neolithic farmers in the Lake District of Cumbria caused the first nutrient enrichment of natural waters when they cleared forests around the lakes about 5000 years ago (Ferguson *et al.*, 1996; Moss, 1996). This brings out the general point that any perturbation of a natural ecosystem is likely to result in nutrient losses, a point that is taken up in terms of entropy in Chapter 14. Arable farming is one of a number of perturbations that can cause such problems, and some loss of nitrate from it is inevitable. This book is concerned with minimizing losses rather than bringing them to an end.

Arable farming and other ecosystems

Two key differences separate arable farming from grassland, forest and other systems of vegetation. The first is that arable farming is unique among these systems in that it leaves the soil bare for part of the year. While the soil is bare, nitrate left in the soil or produced within it is vulnerable to leaching because there are no living roots to capture it. One way of alleviating this problem is to sow the following crop early to minimize the time the soil is left bare. Widdowson *et al.* (1987) showed that sowing a crop of winter wheat early lessened the amount of nitrate at risk in the soil and thereby decreased the risk of leaching (Fig. 5.1).

The length of time for which the soil is bare depends greatly on the sequence of crops. If winter wheat, for example, is followed by another winter cereal, the new crop may be showing above ground within a month of the old crop being harvested and, as Fig. 5.1 shows, by Christmas it can have taken up 25–30 kg N/ha. But potatoes are not usually sown before April in the UK, so if they follow winter wheat the soil is bare for about 8 months.

Rotations and bare soil

Conserving nitrate within the system is not the only issue on the farmer's mind. Avoiding crop diseases is important too. Sowing one cereal too close behind another can provide a bridge between crops for the pathogens that cause disease. Several crops are grown as 'break' crops between cereals to restrict disease problems. These include winter oilseed rape, which can be sown at the same time as a winter cereal and avoids bare soil to a large extent, but they also include potatoes and sugarbeet, which leave the soil bare over winter. Table 5.1 lists some of the more likely crop sequences and the length of time for which the soil is bare. It also shows the length of time after 15 October for which it is bare, and this is discussed later. Table 5.1 shows sequences of two crops, but Smith *et al.* (1997) demonstrated using the SUNDIAL model how much rotations vary in their efficiency of nitrogen use and the potential financial saving.

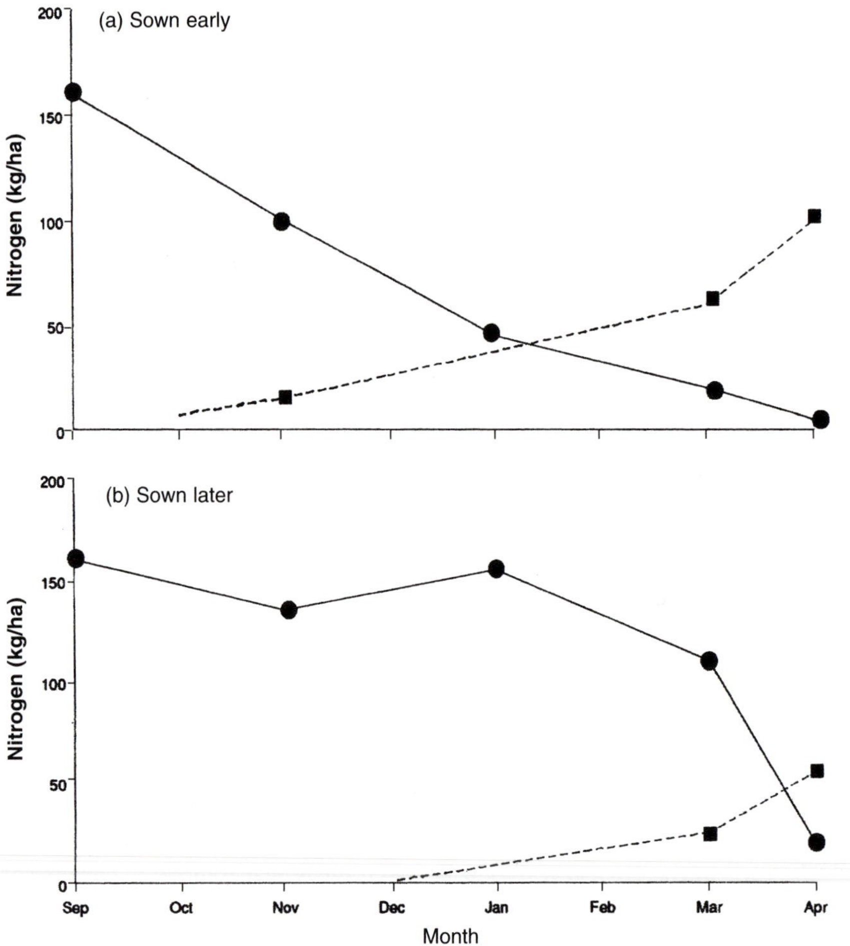

Fig. 5.1. Effect of early sowing on the amount of nitrate taken up by the crop (■) and remaining in the soil (●). (From Widdowson *et al.*, 1987.)

Catch crops

If the soil has to be left bare without a commercial crop for an appreciable length of time, nitrate losses can be restricted by growing a 'catch crop or cover crop', a temporary crop which may have little or no cash value but which takes up nitrate effectively. At one time farmers used to grow turnips during winter for the sheep to graze but, with fewer mixed farms today, a catch crop is usually used to remove and hold back nitrate from risk of leaching. Rape, white mustard and phacelia are among the crops used, and these crops can be very effective, taking up 50 or even about 100 kg/ha of nitrate-N (Christian *et al.*, 1990).

Catch crops do not seem to be attracting as much interest as they did 10 years ago, probably because of problems that have shown up with their use.

Table 5.1. Some common crop sequences and the time for which the soil is bare between crops and after 15 October.[a] Dates are defaults used in SUNDIAL model (Smith *et al.*, 1997) and were based on discussions with the Rothamsted Farm Manager.

	Date of			Time bare (days)	
Sequence	Harvest	Sowing	Emergence[b]	Total	After 15 October
Winter wheat – winter rape	15 August	1 September	22 September	37	0
Winter rape – winter wheat	15 July	1 October	22 October	93	6
Winter barley – winter rape	21 July	1 September	22 September	61	0
Winter rape – winter barley	15 July	15 September	6 October	82	0
Winter wheat – potatoes	15 August	1 April	22 April	249	188
Winter barley – potatoes	21 July	1 April	22 April	273	188
Potatoes – winter cereal	1 October	Winter cereals often sown	late after potatoes	–	15 (?)
Winter wheat - winter beans	15 August	20 October	10 November	86	25
Winter wheat - spring beans	15 August	1 March	22 March	218	157

[a]Drainage from the soil might resume at this date (but date is extremely variable).
[b]Emergence assumed to be 3 weeks after sowing.

The catch crop is ploughed in before the main commercial crop is sown with the intention that it is broken down by the soil population and its nitrogen released to the main crop. However, this release tends to take place too late for that crop and may be delayed until autumn, just when it is not wanted (Catt *et al.*, 1998a). Catch crops may also use water needed by the main crop and they can also carry diseases from one main crop to the next. They can, however, be useful on farms that provide shooting facilities because they provide cover for game birds.

Cultivation

The other factor that differentiates arable farming from other systems is cultivation. In *primary cultivation* the farmer uses a plough to invert the topsoil and thereby bury weeds and weed seeds. The layer inverted, known as the 'plough layer', is usually 23–25 cm deep. *Secondary cultivation* breaks up the plough layer, usually with a harrow, to produce a 'seedbed' of fairly small soil aggregates. The perturbation of the soil in this way tends to stimulate nitrate production by the soil's microbial population and this nitrate is vulnerable to loss if the soil remains bare.

Forty to fifty years ago many farmers did their primary cultivation in autumn and held their secondary cultivation until spring and sowed spring crops. The results were aesthetically quite pleasing, and various artists, Rowland Hilder, for example, painted pictures of neatly ploughed furrows stretching away towards a line of trees in their winter black. This system had several advantages. The frost got into the ploughed soil and broke it up, so that the farmer got some of his secondary cultivation done for free. And while the frost was in the soil it killed of some of the pathogens that caused disease. The big disadvantage of the system was that the soil was bare, and nitrate free to leach from it, for the entire winter. The change to winter crops, notably winter wheat, sown in the autumn was probably driven by the larger yields available from winter crops, but there was a definite benefit in terms of nitrate loss too (Croll and Hayes, 1988), provided farmers did not apply fertilizers in autumn. We shall return to winter nitrate losses later in the chapter. But before we discuss nitrate losses in detail we need to consider the implications of a problem that has attracted increasing attention from soil scientists during the past 10–15 years – that is, scale. This is related to the associated topic of hierarchy.

Scale and Hierarchy

'Scale' is a potentially confusing word because a large-scale map describes only a small area. It may be best to use the word mainly as pointer to a problem and to avoid attaching the words 'large' or 'small' to it. Scale is a problem in modelling studies, to which Hoosbeek and Bryant (1992) responded with their 'Scale Diagram'. As we are not concerned here with modelling, we do not need to consider the diagram *per se*, but the hierarchical levels it

Table 5.2. The hierarchical levels of the Scale Diagram of Hoosbeek and Bryant (1992).

Scale	Unit
$i + 4$	Region
$i + 3$	Interacting catchments
$i + 2$	Catena or catchment
$i + 1$	Field (polypedon)
i	Pedon
$i - 1$	Profile horizon
$i - 2$	Peds, aggregates
$i - 3$	Mixtures
$i - 4$	Molecular

describes (Table 5.2) are a useful guide to thinking about scale. Many of the measurements mentioned in the previous paragraphs are made somewhere between level $i - 2$ and level i, or even level $i - 4$ for molecular biological assays, but we may be looking for answers to questions raised at levels $i + 1$, $i + 2$ or higher. Bridging this gap may be a problem.

A more difficult but interesting point about a change from one hierarchical level to another is that it may involve a change in the level of determinism. At one level a process may operate in a deterministic way, which means that the outcome of a given set of circumstances is entirely predictable. At another level, the outcome may be entirely random. Determinism is discussed in detail in Box 5.1, and the relationship between determinism and scale (or hierarchical level), which is described as decoherence, is explained in Box 5.2.

Discontinuity

The chemical, physical and biological processes discussed in Chapters 2 and 3 are all affected by scale in the form of the problem of discontinuity. We need to know what happens over substantial areas of land but the amount of soil on which the processes are measured is usually quite small. Soil chemical measurements made on a kilogram or less of soil may have to represent a hectare of land and may fail to do so, unless sampling procedures are designed with great care. Soil physical measurements have a similar problem. Back in the 1960s and 1970s, leaching measurements were often made on sieved soil in tubes 10–20 cm long and a few centimetres in diameter. We now know that much larger columns of undisturbed soil are needed and that even those do not tell the whole story. The vessels in which biological measurements are made usually have to be small enough to go into a constant-temperature cabinet, and molecular biological assays are by their nature at a very small scale. How much can these measurements tell us about what is happening at the scale of the field? All these questions arise from the problem of discontinuity.

Box 5.1. Determinism.

Determinism underpins much of our scientific thinking and, indeed, many of our assumptions about life in general. Its origins lie in a statement by the French mathematician and astronomer Laplace, which Stewart (1995) quotes in this way:

> An intellect which at any given moment knew all the forces which animate Nature and the mutual positions of all the beings that comprise it, if this intellect was vast enough to submit its data to analysis, could condense into a single formula the movement of the greatest bodies of the universe and that of the lightest atom: for such an intellect nothing could be uncertain, and the future like the past would be present before its eyes.

Even if we put to one side the theological implications of this statement and our doubts about the vastness of our intellect, we still have two problems. One is that we do not know all the forces that operate within even one of the hierarchical levels listed in Table 5.1. The other is that we are rarely, if ever, able to define the initial state of any of the levels. We usually assume that the systems we are studying behave in a deterministic way, but this comes into question if we cannot resolve these problems. (This is not a problem if we think that the system operates purely randomly.) Our ability to make predictions at any of the hierarchical levels may simply depend on how certain we are as to which are the dominant forces in the system and whether we know the initial state of the system. This could mean, for example, that at the catchment ($i + 2$) and higher hierarchical levels, the dominant force is the weather, which can be measured reliably, whereas at the lower levels the forces which dominate are less easy to establish or measure. It could also mean that a 'broad brush' assessment of the initial state of one of the higher hierarchical levels, based on readily available information, is more reliable than the corresponding assessment at a hierarchical layer less than i, where the necessary detail may not be available. We should not, therefore, be surprised if the models and measurements used at the various levels differ.

Discontinuity is a particular problem in the study of nitrous oxide emissions. Dobbie *et al.* (1999) and Dobbie and Smith (2003) used closed chambers to capture the emissions, cylindrical ones of 0.4 m diameter or square ones with a length of side of 0.4 m, according to the crop. The cylinders covered 1.26×10^{-5} ha and the square chambers 1.6×10^{-5} ha. A 16-ha field is a million times larger than the area covered by a square chamber, but the emission of nitrous oxide from it is most unlikely to be a million times the emission from the chamber. The soil is far too variable a material and the emissions too episodic for so simple a relationship to apply. Dobbie and her colleagues used up to eight replicate chambers to counter this problem, but the log-normal distributions of the emissions they found suggests that more might have been desirable though probably not practicable. We return to the discontinuity problem later when we discuss spatial patterns in nitrate losses.

Box 5.2. Decoherence.

Decoherence (Stewart, 1995) is probably an idea that is not familiar to you. It links the concepts of determinacy and scale. Classical physics was dominated by Newton's Laws and Laplace's ideas about determinism and concerned itself to a large extent with events in the solar system. Modern quantum physics is interested in what happens at the atomic or sub-atomic scale and is virtually indeterminate. The planets obey Newton, but particles within the atoms of which they are built obey the laws of chance. Decoherence is the term used to describe the loss of indeterminacy that seems to occur as small quantum systems are aggregated to make up very much larger ones. Put another way, it expresses changes in determinacy between hierarchical levels.

The concept of decoherence can be applied to soils and landscapes (Addiscott and Mirza, 1998). It suggests that large areas of land should behave in a more determinate and predictable way than small volumes of soil. This may seem counter-intuitive but denitrification provides an excellent example. We saw in Chapter 3 how 'hot spots' arising from fragments of plant residues and other localized anoxic zones in soil samples make denitrification unpredictable in terms of its location and intensity within the sample. The soil samples probably represent the $i - 1$ hierarchical layer. The process remains difficult to predict at the i and $i + 1$ levels (pedon and field), but predictive relationships for denitrification can be established more readily at the $i + 2$ or $i + 3$ level, the scale of the catena or landscape (Groffman and Tiedje, 1989; Corre *et al.*, 1996; Velthof *et al.*, 2000). These relationships depend on the observation that soil wetness, which strongly influences denitrification (Chapter 3), can usually be assessed from the topography. A soil on a spur tends to be relatively dry, but one on a footslope rather wetter. These topographic factors can be combined for predictive purposes with climatic, seasonal and land-use ones. These concepts have been refined by Lark *et al.* (2004a,b) using wavelet theory as described in the main text, and Milne *et al.* (2004) have examined their implications for the evaluation of models, using scale-dependent wavelet correlations to quantify the performance of the model at different scales.

Nitrogen Loss before Harvest

Bare soil without a crop was identified as a key factor in nitrate losses, so to examine these losses in greater detail we need to consider two periods of the year:

- The period of active growth before harvest, when the crop is taking up nitrate vigorously.
- The period after harvest, when the crop has been harvested and the soil is bare apart from stubble until the new crop emerges.

In this section, we discuss losses before harvest.

In some parts of the world the onset of active growth and nitrate uptake is clearly marked – by the spring thaw or the arrival of the rains, for example. In the UK and other temperate zone countries the vagaries of the weather make the event less obvious, but one useful signal is simply the time at which the grass first needs to be mown regularly. The lawn, like nearby crops, is responding to warmer temperature in the soil and the air, longer day lengths and more intense sunlight. Photosynthesis increases and the growth of both lawn and crops accelerates. Root growth accelerates too, taking mineral nitrogen (ammonium and nitrate) out of the soil for the plant to produce the proteins it needs for growth.

Soil microbes also respond to warmer temperatures by increasing their production of mineral nitrogen. In natural ecosystems plants limit their growth according to the amount produced by the microbes, but arable crops other than legumes nearly always need to be supplied with extra nitrogen to obtain a satisfactory yield. This is another way in which arable farming differs from other ecosystems. This extra nitrogen usually comes in mineral form as fertilizer, but it may be supplied as farmyard or poultry manure or slurry. All forms of nitrogen will put nitrate into the soil eventually but some will do so over a much longer period of time than others. We are concerned most with mineral fertilizer, simply because it is by far the most widely used form of nitrogen. Organic farming will be discussed in Chapter 14.

All the main forms of nitrogen fertilizer, except urea, contain ammonium or nitrate or both. Ammonium is converted rapidly by nitrifying bacteria to nitrate (Chapter 3). Urea has to be hydrolysed by urease to ammonium (Chapter 4) but this usually delays matters by only a few days at normal fertilizer application times. What happens to nitrate from fertilizer is central to this chapter. There are four likely ultimate fates for the nitrate (Fig. 5.2):

1. It may be taken up by the roots of the crop and used to produce the proteins needed for growth and yield by the crop.
2. It may become incorporated in the soil's organic matter, where it will

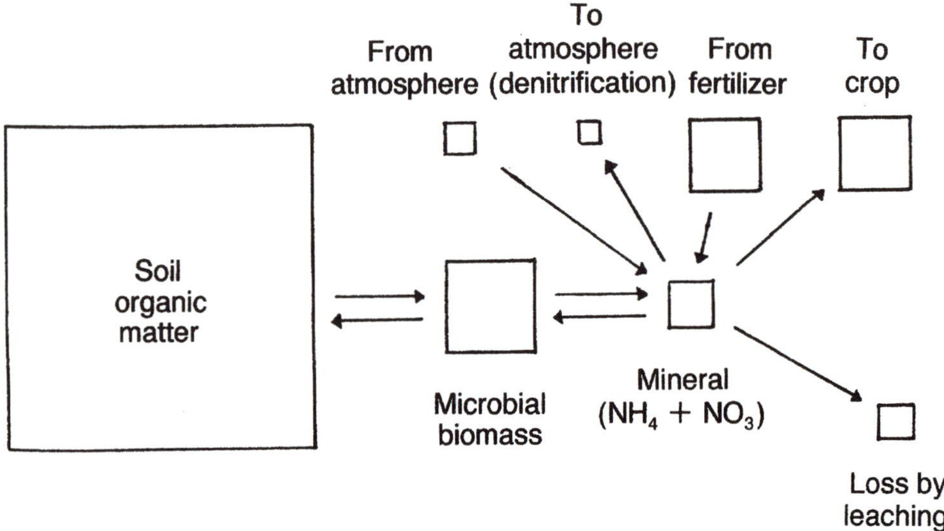

Fig. 5.2. The four ultimate fates of nitrate. The size of the square is proportional to the quantity of nitrogen involved. (From Addiscott *et al.*, 1991.)

cause no problems unless it is released as nitrate by mineralizing and then nitrifying bacteria.

3. It may be leached, or washed out, from the soil and end up in groundwater, fresh water (streams, rivers or lakes) or the sea. The problems it may cause are discussed in Chapters 7 and 8.

4. It may be denitrified (Chapter 3) to dinitrogen N_2 or nitrous oxide N_2O, both of which are gases. Dinitrogen is not a big problem because it makes up 78% of the atmosphere anyway, and if the nitrate came from a fertilizer manufactured by the Haber–Bosch process (Chapter 4) it has simply returned whence it came. There are losses – of the energy expended in its production and of the money the farmer paid for the fertilizer – but nothing worse. Nitrous oxide is much more of a problem because it is a greenhouse gas and is involved in the destruction of ozone in the upper atmosphere. These issues are discussed in Chapter 7.

The first of these outcomes is obviously desirable and the second satisfactory, in that the nitrate has not been lost from the system and no harm has been done to the environment. The loss of nitrate by leaching and the emission of nitrous oxide to the atmosphere are clearly undesirable outcomes from an environmental point of view, and as such they need to be quantified by measurement.

The Fate of ^{15}N-labelled Fertilizers applied to Crops before Harvest

Winter wheat

A crop has the valuable characteristic that it can integrate the processes involved in nitrogen uptake over a range of scales up to the $i + 1$ scale in Table 5.2. The main factors affecting nitrogen uptake are nitrogen supply, crop, soil type and weather. The 'labelling' of fertilizer with ^{15}N was explained in Box 3.2, and using this procedure enables the experimenter to define the initial conditions and the relation between nitrogen supply and nitrogen uptake. He or she can then examine the effects of crop, soil type or weather on nitrogen losses. Another advantage of this approach is that the results it yields are in the form of ratios of some kind – '*70 per cent of the fertilizer was in the crop*' or 'the uptake was *154 kilograms per hectare*'. Ratios are independent of scale. They mean the same at hierarchical level $i + 5$ as at level i and are therefore readily portable. The problem of discontinuity identified above does not arise.

Experimental work with ^{15}N demands substantial resources of time, equipment and skilled manpower, but the commitment of these resources has proved worthwhile. Research of this kind was made throughout the world and provided decisive insights into the nature of the nitrate problem (e.g. Smith *et al.*, 1984; Powlson *et al.*, 1986, 1992; Recous *et al.*, 1988). The results of Powlson *et al.* (1992), for example, were obtained with winter wheat grown at three sites: the flinty silt loam at Rothamsted, a sandy loam

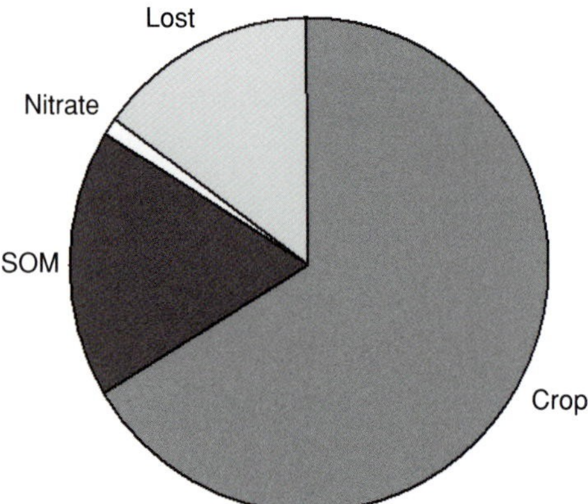

Fig. 5.3. Fate of ^{15}N-labelled fertilizer applied to winter wheat. Likely proportion found at harvest in the crop, in soil organic matter (SOM), as nitrate in soil or lost.

at Woburn in Bedfordshire and a sandy clay over a heavy clay subsoil at Saxmundham in Suffolk (Fig. 5.3). They showed that at harvest 8–35% (average 15%) of the nitrogen applied was missing and had to be presumed lost, but 50–80% of the labelled nitrogen was in the crop and 10–25% was in the soil, almost all of which was in organic matter.

That the nitrogen in the soil was nearly all in the organic matter was a critically important finding. Only about 1% of the labelled nitrogen from the fertilizer was left in the soil at harvest as nitrate. Nearly all the nitrate found was unlabelled and therefore did not come from the fertilizer. Furthermore, the amount of nitrate found in the soil bore no relation to the amount of fertilizer supplied. This seemed to refute the claim that the nitrate problem was caused by nitrogen fertilizer left unused in the soil, but we need to examine all the evidence before we can draw such a strong conclusion. It is also important to remember that these results were for winter wheat, which is recognized as parsimonious with nitrogen, at least in part because it has a well-established root system by the time the fertilizer is applied. Other crops, discussed later, were more profligate than winter wheat.

What happened to the missing labelled fertilizer?

We are concerned here with the 8–35% of labelled fertilizer that went missing between the spring application and the harvest. Much more tended to be missing when the application was followed by above-average rainfall, so Powlson *et al.* (1992) examined the relationship between the loss of ^{15}N and the rainfall during various lengths of time after application. Considering the rainfall in the first 3 weeks gave the regression that accounted for the greatest percentage of the variance in ^{15}N loss (Fig. 5.4), they concluded that this first 3 weeks was the critical period. This relationship is useful as a rough guide to fertilizer losses, but it does not tell us how the fertilizer was lost.

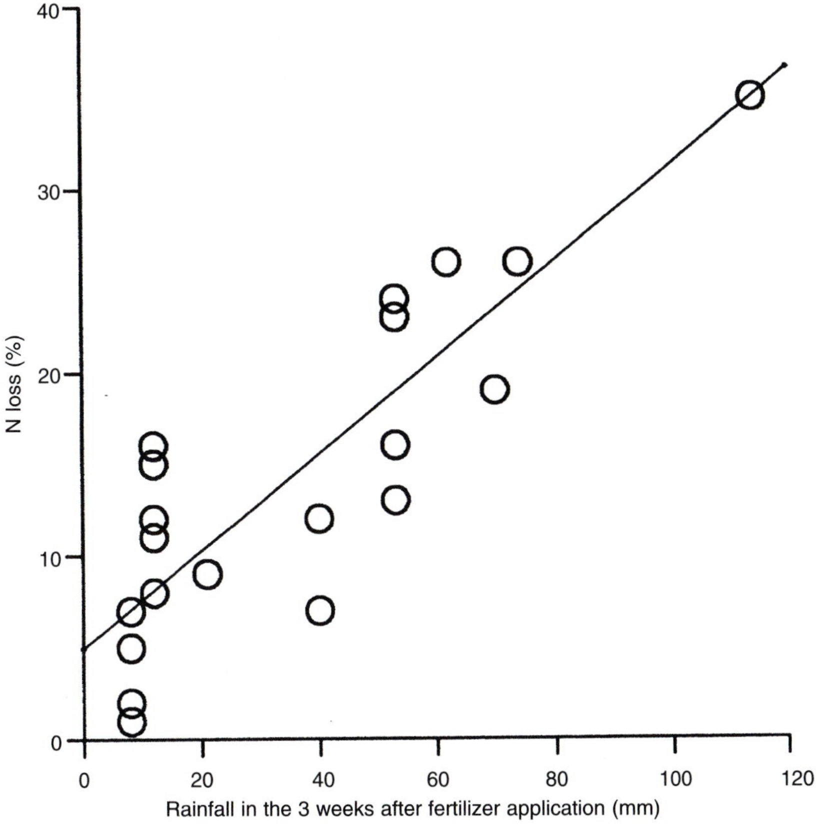

Fig. 5.4. Relationship between ^{15}N lost and rainfall in the 3 weeks after application for the experiments of Powlson *et al.* (1992).

Partitioning the loss between leaching and denitrification

The losses of ^{15}N caused by rain may occur by leaching or denitrification. These processes are likely to occur simultaneously and both are difficult to measure, so partitioning losses between them is not easy. It was not feasible to do the partitioning experimentally with the resources available to the experimenters, so a computer model was used (Addiscott and Powlson, 1992). The SLIM leaching model (Addiscott and Whitmore, 1991) was used in conjunction with models for mineralization and crop uptake to estimate the percentage losses of labelled fertilizer by leaching, and these estimates were subtracted from the total percentage losses of labelled fertilizer to provide estimates of the loss by denitrification. These losses by apparent denitrification could have resulted from an underestimate of leaching, so a check was made to ensure that this was not the case. This check showed that they were better related to the wetness of the soil (as calculated by the model) than they were to the rainfall, suggesting that denitrification was the cause.

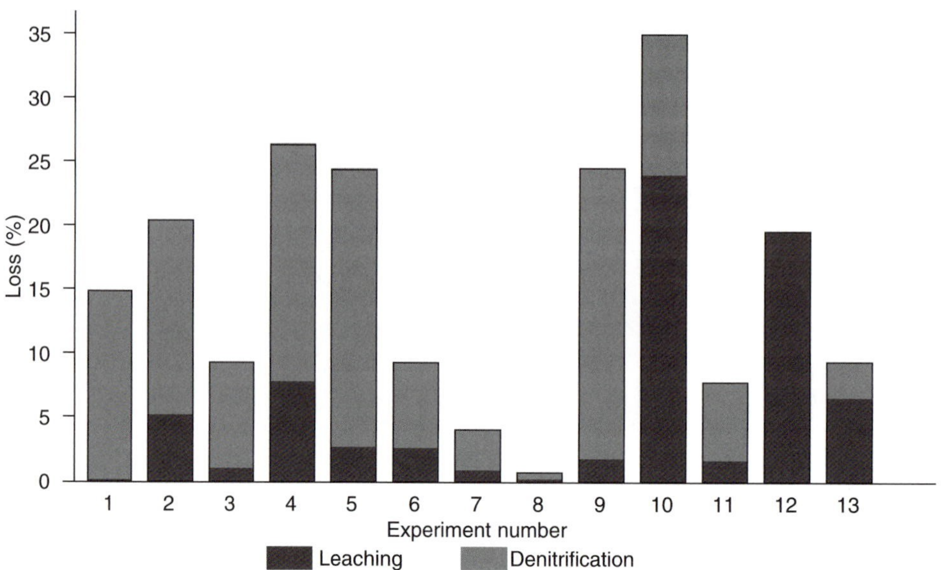

Fig. 5.5. Partitioning of ^{15}N loss in Fig. 5.4 between leaching and denitrification. (From Addiscott and Powlson, 1992.)

If the assumptions implicit in the approach were correct, the total percentage losses of labelled nitrogen could be partitioned between the two pathways of loss as shown in Fig. 5.5. In two of the 13 experiments the loss was totally by leaching, although one of these losses was very small. In another experiment denitrification accounted for all the loss. In the remaining ten experiments both pathways of loss contributed. Overall, nearly two-thirds of the loss of labelled nitrogen was by denitrification. Thus of the average total loss of 15%, leaching contributed only about 5% and denitrification 10%. As only about 1% of the labelled nitrogen was left in the soil at harvest, direct leaching losses from the fertilizer can have been no more than about 6%.

This result differs considerably from the impression given by environmental pressure groups and the media, and we need to seek complementary evidence. This further evidence (Goss *et al.*, 1993) was provided by the Brimstone Experiment, described in Box 5.3 and Fig. 5.6. No isotopes were used in this experiment so, between fertilizer application and harvest, nitrate lost by leaching could either have come from the fertilizer or been produced in the soil. The combined losses can, however, be expressed as percentages of the fertilizer applied, and the results, 0–17% with a mean of 3.3%, compare well with the losses measured in the isotopic experiments. This appears to confirm the very small direct losses from fertilizer, but two cautionary notes are needed.

One such note is that these experiments received much more detailed attention than can be given on an ordinary commercial farm. In particular, the nitrogen was spread more evenly on the plots that were used than

Box 5.3. The Brimstone Experiment.

'The Brimstone Experiment' is the name usually used for an important experiment at Brimstone Farm near Faringdon in Oxfordshire, UK. It is sited on a heavy clay soil of the Denchworth Series (Jarvis, 1973) and has a subsoil that is impermeable to water at depth. The soil is strongly fissured, especially in late summer and early autumn, and provides pathways for preferential flow during much of the year. The 20 plots, established in 1978 (Cannell *et al.*, 1984; Harris *et al.*, 1984), are each 0.24 ha in area. They are isolated hydrologically from each other by polythene barriers inserted to a depth of 1.1 m parallel to the 2% slope and by interceptor drains at a depth of 1 m in a gravel-filled trench along their upslope margins (Fig. 5.6). Eighteen of the plots are now drained by mole channels 0.55 m below the surface, which intersect with gravel-filled trenches along the downslope margins of the plots; these have collector drains at a depth of 0.9 m. Surface runoff and downslope flow at the base of the plough layer are collected together in a deep furrow across the downslope margin of each plot to give what is described as 'cultivated layer flow'. This is usually very small compared with the flow through the drains, and is often not reported. The other two plots remain undrained.

The flow from the collector drains and the cultivated layer flow are measured by V-notch weirs, which have a flow range of 0.0–7.0 l/s and in which the head level in the weir chambers is recorded by float systems connected to rotary potentiometers and a Campbell CR10 datalogger. An Epic 1011T sampling system (Montec, Salford, UK) controlled by the datalogger samples each flow for analysis at the fresh flow end of each collector pipe before it enters the flow-measuring chamber. The rate of sampling is driven by the rate of increase in the flow. The losses of nitrate, phosphate or other ions are computed from the area under the relevant concentration–volume curve using Simpson's rule.

The experiment is representative of the fairly large area of arable land in the UK that can be farmed only when drains are installed to carry away water which cannot otherwise move downwards beyond the depth at which they are installed. Without drains, it is impossible to use agricultural machinery on such soils at times during spring and autumn. The layout described, with mole channels feeding into gravel-filled trenches with collector drains at the bottom, is the usual one. The mole channels are made with a 'mole', a mole-shaped piece of steel which is drawn through the soil by a 'leg', a steel bar attached via a frame to the tractor. The leg not only draws the mole, it also makes a 'leg-slot' in the soil and a network of herring-bone cracks which greatly facilitate the flow of water into the mole channel.

The system can remove water rapidly from the soil when the mole channels are newly drawn. Unfortunately, it removes nitrate and other solutes with equal rapidity. The mole channels and the leg-slots gradually deteriorate and eventually need to be renewed. Some argue that drainage systems should be allowed to deteriorate because the rapid removal of water contributes to flooding, but this obviously depends on the relative importance of other contributors to flooding such as the removal of vegetation or soil that is impermeable because of the way it is managed or because it is covered in tarmac.

would be possible at field scale. Direct losses may be somewhat larger on commercial farms, particularly if the fertilizer spreader is not set accurately or if passes with the spreader overlap.

We also need to recall that these experiments were made with winter

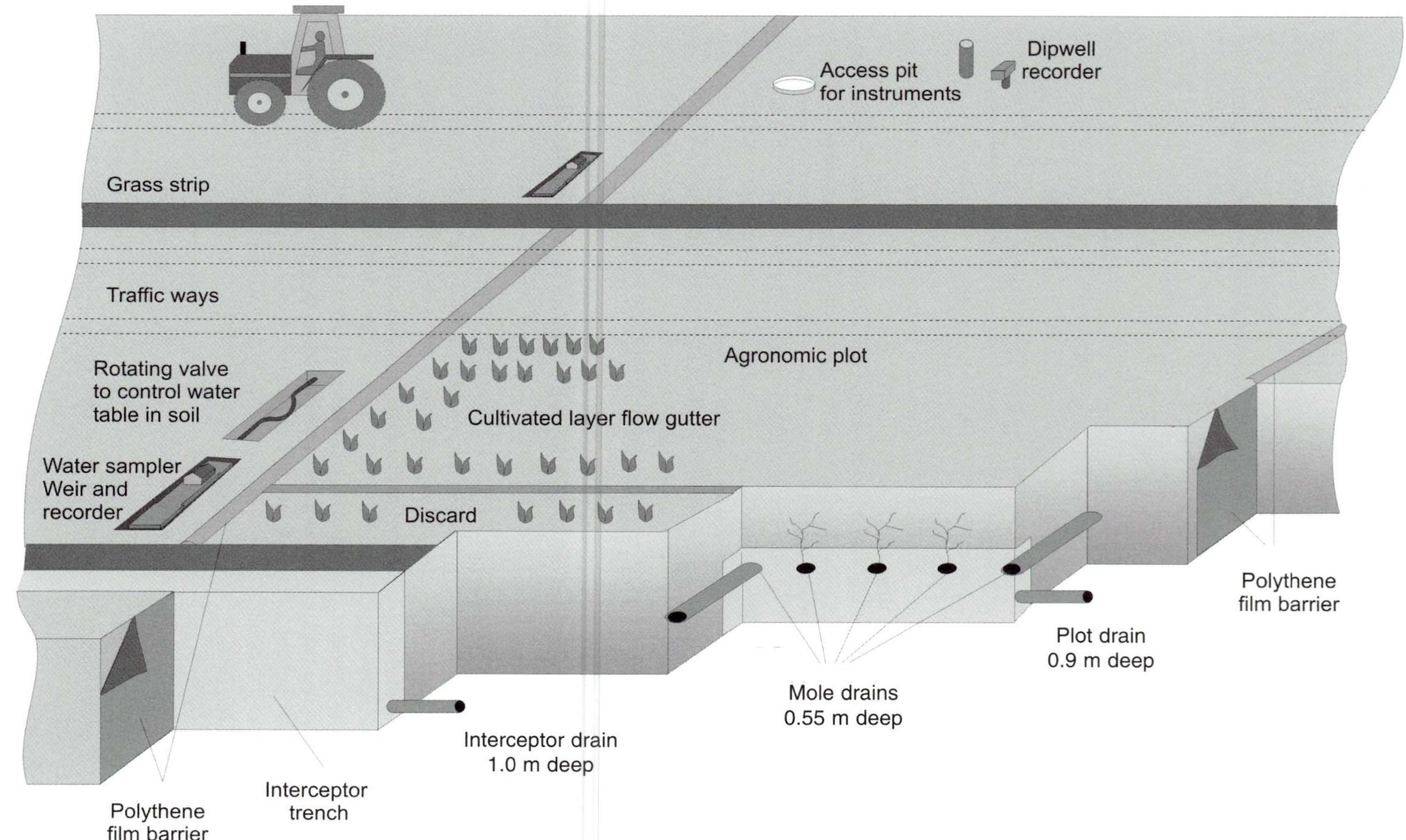

Fig. 5.6. Schematic diagram of a plot on the Brimstone Farm Experiment. (Courtesy of Rothamsted Research.)

wheat, a crop that is recognized to be parsimonious with nitrogen, and ask how much should we be influenced by the results from this one crop? One answer is that about 60% of the arable land in England and Wales grows winter cereal crops, so we need to note these results carefully. The crop is also fairly representative of crops that are sown in the autumn and establish their roots before spring.

Spring-sown cereals might be expected to be less parsimonious with their nitrogen than those sown in the autumn, but an experiment by Dowdell *et al.* (1984) suggests otherwise. When these authors applied ^{15}N-labelled nitrogen fertilizer to spring barley grown on a shallow soil overlying chalk and collected drainage for another 3 years after the year in which the ^{15}N was applied, they found that only 6.3–6.6% of the labelled nitrogen had been lost by leaching in total. Perhaps winter wheat was not exceptional in losing directly by leaching only 6% of the nitrogen it was given.

But we need to look at other crops as well as cereals, particularly those that are grown as 'break crops' to prevent the carry-over of disease from one cereal crop to another.

Break crops

Far more research has been conducted on the fate of ^{15}N applied to cereals than its fate under break crops. The break crop grown depends to some extent on soil type – potatoes, for example, tend to be grown on lighter soils – but other factors such as the market price of the crop come into play. Sugarbeet can be grown only if there is a processing factory within reach. Oilseed rape is the commonest break crop overall. Table 5.3 shows the results of four Rothamsted-based experiments made by Macdonald *et al.*

Table 5.3. Efficiency of nitrogen use by winter oilseed rape, potatoes and sugarbeet compared with that of winter wheat: Criterion 1, % of ^{15}N-labelled nitrogen fertilizer left in the soil at harvest; Criterion 2, % increase N_{min} in the soil at harvest caused by fertilizer application. Field beans also included. (Derived from Macdonald *et al.*, 1990.)

	Winter wheat	Winter rape	Potatoes	Sugarbeet	Field beans
Number of experiments[a]	4	4	2	1	1
N application (kg/ha)	211	235	223	122	–
% ^{15}N in soil at harvest	3.5 (2.6[b])	4.3	7.2	1.1	–
N_{min} in unfertilized soil[c] at harvest (kg/ha)	58	42	40	19	86
% increase in N_{min} from fertilizer	14	69	98	0	–

[a]The data in the table were obtained by averaging across the number of experiments shown.
[b]Omitting one crop which was seriously damaged by disease.
[c]The N_{min} was almost all nitrate.

(1989) in which they compared the fate of under winter oilseed rape with its fate ^{15}N under winter wheat. Two experiments included potatoes and one each sugarbeet and field beans.

Two criteria can be used to assess how well the crops in Table 5.3 used nitrogen: (i) the percentage of the ^{15}N applied that was left in the soil as nitrate; and (ii) the percentage increase in mineral nitrogen, labelled or unlabelled, left in the soil as a result of fertilizer use. The first criterion specifically measures the efficiency of fertilizer use, while the second shows the efficiency with which the crop uses nitrate, regardless of its origin. Both criteria showed oilseed rape to be less efficient in using nitrogen than winter wheat, but this may have resulted in part from the rape crop's habit of shedding leaves before harvest. Potatoes were less efficient still – their roots do not scavenge effectively for nitrate – while sugarbeet was clearly the most efficient crop of all. Field beans are legumes (Chapter 3) and do not need nitrogen fertilizer because bacteria in nodules on their roots obtain nitrogen from the air and supply it to the plant. But they left more nitrate in the soil at harvest than any other crop not given nitrogen. In fact, when given no nitrogen, all the crops left a fair amount of nitrate in the soil (Table 5.3), emphasizing the point that it is not just fertilizer nitrogen that is left at risk in the soil.

Changes in nitrogen use efficiency during 100 years

Winter wheat has not always used nitrogen as efficiently as in the experiments described above. This can be shown by comparing nitrogen leaching measurements made more than a century ago on the Broadbalk Experiment by Lawes *et al.* (1882) and recently by Goulding *et al.* (2000). Fuller details of the Broadbalk Experiment and its soil are given in Chapter 1 and Box 1.1. Tile drains were installed longitudinally in the centre of most of the plots in 1849 (Lawes *et al.*, 1881, 1882) and still run intermittently. The drains on one section of Broadbalk were replaced in the autumn of 1993. The subsoil of the field cracks sufficiently to allow free drainage and the field is not one that would necessarily require drainage to ensure trafficability. Some drainage will have occurred through the subsoil as well as through the drains, so it is not clear what proportion of each plot will have contributed to the flow through the drains. But the results are still very useful.

Goulding *et al.* (2000) compared early nitrogen losses from the drains (1878–1883) with losses they measured between 1990 and 1998. As these authors pointed out, there can be no exact comparison because of the weather differences between the two periods ('average' years rarely occur) and the difference in varieties. They were able to make some allowance for drainage differences through calculations that made allowance for the fact that the average annual drainage in 1878–1883 was 10% greater than that in the recent years. The crops made their own compensation, in that the smaller grain yields of the early years were offset by larger straw yields.

The results for 1878–1883 showed (Fig. 5.7) that there was a measurable

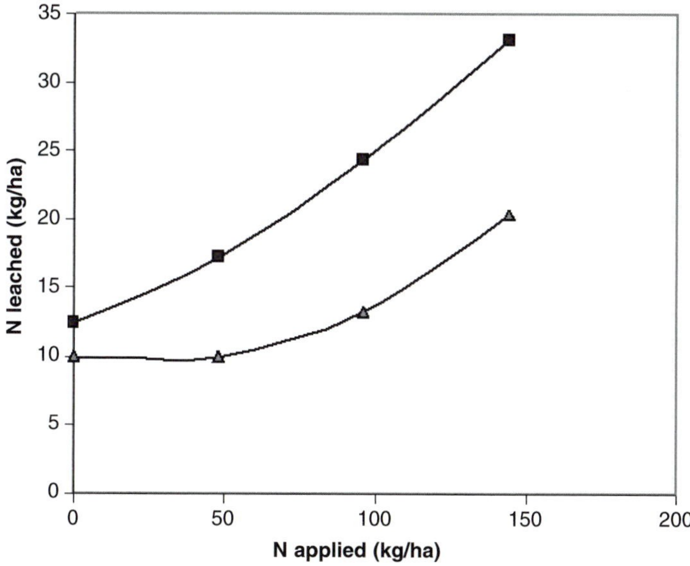

Fig. 5.7. Comparison of losses of nitrate in drainage in 1878–1883 (■) with those in 1990–1998 (▲) for Broadbalk plots 5, 6, 7 and 8 (continuous wheat given 0, 48, 96 or 144 kg N/ha). (From Goulding *et al.*, 2000.)

loss of nitrogen by leaching even when no fertilizer was given. Subsequently, as might have been expected, the losses increased steadily with the amount of nitrogen fertilizer given. The increase was curvilinear, implying that, as more fertilizer was given, the nitrogen loss increased as a proportion of the nitrogen given as fertilizer.

The results for the recent years, 1990–1998, were interesting in two respects (Fig. 5.7). The overall annual losses were 43% less than in 1878–1883, and the pattern was different. The nitrogen loss from 48 kg/ha of nitrogen fertilizer was no bigger than that when no fertilizer was given at all, and the loss from 96 kg/ha was only marginally (and not significantly) greater. That from 144 kg/ha was somewhat greater but not significantly so. The changes in overall loss and the pattern of loss can be attributed to the greater grain yield, by a factor of two to three, of modern varieties of wheat and their more efficient uptake and use of nitrogen, together with better methods of crop protection.

Nitrogen Loss after Harvest

Harvest is the point at which arable land becomes most different from grassland and natural systems of vegetation, and this is one good reason for looking for the roots of the nitrate problem in the postharvest period. What happens to water and nitrate in the soil during this period?

Winter often feels wetter than the summer in the UK, but this is not necessarily because more rain falls. In fact, the average daily rainfall varies little

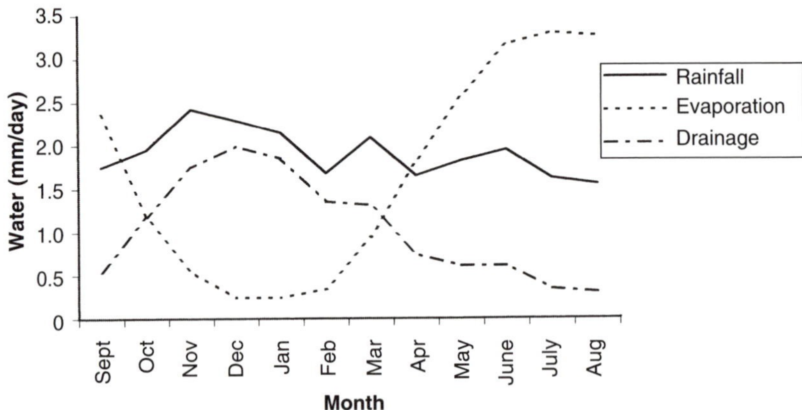

Fig. 5.8. Changes in the course of a year in rainfall, evaporation and drainage through 0.5 m soil at Rothamsted, expressed as mm/day. Average mm/day for each month.

between the months of the year (Fig. 5.8). The amount of evaporation varies far more. During autumn, the radiation energy capable of evaporating water from the soil becomes less as the days become shorter and the sun's radiation is attenuated by the increasing distance it has to travel through the atmosphere.

At harvest, the soil is usually dry because of direct evaporation of water and transpiration by the crop. As autumn progresses, rainfall overtakes evaporation and the soil becomes moister until water begins to flow downwards and then out of the soil. Drainage has begun. There is a critical period of time during which drainage occurs but there is no crop to capture the nitrate in the soil. This is a particular danger of nitrate leaching, whether the nitrate was left from the fertilizer application or produced by microbes. Table 5.1 shows the danger period as being between 15 October and the emergence of the crop. This date is a likely time for the resumption of drainage, but the actual date varies enormously from year to year. The length of the danger period differs greatly between autumn- and spring-sown crops, but there are also considerable differences within these categories.

Production of nitrate by mineralization in the autumn

The ^{15}N experiments suggest that only about 6% of the nitrogen fertilizer is lost directly by leaching as nitrate. But 6% of an average application of nitrogen to winter wheat of 180 kg/ha is 11 kg/ha, and we know that far more is lost in practice. From where does it come?

In Chapter 3 we saw how microbes and the other members of the soil population play a vital role in ecosystems by recycling plant remains. They release as ammonium and nitrate the nitrogen entombed in large, redundant macromolecules and thereby enable the next generation of plants to use them for growth. The soil population is active when the conditions suit it, and it certainly does not cease work when the soil is bare. The state of the soil after harvest in early autumn, when the soil is still warm from summer but is becoming moister as rainfall overcomes evaporation, suits it very well indeed and it produces nitrate vigorously. This nitrate is helpful if there is an actively growing crop in the soil, but it becomes part of the nitrate problem if there is not, particularly if the resumption of drainage is approaching or has occurred.

We saw in Chapter 2 that nitrate produced in the soil has exactly the same structure and properties as nitrate from nitrogen fertilizer. The nitrate produced by the soil population is therefore just as untimely as nitrate left from fertilizer. This entirely natural nitrate is the greatest cause of the nitrate problem – responsible for a far greater proportion of nitrate losses to natural waters than direct losses from fertilizer. This important conclusion was reached in the 1980s, but it receives strong support from an experiment made 100 years earlier.

Nitrogen Loss from Organic Matter: Evidence from the Rothamsted Drain Gauges

Lawes *et al.* (1881) constructed the Drain Gauges at Rothamsted in 1870. There were three gauges, each comprising a natural block of soil of 1/1000 of an acre (4 m^2) surface area. Their depths of soil were 20, 40 and 60 inches (0.5, 1.0 and 1.5 m). The blocks of soil were isolated by brick walls and undermined to enable drainage-collection systems to be installed (Fig. 5.9). This method of construction left the soil almost undisturbed. The gauges were designed to measure the amount of water passing through the soil. But nitrate was noticed in the drainage and from 1877 until 1915 its concentration was measured daily. The measurements were used by Addiscott (1988a) to derive information about long-term losses of nitrate from soil organic matter. The results given here are from the deepest gauge, considered the most reliable. Those from the other two gauges were broadly similar.

During the first 7 years of these measurements the soil lost an average of 45 kg/ha of nitrate-N each year. These losses were noteworthy in several respects.

- The soil had not been seriously disturbed during construction – any serious disturbance would have encouraged the soil population to produce nitrate.
- The soil had already had 7 years to settle down before the measurements started.
- No crop was grown and the only form of cultivation was hand-weeding.
- After 1868 neither fertilizer nor manure was applied, so the measurements were made between the ninth and the 16th years since any previous application.

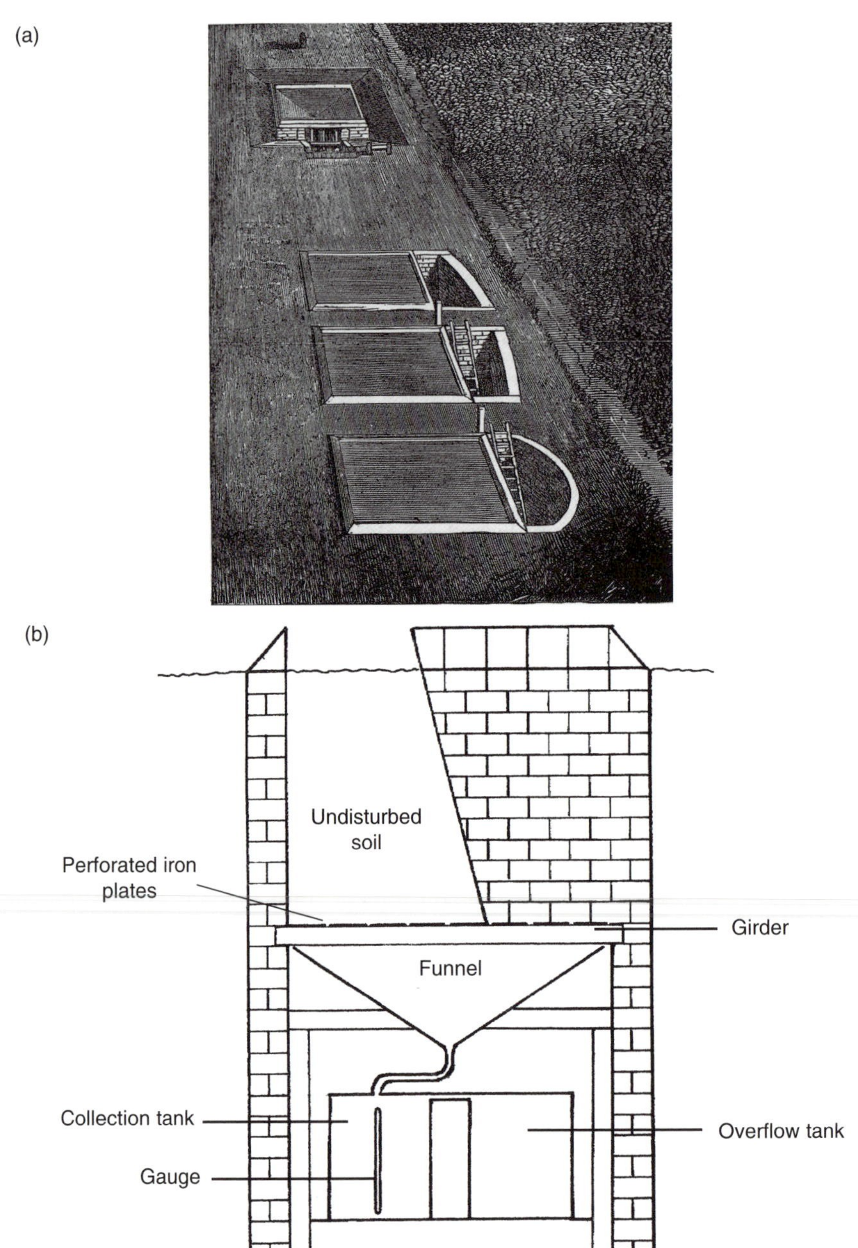

Fig. 5.9. (a) The Rothamsted Drain Gauges as depicted by Lawes *et al.* (1882). There are three Drain Gauges in the foreground; the structure in the background is the 1/1000 acre rain gauge. (b) Details of an individual gauge (from Addiscott *et al.*, 1991). Lawes *et al.* constructed three gauges, each with a surface area of 1/1000 acre (about 4 m^2), in 1870 to measure the quantities of water draining through 20, 40 and 60 inches (0.5, 1.0 and 1.5 m) of soil. These comprised natural blocks of undisturbed soil that were isolated by brick walls and undermined such that collectors for the drainage could be installed. The soil in the gauges carried no crop, received no fertilizer and was cultivated only minimally to kill weeds.

During the 7 years only 3–5 kg/ha was brought in by rain each year, so practically all of the remaining 40–42 kg/ha must have come from the soil.

During the whole 38 years of the measurements about 1450 kg/ha of nitrate-N leaked from the soil in the gauge. This amount was almost exactly the same as the measured decrease in the soil nitrogen in the gauge, about 1410 kg/ha. Mineral nitrogen is only 1–2% of the total nitrogen in the soil, so virtually all the nitrate that leaked from the soil must have been mineralized from nitrogen in the soil organic matter. Put another way, the organic matter was the source of nearly 1.5 t of nitrate-N – that is, more than 6 t of nitrate – lost per hectare during the 38 years. There can be no doubt about the role played by mineralization in nitrate losses from the soil.

The leakage was not only large, it went on for a long time (Fig. 5.10) and had a half-life of 41 years. That is to say, the rate of leakage took 41 years to fall to half its initial value. It would probably have taken 130–140 years to fall to a tenth of its initial value. This was a slow and inexorable loss of a very large pool of organic nitrogen which, incidentally, showed no respect for European Community (EC) legislation. The water draining from this unfertilized soil in the 1870s initially had a nitrate concentration that exceeded the EC limit of 50 mg nitrate/l. It was 7 years before the drainage water became officially potable.

What the crop leaves behind: memory effects and the surplus nitrate curve

The slow and inexorable loss of nitrate described above happened without additions of fertilizer nitrogen, but surely such additions must increase the

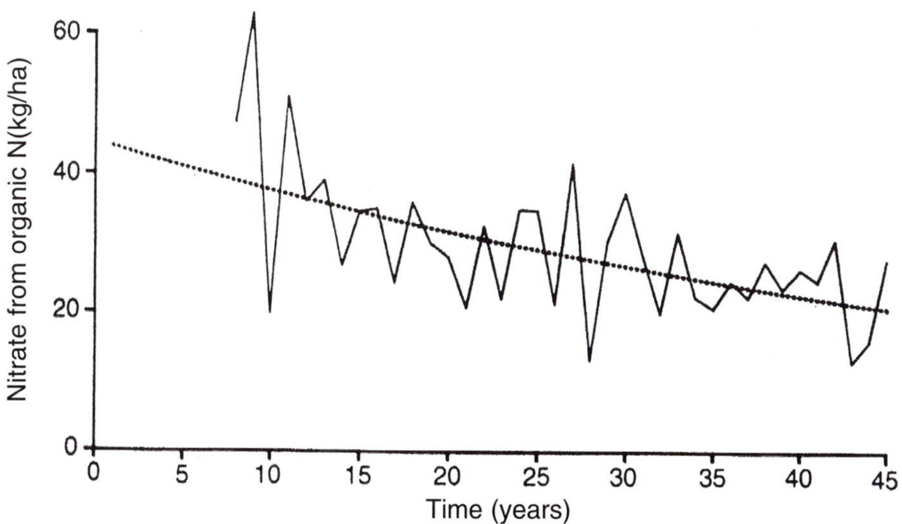

Fig. 5.10. The leakage of nitrate derived from nitrogen in soil organic matter in the 60 inch (1.5 m) Drain Gauge at Rothamsted between 1877 and 1915. The broken line is the result of removing the effect of fluctuations in rainfall. (From Addiscott, 1988a, and Addiscott *et al.*, 1991.)

loss? Direct losses from nitrogen fertilizer *before* harvest are small, but what is the effect of the fertilizer on losses of nitrate mineralized *after* harvest? The experiments with ^{15}N-labelled fertilizer applied to winter wheat showed that 10–25% of the fertilizer nitrogen was found in organic matter in the soil at harvest. Will this 'new' organic matter, most of which comes from the residues of the crop just grown, be broken down more readily by the soil population than older existing organic matter, some of which is inert? This effect of fertilizer on subsequent mineralization was described by Addiscott *et al.* (1991) as the 'memory effect'. Memory effects obviously depend on the crop and the number of years for which it is grown in the same place.

Research at Rothamsted suggests that nitrogen fertilizer applied to winter wheat grown for 1 year does not leave a memory effect. The amount of nitrate produced by the soil population does not seem significantly greater with fertilizer than without. There is, however, a definite memory effect on the Broadbalk Experiment at Rothamsted (Powlson *et al.*, 1986; Glendining *et al.*, 1992). Winter wheat has been grown for 160 years and the production of nitrate produced by mineralization increases with the annual nitrogen application. Figure 5.11 shows this through the uptake of unlabelled nitrogen when ^{15}N-labelled fertilizer is applied. (The 42 kg/ha uptake with no applied nitrogen reflects the deposition of nitrogen oxides and ammonia from the atmosphere, estimated in Chapter 7 as between 37 kg/ha for bare soil and 48 kg/ha for a well-established winter wheat crop. This was a thin crop.)

It is not yet clear how long such memory effects take to build up for winter wheat, but nitrogen applied to some crops, including oilseed rape, seems to leave a memory effect after one season. Researchers at ADAS found that mineralization of nitrate in the soil after the harvest of a rape crop increased with the amount of fertilizer given to the crop (R. Sylvester-

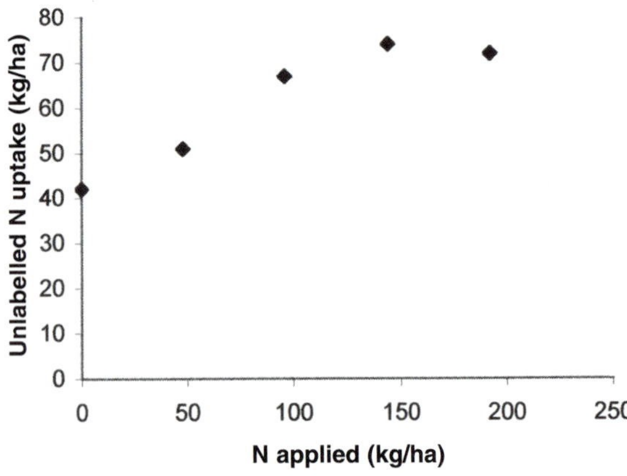

Fig. 5.11. The 'memory effect' on the Broadbalk Experiment. Effect of the amount of nitrogen fertilizer given long-term on the rate of nitrogen mineralization, measured as the uptake of unlabelled nitrogen when ^{15}N-labelled fertilizer had been applied. (From Powlson *et al.*, 1986.)

Bradley, Cambridge, 1990, personal communication). The rapid development of the memory effect with oilseed rape implies that its residues decompose quickly in the soil. This is probably because the crop has a habit of shedding its leaves while the pods are ripening, which leaves the residues vulnerable to attack by the soil population before harvest. Other crops that shed their leaves early may also show rapid memory effects. Memory effects for potatoes and sugarbeet must be dominated by the way in which the haulm and leaves are treated. They could give definite memory effects if ploughed into the soil in the autumn.

In the experiments with ^{15}N-labelled nitrogen fertilizer, winter wheat left only about 1% of the fertilizer nitrogen in the soil as nitrate at harvest. Part of the reason for this small proportion was that nearly all of the nitrogen was taken up by the crop. If too much fertilizer had been supplied, the proportion left in the soil would have been much greater. It may help to think of a 'surplus nitrate curve' relating the amount of nitrogen left in the soil to the amount given as fertilizer (Fig. 5.12). This shows the amount remaining to remain constant and relatively small until more fertilizer nitrogen has been supplied than the crop can use, when the curve turns sharply upwards. The relation of this 'upturn point' to N_{opt}, the optimum fertilizer application estimated by experiment, is discussed in Chapter 10. Chaney (1990) provided experimental evidence for the surplus nitrate curve.

Memory effects become somewhat academic if the farmer strays too far

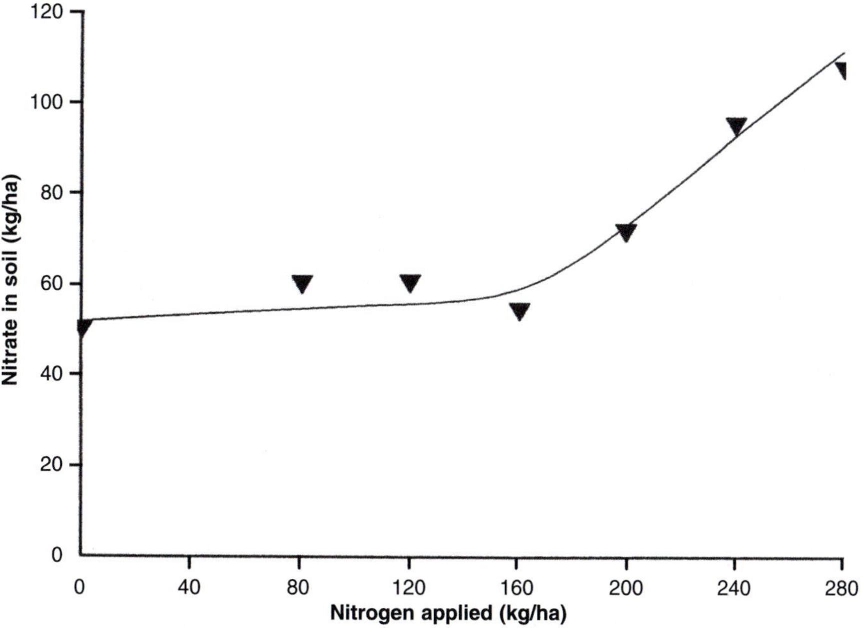

Fig. 5.12. The 'surplus nitrate' curve. Relationship between the amount of fertilizer supplied and the amount of nitrate remaining unused in the soil after winter wheat was harvested. (From Chaney, 1990.)

along the surplus nitrate curve. Chaney's (1990) results (Fig. 5.12) suggest that the upturn point, which marks the safe limit, is about 180 kg/ha of nitrogen. Larger applications are quite common, but they can be justified only if the crop uses them. A useful rule of thumb says that winter wheat takes up about 23 kg/ha of nitrogen for each tonne of grain produced. Wheat yields of 10 t/ha are now not uncommon, and farmers who achieve them can make a case for applying 230 kg/ha of nitrogen. Yields of 13 t/ha are not at all common, and farmers who apply 300 kg/ha have probably gone too far down the surplus nitrate curve. Even a more modest target can be a problem. A farmer who aims for 10 t/ha and achieves only 7 t/ha, perhaps because of a crop disease, could leave 60–70 kg/ha of fertilizer nitrogen in the soil at harvest. This is not just a waste of money. If the drainage the following winter is not very great, that unused fertilizer could *on its own* take the nitrate concentration in that drainage close to the EC limit of 50 mg/l. Farmers need to think carefully before they apply more than 200 kg/ha of nitrogen.

Many farmers in England now find themselves in Nitrate Vulnerable Zones (NVZs), areas in which farming activities are subject to a set of rules (Box 5.4). NVZs are now closely defined by maps that can be accessed by computer and which allow the farmer to find out why a particular piece of land is in an NVZ. One reason will often be that the soil is 'leaky' with respect to nitrate. The NVZs that were designated in 1996 were localized and occupied a fairly small part of the landscape, but new NVZs designated recently cover large areas of central and Eastern England.

Application of nitrogen fertilizer in autumn

Nearly all nitrogen fertilizer is now applied in spring. But for many years autumn-sown crops were assumed to need a small amount of nitrogen fertilizer in the seedbed to help them through the rigours of winter. The Lawes Chemical Company, for example, used to produce a 6–15–15 compound (containing 6% of N and 15% each of P_2O_5 and K_2O) for use in the autumn. This practice was brought into question when it was realized how much nitrate was produced by the soil population in autumn and that the fertilizer nitrogen was just as much at risk as the naturally produced nitrate. It was shown clearly to be wasteful by Goss *et al.* (1993) in work on the Brimstone Experiment. They showed that for every kilogram of nitrogen applied as fertilizer an extra kilogram of nitrogen was leached as nitrate. It was probably not the fertilizer nitrogen that was leached but soil nitrate that the crop would have taken up if the fertilizer had not been applied.

Not many farmers now use nitrogen in the autumn, but the reasons for which it might still be used reflect the complexity of the nitrate issue. The burning of straw is now banned, and farmers have to plough it into the soil, sell it or find a use for it. Many plough it into the soil, but when they do so the straw immobilizes mineral nitrogen in the soil. This is because straw has an N:C ratio that is too low for the soil population to be able to metabolize

Box 5.4. A summary of the rules for Nitrate Vulnerable Zones.

The rules for Nitrate Vulnerable Zones (NVZs) apply in different ways to the arable land discussed in this chapter and the grassland discussed in Chapter 6. Both types of land are included in this box, but please note too the code for managing manures and slurries (Box 6.1). These rules are concerned with closed periods, when no applications may be made, nitrogen limits on applications, spreading controls, slurry storage and record keeping. (Based on the Department for Environment, Food and Rural Affairs' summary of NVZ rules.)

Closed periods and nitrogen limits

Closed periods	Arable	Grassland
N fertilizer (all soil types)	1 September–1 February	15 September–1 February
Organic manures (containing much available N, e.g. slurry and poultry manure, on sandy and shallow soils)	1 August–1 November (with no autumn-sown crop)	1 September–1 November (includes arable with autumn-sown crop)

Nitrogen limits	Arable	Grassland
N fertilizer	Do not exceed crop requirement	Do not exceed crop requirement
Organic manures		
(i) Whole farm within NVZ (including grazing deposition)	210 kg/ha[a] total N	250 kg/ha total N
(ii) Field limit[b] (excluding grazing deposition)	250 kg/ha total N	250 kg/ha total N

[a]Reduces to 170 kg/ha after first 4 years of Action Programme in NVZ.
[b]Available N from organic manures must not exceed crop requirement.

Spreading controls

- Do not apply nitrogen fertilizer or organic manures when the soil is waterlogged, flooded, frozen hard or covered by snow.
- Do not apply nitrogen fertilizer or organic manures to steeply sloping fields.
- Spread nitrogen fertilizer and organic manures evenly and accurately.
- Do not apply nitrogen fertilizers in a way that contaminates water courses.
- Do not apply organic manures within 10 m of water courses.

Slurry storage

- There must be sufficient storage to meet the autumn closed period for spreading slurry.
- All new or substantially enlarged/reconstructed storage facilities must comply with the relevant regulations.

Record keeping

Adequate farm records must be kept for at least 5 years for cropping, livestock numbers and the use of nitrogen fertilizers and organic manures.

it properly (Fig. 3.2) and they take mineral nitrogen from the soil to do so. The amounts of nitrogen involved are not large – 1 t of straw immobilizes about 10 kg of nitrogen. But one or two farmers might be sufficiently sensitive to yellowing in their crops to apply some nitrogen.

Autumn nitrogen applications are not mentioned in the 2002 Fertilizer Review (Fertilizer Manufacturers' Association, 2003) but the most recent version of RB209, the fertilizer recommendation booklet (MAFF, 2000), suggests 30 kg/ha of nitrogen in the seedbed for autumn-sown oilseed rape grown in soils with the two lowest categories of nitrogen availability. Curiously enough, this might have to do with pigeons. Young oilseed rape plants are very attractive to hungry pigeons, and a large flock can devastate a crop. But pigeons, like aircraft pilots, do not like to land in heavy vegetation in case they cannot take off again. That 30 kg/ha of nitrogen may help the crop to grow rapidly to a height at which it signals danger to the pigeons, but it probably has no other benefit to yield.

Fate of nitrogen applications that exceeded uptake by the crop

Calculations made by ADAS (Sylvester-Bradley *et al.*, 1987) showed the following trend. Before 1976 farmers nationally applied less nitrogen to winter wheat than the crop took up. In 1976 the amounts supplied and taken up were in balance, and from 1977 onwards more was applied than was taken up. By 1986 the cumulative excess was more than 300 kg/ha. Addiscott (1996) was curious to know what had happened to the excess nitrogen and attempted an assessment based on the following rules.

1. The fate of the fertilizer nitrogen at harvest was assumed to be similar to the fate of the ^{15}N in the experiments of Powlson *et al.* (1992) at Rothamsted, discussed earlier in the chapter. Taking the average figures: 65% in the crop, 18% in soil organic matter, 1% as mineral nitrogen in the soil, 10% denitrified and 6% leached.

2. The excess was assumed to be made up by the 35% of the nitrogen that had not gone into the crop. Of this, 51% was in the soil organic matter, 29% denitrified and 20% leached. (The leached component included the 1% of the original application that was in the soil at harvest.)

3. The fertilizer nitrogen that went into the soil organic matter was assumed to be remineralized as follows: 10% in the first year, 3% of the remainder in the second year and 1% of the remainder in each subsequent year.

4. One-third of the nitrogen that was remineralized was leached and the remainder taken up by the crop.

These rules were applied to the excess of nitrogen applied over nitrogen taken up read for each year from Fig. 4 of Sylvester-Bradley *et al.* (1987), giving the results shown in Table 5.4. The nitrification and denitrification losses are the cumulative values for 10 years and should be divided by 10 to give an average annual loss. If the assumptions underlying the calculations are correct, 49% of the excess nitrogen was lost in the year in which it was

Table 5.4. Fate of nitrogen fertilizer calculated by Sylvester-Bradley *et al.* (1987) to have been applied in excess of crop uptake in England and Wales, 1977–1986. Based on assumptions listed in text. (Adapted from Addiscott, 1996.)

Fate of nitrogen	Quantity of N	
	As kg/ha	As % of total excess
Leached during year of application	65	20
Denitrified during year of application	95	29
Re-mineralized and then leached[a]	7	2
Re-mineralized and taken up by crop[b]	13	4
Left in organic matter at end	146	45
Total	326	100

[a]Corresponds to 0.7 kg/ha per annum.
[b]Corresponds to 1.3 kg/ha per annum.

applied and 45% of it was in the soil organic matter at the end of the 10 years. A small annual release of mineral nitrogen (nearly all nitrate) from the excess was to continue indefinitely. This release is the source of the greatest uncertainty in the results because of the fairly arbitrary assumptions that led to it. Two things, however, can be predicted. The first is that, if the excess of nitrogen applied over nitrogen taken up has continued, the release will have increased, but only slightly. When the excess ceases, the release will decrease only very slowly, with a half-life comparable to that of the loss of nitrogen in the Drain Gauges.

Eighteen years have elapsed since 1986 and much has changed. Applications of nitrogen fertilizer peaked in the UK in 1985 and have been on a downward trend since then (Greenland, 2000). Crop uptake, however, has increased, leading the Fertilizer Manufacturers' Association (2002) to comment in its most recent report that, '... continually increasing yields and nutrient offtake with reducing fertilizer use cannot continue indefinitely without adversely affecting crop quality and soil fertility'. The ADAS calculations have not been repeated so far as I am aware, but, if they were, the excess of fertilizer application over uptake would probably not be found to have accumulated after 1986 at anything near the rate it did between 1977 and 1986.

Spatial Patterns in Losses of Nitrate from Arable Land

Spatial patterns in nitrate losses are important. If the farmer knows where losses are most likely to occur, he can use precision farming techniques to try to avoid them. Environmentalists are interested in spatial patterns because they are essential in scaling up from localized measurements at the $i - 1$ or i scale to provide assessments of losses at the catchment, regional or national scale ($i + 2$ to $i + 4$) for policy purposes. The problem of discontinuity implies that we cannot simply multiply up from small-scale measurements. An emission of 10 g from 1 m^2 does not necessarily imply an emission of 10 kg from 1000 m^2, because we do not know the whereabouts of the 1 m^2 in the overall spatial pattern. We are concerned here with both of the main environmental nitrate issues, nitrate leaching to surface waters and emissions of nitrous oxide to the atmosphere.

The number of individual measurements that are necessary arguably makes it more difficult to assess the spatial pattern of leaching losses from the base of the soil profile than that of nitrous oxide emissions from the soil surface, but neither is easy. In many cases, the patterns for both will depend on the topology and hydrology of the landscape as shown by Farrell *et al.* (1996). Nitrate that is lost upwards in gaseous form obviously does not contribute to the nitrate concentration in drainage water, and these authors found that, within a landscape, there was a negative correlation between denitrification activity and the concentration of nitrate leached to depth. These are patterns that become apparent only the scale of the landscape ($i + 2$ to $i + 3$) and provide an example of decoherence (Box 5.2). Interestingly, the authors also found a negative correlation between the concentration of deep-leached nitrate and its $\Delta^{15}N$ value (the abundance of ^{15}N relative to the natural abundance of 0.3663 atom%). They interpreted this in terms of isotopic discrimination (or fractionation) during denitrification (Box 5.5).

Box 5.5. Isotopic discrimination.

We saw in Box 3.2 that isotopes of an element are atoms that have the same number of protons but differing numbers of neutrons and that some elements have only one isotope, while others have more. Nitrogen is the seventh atom in the periodic table and has an atomic number of 7. Its commonest isotope ^{14}N has (as the superscript implies) an atomic weight of 14, but measurements of the atomic weight of nitrogen give a value of 14.008, implying that nitrogen must have a heavier isotope. In fact, nitrogen has both a heavier and a lighter isotope, ^{15}N and ^{13}N , but the former is more common.

All the isotopes of an element have the same chemical properties and the only thing that can make them differ in behaviour is the difference in weight. For example, ^{15}N is 7% heavier than ^{14}N and 15% heavier than ^{13}N . These differences are not particularly large but they are enough to make ^{15}N the most inert in certain processes, particularly those involving the evolution of gases. The dinitrogen and nitric oxide lost during denitrification will be slightly richer in ^{13}N and ^{14}N than in ^{15}N , while the nitrate left will be richer in ^{15}N. These differences are not large but, as the experiments of Farrell *et al.* (1996) show, they are clearly detectable.

We saw in Chapters 3 and 5 that nitrogen enriched with ^{15}N is important in scientific research on the fate of nitrogen fertilizer, and isotopic discrimination during movement on a separation column provides one method of obtaining enriched material.

Spatial patterns in nitrate leaching

There do not seem to be many reports of spatial patterns in measurements of nitrate leaching *per se*, but this is not surprising because it would involve a substantial number of lysimeters or porous ceramic cups, the installation of which could well disrupt any pattern.

Nitrate concentrations measured in the soil provide an indication of potential leaching but, as the study of Farrell *et al.* (1996) showed, potential leaching may not become actual leaching. Patterns in cropped land depend on the nature and spatial distribution of the crop. Cereals are sown close together and their roots take nitrate from the whole volume of the soil, so no clear patterns may occur. Land at Rothamsted which had previously carried cereals showed little spatial pattern in nitrate concentration (Whitmore *et al.*, 1983). A much clearer pattern was found by Schroth *et al.* (2000) in an oil palm plantation. The roots had not exploited the whole soil volume, and the nitrate concentration increased with distance from the palms. The authors suggested that broadcasting fertilizer rather than placing it at the base of each tree might have favoured lateral root growth and given better recovery of nitrate.

Box 5.1 suggests that the larger the area of land, the clearer the spatial patterns of leaching will be, because there are more likely to be changes in topography, soil type and vegetation.

Spatial patterns in nitrate leaching can be investigated with rather less effort by computer modelling. Very large areas, such as the central USA, can be simulated (Wu and Babcock, 1999). But there are potential traps for the unwary. One arises if there is non-linearity in the model – that is, when an output from the model shows a non-linear relationship with one of the model's parameters. The interaction between the non-linearity and the spatial pattern has the potential to give misleading results (Addiscott and Bailey, 1990; Stein *et al.*, 1992), and is part of the general problem of error propagation by models, the transmission of error through the model from parameter to output. The error is usually error in the statistical sense of variation, but it may take other forms. Error propagation is a particular problem when models are used in conjunction with GIS, and Heuvelink's (1998) monograph on the topic provides a warning of the possible consequences, which can be serious.

Spatial patterns in nitrous oxide emissions

Gaseous losses have long been recognized as a spatially variable process (e.g. Foloronso and Rolston, 1985), but certain aspects of the process add to the complexity. Discontinuity is one problem. Another is the likelihood that the assumption of stationarity that underlies such studies is not valid (Lark *et al.*, 2004a). (This assumption implies that the data can be treated as if they were generated by a statistical process that is uniform at all points in space. This assumption becomes untenable if the variability of an item in the data changes markedly from one part of the landscape to another.) Also, losses of nitrous oxide from the soil result from two processes, nitrification and incomplete denitrification, that have different substrates (ammonium and nitrate) and are not subject to the same controls. Lark *et al.* (2004a,b) have addressed these problems through the application of Wavelet Theory, a new mathematical technique for describing spatial processes that does not depend on assumptions of stationarity. It has recently been extended from signal processing and geophysics to soil processes (Lark and Webster, 1999).

Lark *et al.* measured the emissions of nitrous oxide at 256 points along a 1024 m transect of arable land using the soil core method of Webster and Goulding (1989). (They would have used closed chambers to measure the emissions, but making 256 such measurements on the same day with the staff available was not feasible. The emissions are temporally as well as spatially variable (Smith and Dobbie, 2001), and the two forms of variation would have been confounded if the measurements had been split over the 8 or so days probably needed for measurements using chambers.)

The wavelet transform used by Lark *et al.* (2004a) enabled them to successfully describe the complex spatial pattern in emissions of nitrous oxide along the transect. Their analysis showed intermittency in the variation of nitrous oxide emission at different spatial scales, with changes in variability that were clearly associated with changes in topography and parent material. The implication of this is that geostatistical techniques may not be appropriate for the spatial analysis of nitrous oxide emissions from the soil.

Examining covariations of nitrous oxide emissions with soil properties (Lark *et al.*, 2004b) revealed further complexity. Not only do nitrous oxide emissions vary spatially, but also the factor that most limits the emissions varies spatially too, and the factors that are most important differ between spatial scales. For example, in the first 50 sampling points on the transect, the emissions were best correlated at the finest scale with soluble organic carbon, at the next scale with water content, and at the coarsest with soil pH. The pattern differed considerably on much of the rest of the transect and the strongest correlation was with pH, reflecting the influence of the parent material. Overall, the results could be summarized in the words of Wagenet (1998), '... at different scales, different variables are often needed to describe similar processes'. Box 5.1 suggests that this should not come as a surprise.

6 Losses of Nitrogen from Grassland

Nitrate losses and emissions of nitrous oxide from grassland differ from those from arable land in two ways.

- Arable cropping always leaves the soil without a crop for part of the year. During this time, there are no roots to retrieve nitrate before it is lost from the soil by leaching or denitrification. Under grassland, roots are there all the time when the grass is permanent and usually for at least 3 years when it is temporary, so losses of nitrate from it are inherently smaller.
- Nitrate management is complicated in arable cropping by the activities of the soil's population of microbes and soil animals, which release nitrate by mineralization and destroy it by denitrification. Grassland farmers encounter not only these complications but also an additional set of complications arising from excretions of dung and urine by much larger animals, which result in large losses of nitrate. This chapter mentions only cattle and sheep but other animals may be involved. Remember, from Chapter 3, that when you see a field of sheep, the billions of microbes within the soil of the field have roughly the same collective body mass as the sheep (Jenkinson, 1977).

There is obviously a difference between grassland which is grazed by cows or sheep and ungrazed grassland. There is a further distinction between intensively managed grassland, which receives substantial inputs of nitrogen fertilizers, and rough grazing, often on hills, which usually receives none. In the UK, roughly speaking, agricultural land is split three ways: one-third each to arable cropping, managed grassland and rough grazing. The rough grazing contributes little to nitrate leaching and only slightly to gaseous losses of nitrous oxide, and needs no further specific mention. The chapter is therefore mainly concerned with grazed or ungrazed managed grassland. The complications introduced by the animals are considerable, and research on systems that include them is usually appreciably more difficult than that on arable systems, but very interesting progress has nevertheless been made.

Because grassland always has a well-established root system to retrieve nitrate, it can be given substantial applications of nitrogen fertilizer, up to about 400 kg/ha per year, without appreciable nitrate losses from the soil (Prins *et al.*, 1988; ten Berge *et al.*, 2002). Figure 6.1 shows the surplus nitrate curve for ungrazed grass (a) in relation to the yield response curve (b) (cf. Chapter 5). This curve applies until you want to use the grass for feeding, most often of cattle. You can either feed it *in situ* by allowing the animals to

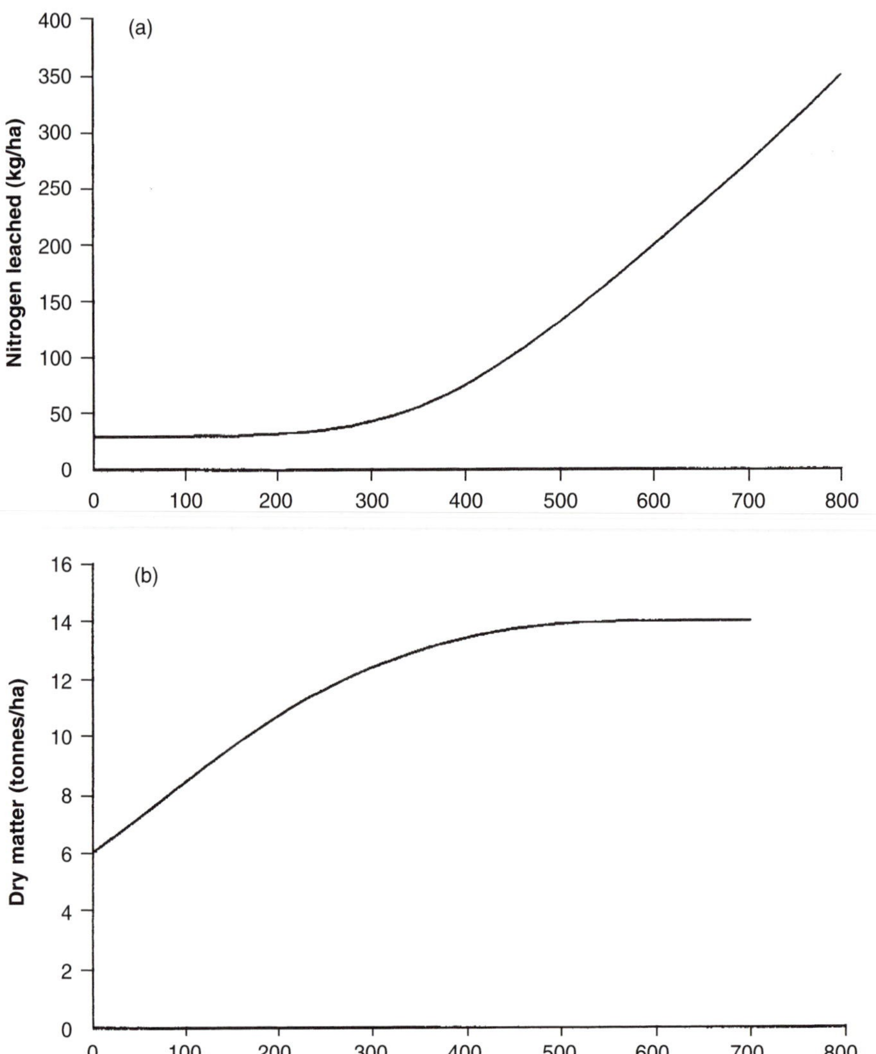

Fig. 6.1. (a) Surplus nitrogen curve for grass without cows. (b) Response of grass to applied nitrogen. (Adapted from Prins *et al.*, 1988.)

graze it or you can cut it and feed it to the animals elsewhere. Either way you get problems with nitrogen-containing material excreted as urine or dung.

Grazed Grassland

Until 1984 it was generally assumed that, because grassland maintained a root system all the year round, leaching of nitrate from it was not a problem. That assumption had to be abandoned when Ryden *et al.* (1984) showed unequivocally that leaching from grassland occurred because of the non-uniform deposition of urine and dung by cattle and sheep. A cow urinates about 2 dm^3 of urine on to an area of roughly 0.4 m^2, corresponding to a nitrogen application of 400–1200 kg/ha to that area. The urea in the urine is hydrolysed by the enzyme urease to ammonium, which is soon nitrified in all but the coldest temperatures. The resulting nitrate is very vulnerable to leaching because the grass simply cannot use so much nitrate. The safe application of 400 kg/ha is split into several aliquots to ensure even growth during the season and becomes irrelevant when so much nitrogen arrives in urine all at once. The problem is exacerbated by the gregarious habits of cows and the preference they seem to have for urinating in a particular part of a field. This may result in only about 15–20% of the field being fertilized in this way. Depositions of nitrogen in dung are smaller, about half those in urine, but they are spread on an even smaller area, maybe 7–10% of the field. The dung and urine patches also lose significant amounts of nitrous oxide and ammonia to the atmosphere.

The problem that underlies these losses is that ruminant animals are not efficient at converting nitrogen into useful products. Of every 100 kg of nitrogen applied as fertilizer, as little as 10–20 kg may be found in the meat, milk, wool or other products for which the animals are kept. Much of the rest, perhaps 80% of the nitrogen consumed by the animal, is excreted and, because of its concentration on small areas, a substantial part of it is lost. This inherent inefficiency implies that we need to know how the efficiency of nitrogen use changes as more nitrogen is applied. Using more nitrogen means that more grass is produced and more animals can be kept (van Burg *et al.*, 1981). The productivity per hectare increases, but the productivity per animal shows a sharp, almost linear decrease. More nitrogen means not only more animals excreting but also more nitrogen excreted per animal. This may be because the extra nitrogen encourages the grass to take up more nitrate than it needs for growth and to store some in its tissues. Farmers are careful, incidentally, not to allow their cattle to graze fields for about 10 days after application of nitrogen fertilizer. This is because the pH of the rumen is much less acid than that of the human stomach, making the animals vulnerable to methaemoglobinaemia. (The occurrence of this condition in infants is discussed in Chapter 10.)

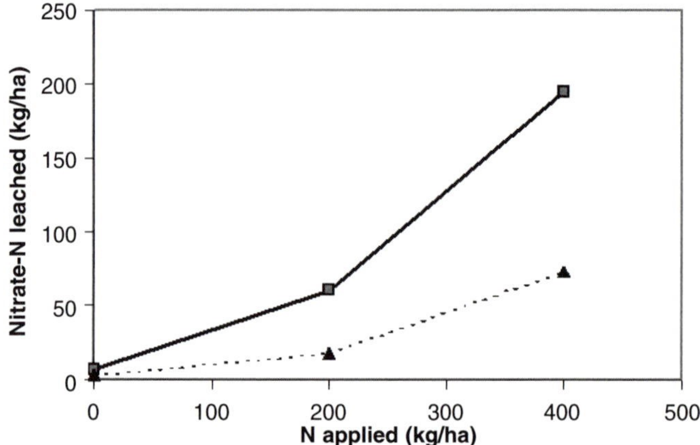

Fig. 6.2. Losses of nitrate by leaching from drained (■) and undrained (▲) grazed grassland. (Redrawn from data of Scholefield *et al.*, 1993.)

Quantifying leaching losses from grazed systems

The pattern of efficiency outlined above is reflected clearly in measurements of nitrate leaching from grazed grassland on a silty clay loam in south-west England (Scholefield *et al.*, 1993). Plots with no fertilizer lost virtually no nitrate. Increasing the nitrogen application from 0 to 200 kg/ha resulted in a modest increase in nitrate loss, but increasing it from 200 to 400 kg/ha gave a much greater increase (Fig. 6.2).

This experiment also examined the side-effects of installing an efficient field drainage system. Such a system is desirable because it alleviates the damage done when heavy-footed cattle trample a water-logged grass sward (poaching). But, when the drainage lessened water-logging, it also improved conditions for aerobic, particularly nitrifying, bacteria with the result that more nitrate was produced. This and the freer movement of water in the soil led to a threefold increase in losses of nitrate from the soil (Fig. 6.2). The losses also increased with the age of the sward.

Cutting losses of nitrate from grazed grassland

Using more nitrogen means more grass is produced and more animals can be kept. But each animal excretes more nitrogen, so losses of nitrate from grazed systems become disproportionately greater as more fertilizer is applied. Losses can only be restrained by keeping fertilizer application within safe limits. How much nitrogen is safe? This depends on the age of the sward, the type of soil and how well drained it is, and the time the cattle are taken off the land. The experiments of Scholefield *et al.* (1993) and others suggest that about 250 kg/ha is probably the most nitrogen that should be applied to grazed grassland, but Jarvis (1999) found that the typical input to a dairy farm in the UK is 281 kg/ha. The extra 31 kg/ha does not sound a particularly large difference until you note that the typical offtake of nitrogen from the farm was only 36 kg/ha. Its potential cumulative effect is considerable.

The largest safe application suggested for grazed grassland, 250 kg/ha, does not differ greatly from the safe application to winter wheat (Chapter 5), which was suggested to be 200 kg/ha or 230 kg/ha if the farmer had reasonable expectations of a 10 t/ha yield. But we need to remember that the safe application also depends on soil type, and grassland soils are often heavier than arable soils.

Quantifying emissions of nitrous oxide from grazed systems

Estimates of emissions from grazed systems are needed at various scales. Working at the scale of the field plot, Yamulki *et al.* (1998) measured the emission of nitrous oxide from the urine and dung of cattle applied to experimental areas using closed steel chambers pushed into the soil to a depth of about 5 cm. Each chamber was 40 cm in diameter, but the size in relation to soil variability was probably not an issue because the emissions were dominated by the patch of dung or urine and the chamber covered the patch. The losses of nitrous oxide from the excreta were significant but were not correlated with rainfall or mineral nitrogen in the soil, possibly implying that they were of a somewhat different nature from losses from arable soils. Nitrogen lost as nitrous oxide from dung and urine patches during 100 days was about 0.5% and 1.0% respectively of the nitrogen excreted, and the annual emission was roughly five times greater from the urine patches than from the dung patches. Fluxes from the excreta were greater during the wet conditions occurring in autumn than in drier summer conditions. In principle, scaling up from such measurements is not a problem if the likely nitrogen content and number of patches per hectare is known. Doing so suggested that emissions of nitrogen as nitrous oxide from dung and urine deposited by grazing animals in the UK may amount to 20 Gg/year and contribute up to 22% of total nitrous oxide emissions from grassland in the UK, but this figure is subject to enormous spatial and temporal variability.

Kelliher *et al.* (2002) recently developed an interesting approach to assessing emissions at the field scale. During calm clear nights a shallow inversion layer develops in the atmosphere above the soil surface. This effectively traps nitrous oxide, carbon dioxide and other gases emitted from the soil, and defines a volume within which they are held. Kelliher and his colleagues used a method that measures the absorption of infra-red radiation by the gas molecules (open path Fourier-transform infra-red spectroscopy) to measure concentrations of nitrous oxide and carbon dioxide along a path of 97 m at a height of 3 m in this layer. This approach not only provided a means of estimating nitrous oxide emissions but also showed a very strong correlation between these emissions and those of carbon dioxide. This approach is useful because it can integrate nitrous oxide emissions over a larger area than other methods, but it depends greatly on the number of calm clear nights in the area under study.

Cutting losses of nitrous oxide from grazed systems

Nitrous oxide emissions, like losses of nitrate by leaching, result from untimely nitrate, nitrate that is in the soil at the wrong time. They can be restricted, as can leaching losses, by giving no more than 250 kg/ha of nitrogen fertilizer. The improved soil aeration that resulted from field drainage encouraged aerobic nitrifying bacteria and increased nitrate losses by leaching (Scholefield *et al.*, 1993), but the improved aeration should lessen emissions of nitrous oxide from denitrification. Unfortunately, however, work by Yamulki *et al.* showed that drainage increased emissions of nitrous oxide. This was presumably nitrous oxide emitted during the enhanced nitrification.

Ungrazed Grassland: Cutting the Grass and Feeding it Elsewhere

Grass is a safe crop that does not lose nitrate, provided you keep the cows off it. It may therefore be better (e.g. Stout *et al.*, 2000) to grow the grass using nitrogen fertilizer up to the safe limit (without cows) of 400 kg/ha and then store the grass as hay or silage (cut grass stored under anaerobic conditions in a 'clamp', often with additives such as molasses). The stored grass can then be fed to the cows somewhere where the urine and dung can be collected and used to fertilize the soil. Three materials are collected, all of which lose ammonia to the atmosphere:

- Farmyard manure, a largely solid material which results from keeping cattle on straw on to which they deposit urine and dung. This is stored in piles to 'rot down', a process in which microbes use nitrogen from the urine and dung to break down the straw. Well-rotted farmyard manure is an excellent fertilizer, much used by gardeners as well as farmers, but difficult to apply evenly.
- Slurry is the mixture of urine and dung that accumulates when cattle are kept over a solid surface such as concrete and is stored on the farm in large tanks. Cows brought into the milking parlour from the field also generate slurry.
- There is also a category known as dirty water, which has no appreciable value but can be used for irrigation if needed and may even have a small nutrient value. It may be spread through irrigation equipment for disposal even if not needed.

Value of manure and slurry

Manure and slurry are potentially valuable. Chambers *et al.* (2000) estimated that soils that receive manure and slurry get on average 120 kg/ha of nitrogen, 30 kg/ha of phosphorus and 90 kg/ha of potassium. These authors put their potential fertilizer value at more than £150 million in England and Wales. But there are practical difficulties in using them as fertilizers. In the UK, for example, most of the farmyard manure and slurry is produced in the west of the country while the greatest demand for fertilizer is in the east. Both materials are bulky and awkward to handle and it is not economically feasible to transport them more than a short distance.

Storage problems

There are several points at which nitrate or other nitrogen-containing material can escape from the system when grass is stored and dung and urine are collected and used. Ammonia emission is always a problem, and there are liquid losses too.

- An effluent can escape from a silage clamp and cause problems in surface waters or watercourses. Silage also has a distinctive smell, which I like but some others do not.
- Another noxious effluent can escape from a pile of farmyard manure and cause problems in nearby water. Farmyard manure also has a distinctive smell.
- Slurry storage tanks are not infallible. Leaks do occur and cause potentially severe problems in watercourses. Slurry has a distinctive and rather unpleasant smell, but its smell is not as bad as that of gas liquor (p. 49).

Problems when manure and slurry are spread

Farmyard manure and slurry are valuable sources of nitrogen and phosphorus for grass and other crops, but they can be washed off the land into watercourses once they are spread, particularly if the land is sloping (Parkinson *et al.*, 2000). They also lose ammonia and nitrous oxide to the atmosphere when spread (e.g. Rochette *et al.*, 2001; Velthof *et al.*, 2003). Therefore they need to be applied with great care. The Department for Environment, Food and Rural Affairs has produced codes of good agricultural practice to help farmers who produce or use slurry and manure to avoid polluting watercourses or the air or harming the soil which is their main resource (Box 6.1). These codes also provide an insight into the assortment of problems faced by livestock farmers who have to protect the environment while trying to make a living.

Box 6.1. Codes of practice for managing manures and slurry.

The codes of good agricultural practice are designed to help farmers to avoid polluting water and the air and to protect their soil. They are published by the Department for Environment, Food and Rural Affairs and are based on research and organized common sense. They comprise a series of key messages, and those relating to manure and slurry are given here. This box is based on the summary. The wording but not the meaning may have been changed.

The key overall message is that you decrease the risk of pollution by minimizing the quantities of materials handled, stored, spread to land or in need of disposal. This is similar in principle to the comment made about nitrate in Chapter 2, that the less you have in the soil, the less you lose. Other key points include the following.

Manure management planning

- Draw up a manure management plan to aid decisions about when, where and at what rate (of application) to spread manure, slurry and dirty water on the farm.
- Avoid spreading within at least 10 m of a ditch or watercourse and within 50 m of a spring well or borehole.
- These measures will reduce the risk of polluting water not only with nitrate or other chemical species but also with pathogens transmitted from the livestock via the manures.

Slurry and manure spreading

- Use a band spreader or injector to apply slurry if possible. If the manure or slurry is applied to bare land, incorporate (plough in) the material as soon as possible to decrease odours and ammonia loss. Where this is not possible, decrease odour by using a slurry spreader that gives a low trajectory and large droplets.
- Avoid applying more than 50 m^3/ha (50 t/ha) of manure at any one time to decrease the risk of runoff or odours. Adjust the rate of application so that no more than 250 kg/ha of nitrogen is applied per year.

Slurry storage

- Provide enough storage capacity for slurry to be managed safely. Keep this storage in good repair.
- Mix (stir) slurry only when the store is to be emptied. Otherwise mixing should be necessary only to break up a surface crust or to remove sediment.
- Do not add waste milk, whey or silage effluent to stores for slurry or dirty water if this is likely to cause an odour nuisance or if the contents of the store are to be applied to land.

Storage of solid manure

- Make sure that runoff from manure heaps in fields does not pollute water. Runoff from stores on concrete bases should be collected and contained.
- If poultry manure and broiler litter are stored in the open, construct narrow A-shaped heaps that shed rainwater.

Dealing with dirty water

- Produce as little dirty water as possible, and try to keep it separate from clean water. Provide enough storage capacity for dirty water to be managed safely. Keep storage and irrigation equipment in good order.
- Check irrigation systems regularly and make sure that warning devices and automatic cut-offs are functioning.

Livestock housing

- If possible, gather up slurry every day and transfer it into the storage facility.
- Use enough bedding, when needed, to keep livestock clean and keep the manure dry. Avoid overflow or spillage from drinking systems.
- Keep concrete areas around buildings clean and free from accumulations of manure and slurry.

Managing silage liquor

- Minimize the amount of liquor by wilting grass to 25% dry matter before storing. Provide storage capacity sufficient for the liquor to be fully contained. In particular, do not allow any to enter a watercourse. Even a small amount can kill fish and other life in the water.

Quantifying nitrate losses from soils receiving farmyard manure or slurry

Modern fertilizers are easy to store and they ought to be fairly easy to apply uniformly, although that is not always the case. Manure and slurry are not at all easy to apply with any degree of uniformity, and slurries in particular can be a storage problem for the farmer. Slurry storage containers have a limited volume, which can mean that slurry tends to get applied at less than optimal times of year. Planning its application is essential (Box 6.1).

Losses of nitrate from manures and slurries depend on:

- The quantity of vulnerable nitrogen applied per hectare in the manure or slurry. This depends on the proportions of vulnerable nitrogen (nitrate and nitrogenous material readily converted to nitrate) in the material as well as the total amount applied. Nitrogen in slurry or pig or poultry manure is usually more readily leached than nitrogen in farmyard manure.
- The timing of applications, particularly in relation to the amount of subsequent rainfall. The time at which much of the nitrogen becomes vulnerable to leaching depends on the rate at which soil microbes convert non-soluble nitrogenous material to nitrate.

Cutting nitrate leaching losses

Chambers *et al.* (2000) presented an overview of experiments made between

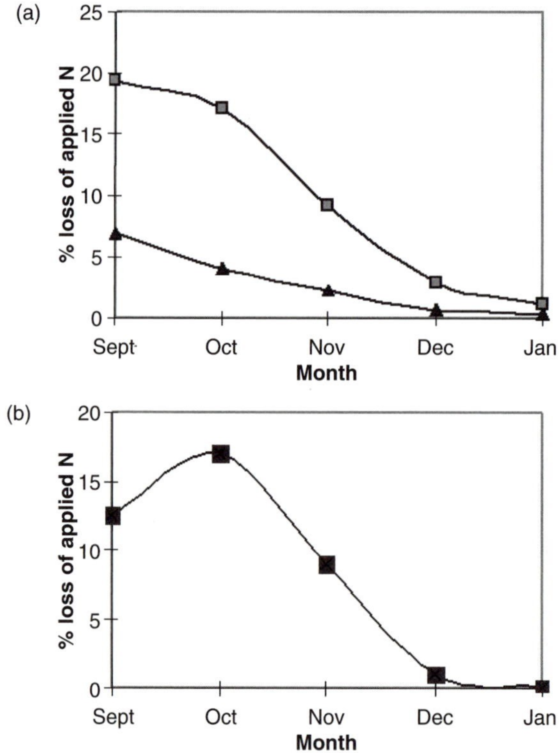

Fig. 6.3. Effect of time of application on losses of nitrate by leaching from manures and slurries applied to (a) arable land (■, slurry or pig manure; ▲, farmyard manure) and (b) grassland (■, slurry). (Redrawn from data of Chambers *et al.*, 2000.)

1990/91 and 1993/94 which illustrates the points in the previous paragraph. The experiments involved the application of various forms of manure and slurry, each supplying about 250 kg/ha of nitrogen to both arable and grassland soils that were considered 'leaky' and expected to allow nitrate leaching. The study on arable sites showed that slurries and poultry manure, which contain a larger proportion of vulnerable nitrogen than farmyard manure, lost more nitrogen (Fig. 6.3a).

Time of application

Figure 6.3, from Chambers *et al.* (2000), illustrates the importance of timely applications. Manure or slurry spread on arable land is at least as much at risk as fertilizer when the soil is bare. Losses of nitrate were greatest when they were spread in September (Fig. 6.3a). Slurry applied to grassland soils showed a slightly different pattern, with the greatest loss from applications in October (Fig. 6.3b). The results in the figure are for 3 years at three sites and may be modified somewhat by further experiments. The results were from the UK and would obviously have been somewhat different in other climates. They should be seen as an example of the kind of effect to be expected.

Amount applied

It is notoriously difficult to analyse a heap of manure or a tank of slurry because of the variability of both the concentration and the form of the nitrogen in both materials. Assessing the amount of nitrogen applied in these materials is therefore difficult, but Chambers *et al.* (2000) reported that an N meter is now available for measuring the ammonium content of slurries. How well it copes with the variability seems uncertain. The analysis of the material is only part of the problem – the farmer has also to spread it uniformly. Smith and Baldwin (1999) (cited by Chambers *et al.*, 2000) found that most manure and slurry spreaders performed badly. But their performance could be improved 'dramatically' by setting the spreading width correctly. (The same principle applied to fertilizer spreaders, p. 75.)

Quantifying nitrous oxide emissions from manures and slurries

As we saw above, manures and slurries show appreciable variations in their nitrogen content and composition both between and within the various types of manure and slurry. Velthof *et al.* (2003) showed how emissions of nitrous oxide reflected these differences in two experiments made with soil in jars. The first compared the emissions of nitrous oxide from manure from various types of livestock and showed (Table 6.1) that the trend in emissions was:

Table 6.1. Effects of the composition of fertilizer and manures/slurries from pigs, cattle and poultry on losses of nitrous oxide from them. Materials applied at 100 mg N/kg soil. (Based on Tables 1 and 2 of Velthof *et al.*, 2003.)

Material	Ammonium-N (as % total N)	C:N ratio (organic matter)	pH	Nitrous oxide emission	
				mg N/kg soil	% of applied N
Control	–	–	–	0.6	–
$(NH_4)_2SO_4$	100	–	–	4.6	4.0
NH_4NO_3	50	–	–	2.7	2.1
Liquid pig manure (C)	60	7	7.23	7.9	7.3
Liquid pig manure (O)	30	16	6.26	8.1	7.5
Liquid sow manure	52	7	7.16	14.5	13.9
Cattle slurry (C)	31	14	6.67	3.6	3.0
Cattle slurry (O)	36	18	6.75	2.4	1.8
Young cattle slurry	43	13	7.71	2.5	1.9
Layer manure	9	9	5.73	2.5	1.9
Broiler manure	8	5	6.22	1.1	0.5
Duck manure	21	15	7.73	1.2	0.6

C, product from conventional farming system; O, product from organic farming system.

Liquid pig manure > cattle slurry > poultry manure

The experiment included pig manure and cattle slurry from conventional and organic farming systems, but the emissions from them did not differ significantly. Regression showed that ammonium-N, expressed as a percentage of total N, accounted for nearly half (44%) of the variance in the emissions from the manures. This suggests that a substantial part of the nitrous oxide *from the manures* was emitted during nitrification rather than denitrification, particularly as none of the manures contained any nitrate. But the results show a large initial flux of nitrous oxide at the beginning of the experiment, which the authors attribute to denitrification of *soil nitrate* when the manures released readily decomposable organic substrates, usable by denitrifying bacteria, into the soil. These substrates could be volatile fatty acids.

The second experiment made by Velthof *et al.* (2003) measured the nitrous oxide emissions from 0 to 200 mg/kg additions of nitrogen to the soil made either as ammonium nitrate or liquid pig manure. The results (Fig. 6.4) showed that the emissions were clearly related to the amount of nitrogen given, but the slope calculated by regression was almost three times greater for the pig manure than for the ammonium nitrate. These nitrous oxide emissions are explicable in terms of the processes described in the previous paragraph. These suggest that the emissions from the ammonium nitrate came from the nitrification of the ammonium and possibly some denitrification of the nitrate. The much larger emissions from the (nitrate-free) pig manure came from denitrification of soil nitrate stimulated by the volatile fatty acids and other substrates released from the manure into the soil.

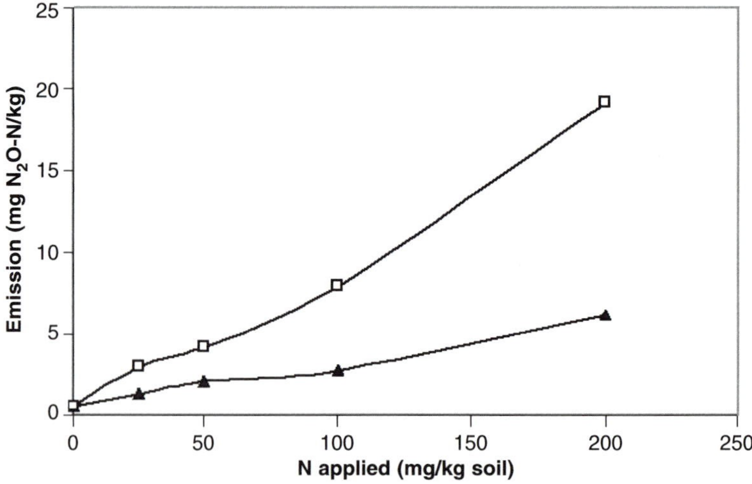

Fig. 6.4. Nitrous oxide emissions from applications of ammonium nitrate (▲) and liquid pig manure (□) plotted against the amount of nitrogen supplied. (Based on the tabulated data of Velthof *et al.*, 2003.)

Velthof *et al.* (2003) also found that the effect of incorporating manure into the soil was not as beneficial as you might expect from its effect on volatilization processes. The smallest emissions of nitrous oxide occurred when liquid pig manure or ammonium nitrate was simply applied to the soil surface. Emissions were greater when either material was mixed homogeneously into the soil or placed at 5 or 10 cm depth. Mixing or placing would have made the materials more accessible to nitrifying bacteria or put them closer to 'denitrification hot spots' (Chapter 3).

Cutting nitrous oxide emissions from manures and slurries

The study by Velthof *et al.* (2003) cited above emphasizes that farmers need to be aware of the composition of the materials they use, particularly the amounts of ammonium and readily mineralized nitrogen they contain. Many of the measures used to lessen untimely nitrogen and cut nitrate losses in leaching are also relevant to cutting nitrous oxide emissions, but the results of these authors point to one conflict. Ploughing in manures helps to restrict losses of nitrate and phosphate by leaching or surface runoff, but now seems likely to increase emissions of nitrous oxide. As in grazed grassland, discouraging denitrification by keeping the soil drained and aerated will cut these emissions. Improved storage facilities for slurry and better and more accurate methods for applying it will also help (Box 6.1).

Total nitrous oxide emissions

The global nature of the nitrous oxide problem requires that we draw up estimates of emissions at a national scale, taking account of both arable and grassland systems and, within the latter, losses during grazing and losses from manures and slurries. Brown *et al.* (2002) estimated that the annual emission of nitrogen as nitrous oxide from the whole of UK agriculture was 87 Gg; F.M. Kelliher (Rothamsted, 2003, personal communication) gave a median estimate of 31 Gg for New Zealand.

Obtaining these estimates cannot have been easy, and one reason for this lies in the essential difference between emissions of nitrous oxide from arable and grassland systems. In arable land, nitrous oxide is a diffuse pollution problem. The transect study in Chapter 5 needed sophisticated mathematics to establish which factors controlled emissions at each scale. In grassland the problem is a mixture of diffuse and point-source pollution, with the point sources having a diversity of sizes. The mathematics were presumably no easier.

Ammonia loss

Nitrous oxide is not the only gas emitted from manures and slurries. The smell of manure and slurry leaves no doubt that ammonia is emitted too, and there is a general consensus (e.g. Jarvis and Ledgard, 2002) that much of the ammonia in the atmosphere comes from dairy farms. But ammonia is not a global problem in the way that nitrous oxide is because ammonia emissions do not travel far so far horizontally or vertically. In wind-tunnel experiments (Ross and Jarvis, 2001), 20–60% of the ammonia from applied urine was deposited on grassland within 2 m.

Ammonia generated on one farm does not usually travel much beyond the neighbouring farm. However, there can be serious problems at the regional level with intensive animal production. Ammonia is among the gases listed in Chapter 7 as contributing to nitrogen saturation of ecosystems, and this is reported to be causing problems in heathland ecosystems in parts of The Netherlands where large numbers of animals are raised. There seems to be some uncertainty about the total quantity of ammonia emitted from the UK (United Kingdom Review Group on Impacts of Atmospheric Nitrogen, 1994), with estimates ranging from about 200 to 400 Gg of nitrogen as ammonia per year. The UK experiences wind from the south-west predominantly, so most of these emissions travel to Scandinavia.

How can ammonia emissions be lessened? Jarvis and Ledgard (2002) published an interesting 'desk study' comparing ammonia losses from dairy farms in the UK and New Zealand. Total nitrogen input was 1.7 times greater in the UK farm, but the offtake of nitrogen was only 1.2 times greater so that the surplus was 1.8 times greater. This difference seemed to arise while the animals were housed and to be associated with manure production. Total losses of ammonia-N were 57 kg/ha from the UK farm but only 24 kg/ha from its counterpart in New Zealand. Interestingly, these losses represented similar proportions of the nitrogen inputs on the two farms, but the New Zealand farm was more productive of milk.

The authors concluded that ammonia emissions from the UK farm could best be lessened through improved storage and application of slurry, and better grazing patterns and use of nitrogen fertilizer. The New Zealand farm needed more effective fertilizer application. On both farms, emissions could be decreased by practices that improved the efficiency with which nitrogen was utilized or improved the C:N balance for productive cows.

Nitrogen balances on farms

Leaching losses of nitrate and emissions of nitrous oxide and ammonia are difficult to estimate for all but small areas of soil. Quantifying losses and emissions reliably for an entire farm is very difficult indeed, and it may be helpful to assess the total amounts of nitrogen brought on to the farm and exported or lost from it. This may at least quantify the gaps in knowledge, and Goss *et al.* (1995) suggested that such studies could form the basis of a

Table 6.2. Nitrogen inputs to and outputs and losses of nitrogen from a 'model' dairy farm. (From Jarvis, 1993.)

Source	Total for farm (kg)	Kg/ha
Nitrogen inputs		
Atmosphere	1,990	25
Fertilizers	19,000	250
Biological fixation	760	10
Feeds	3,746	52
Straw	182	
Total inputs	25,588	337
Nitrogen outputs		
Milk	2,940	
Protein	2,160	
Total outputs	5,100	67
Outputs as % of inputs	20	20
Nitrogen losses		
Leaching	4,272	56
Ammonia volatilization	3,483	46
Denitrification (as N_2O)[a]	4,197 (1,049)	55 (14)
Total losses	11,952	157
Balance not accounted for	8,536	113
Balance as % of inputs	33	33

[a]Assuming that the ratio of dinitrogen to nitrous oxide was 3:1 in the products of denitrification.

systems approach. Two such studies made in the UK are described here and provide interesting comparisons. One (Jarvis, 1993) was a 'desk study' using information from measurements and models to characterize a 'model' dairy farm operating under an ADAS consultancy scheme. The other comprised measurements made at Coates Farm, a mixed farm in the Cotswold Hills (Allingham *et al.*, 2002).

Jarvis's (1993) study showed (Table 6.2) that the nitrogen in the outputs (milk and protein) from the farm accounted for 20% of the quantity of nitrogen coming on to the farm. The losses of nitrogen by leaching, denitrification and volatilization of ammonia accounted for a further 47% of the inputs. One-third (33%) of the inputs were not accounted for. Some of the missing nitrogen must have been incorporated into soil organic matter, but the rest, probably most of it, is likely to have been lost by denitrification, some as dinitrogen and some as nitrous oxide. Exactly what happened to this missing nitrogen is quite important. If (say) half of it was denitrified, and if the dinitrogen:nitrous oxide ratio in the products was 3:1 as is often

Table 6.3. Nitrogen inputs to and outputs and losses of nitrogen from a mixed farm, Coates Farm. (Based on Table 2 of Allingham *et al.*, 2002.)

Source	Mean for 4 years (kg N/ha)
Nitrogen inputs	
Atmosphere	21
Fertilizers	164
Biological fixation	21
Animal feeds	48
Seed	2
Total inputs	256
Nitrogen outputs	
Saleable produce	114
Outputs as % of inputs	44
Nitrogen losses	
Leaching	65
Ammonia volatilization	18
Denitrification (as N_2O)[a]	16 (4)
Total losses	99
Balance not accounted for	43
Balance as % of inputs	17

[a]Assuming that the ratio of dinitrogen to nitrous oxide was 3:1 in the products of denitrification.

assumed, the estimate of nitrous oxide emission in Table 6.2 needs to be more than doubled.

The Coates Farm project (Allingham *et al.*, 2002) studied mixed farming systems involving dairy, sheep and arable systems and operating under 'best management practice'. An interesting feature of the study was that it used 'farmlets' (Laws *et al.*, 2000), 120 m × 20 m areas of larger fields, to represent farming systems, in which farmyard manure and slurry moved between different parts of the system. This and the fact that the study lasted for 6 years enabled the effects of rotations on nitrate losses to be evaluated. The farm was on a shallow, freely draining soil over limestone and most of the losses were by leaching, which was estimated using porous ceramic probes.

Balances between inputs, outputs and losses were calculated for the whole farm for 4 years, and the means are shown in Table 6.3, which is derived from Table 2 of Allingham *et al.* (2002). The nitrogen output:input ratio for this mixed farm was more than twice as large as it was on the dairy farm, perhaps reflecting the greater opportunities for using manure and slurry within the farm. The balance not accounted for, expressed as a percentage of the inputs, was half as large on this mixed farm as for the dairy farm, possibly for the same reason. But making the same assumptions about the fate of the balance for the mixed as for the dairy farm – that is, that half of it was denitrified and the dinitrogen:nitrous oxide ratio in the products was 3:1 – again more than doubled the estimate of nitrous oxide emission.

We have gone a long way towards quantifying the 'enigma' in soil nitrogen balance sheets on which Allison (1955) commented nearly 50 years ago, but the enigma is not dead yet, particularly where farm animals are involved.

7 Nitrate in Fresh Water and Nitrous Oxide in the Atmosphere

WITH A CONTRIBUTION FROM ARTHUR J. GOLD AND CANDACE A. OVIATT, UNIVERSITY OF RHODE ISLAND

We saw earlier that nitrate is lost from the soil by two main processes, leaching and denitrification. Broadly speaking, leaching causes problems in the water environment, and denitrification in the atmosphere. This chapter tackles problems in fresh water and the atmosphere, while the interesting but complex effects of nitrate and other ions in coastal and estuarine waters and the sea are discussed in Chapter 8.

Nitrate in Fresh Water

Nitrate lost from the soil by leaching makes its way into groundwater or into surface waters – that is, streams, rivers and lakes – from which it may eventually reach the sea. Only a small proportion of this nitrate may have come from fertilizer. The rest will have come from mineralization in the soil (Chapter 5), some from grazing animals or the application of manures (Chapter 6), some from the ploughing up of old grassland (Chapter 10) and some, as we shall see later in this chapter, from the deposition of various forms of nitrogen from the atmosphere. The issues involved differ somewhat between ground- and surface waters, but the two categories can interact. Many streams, for example, begin in a boggy area where groundwater seeps out from an exposed aquifer.

Groundwater

Water that is drawn up or pumped from wells from sources beneath the ground is groundwater. Wells can be an unsatisfactory source of water if badly maintained or positioned (Chapter 10), and public groundwater supplies are taken mainly from several large rock formations that are aquifers. (The word aquifer derives from the Latin and simply means water-bearer.) The main geological formations that act as aquifers in the UK are chalk, limestone and sandstone. These rocks are porous and can hold water in up to about half their total volume.

Each aquifer has a saturated and an unsaturated zone (Fig. 7.1). The top of the aquifer may have an interface with the soil, or there may be clay or other material between them. The unsaturated zone then lies between the top of the aquifer and the water table, which is the surface of the saturated zone, the usable water in the aquifer. In the saturated zone, water fills virtually the whole porosity of the rock, and it is this zone from which water is pumped for public consumption.

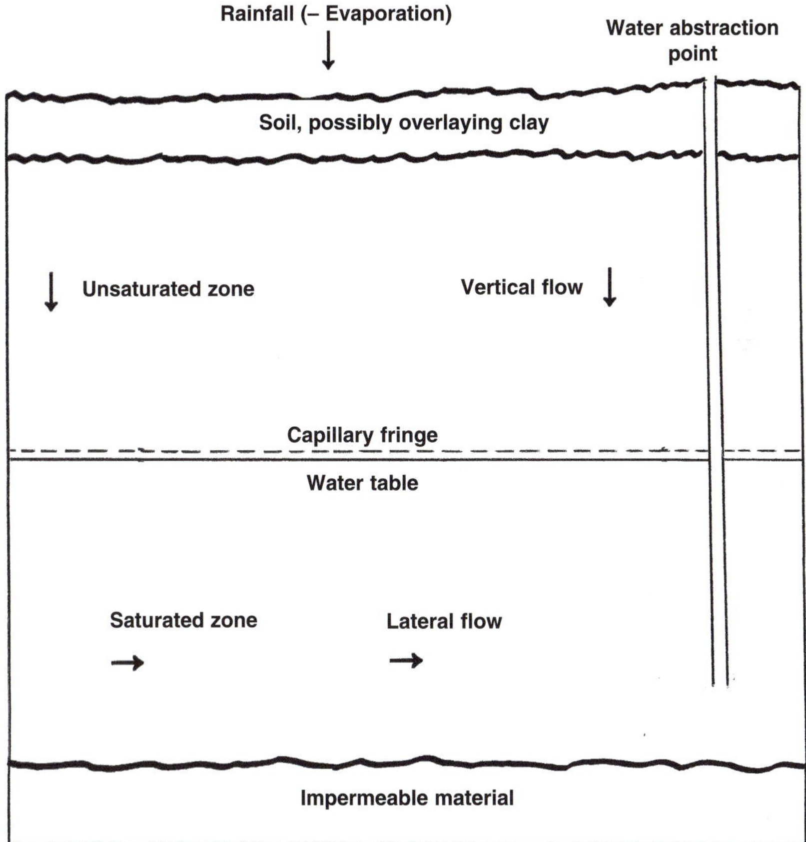

Fig. 7.1. The main features of an aquifer.

Aquifers are an important source of water for human consumption, sometimes the only source, so there is understandable concern about possible pollution of the water contained in them. This concern is strengthened by the long periods of time involved in aquifer processes. Water passing through the unsaturated zone of a chalk aquifer moves downward at about 0.8 m/year (Young *et al.*, 1976). This implies that if a chalk aquifer has a water table 40 m below the top, a not unlikely depth, any water entering the top of the aquifer now will reach the saturated zone, from which it may be pumped, in 2053.

Farming activities above aquifers

Most of the water that enters the unsaturated zone of an aquifer has passed through the soil previously, so there is considerable interest in any farming activity on the land above the aquifer, particularly if it leads to losses of nitrate (discussed in Chapter 5). All forms of farming activity lose nitrate, but the ploughing up of old grassland was particularly interesting to Young *et al.* (1976) because it resulted in bands of nitrate moving down through the unsaturated zone, the rate of whose progress could be measured. This ploughing, the reasons for it and its consequences are discussed in Chapter 10.

Water from aquifers is used for human consumption, so the main issue raised by nitrate in aquifer water is the health issue. The evidence discussed in Chapters 9 and 10 suggests that nitrate is not only not hazardous at concentrations up to twice the European Community (EC) limit, but actually is beneficial, in that it is a key part of the body's defence system against bacteria that cause gastroenteritis and may also protect against heart disease. There are also no environmental issues that involve aquifer water unless it flows out of the aquifer into surface waters. This means that the main constraint on nitrate concentrations in aquifer water is a legislative one, in the form of the EC nitrate limit of 50 mg/l discussed in Chapter 10.

Nitrate removal from groundwater?

Denitrification can help to restrict nitrate concentrations by converting it to gases. The process is not restricted to the soil and can occur in groundwater, provided both the microbes and a suitable carbon substrate are available. The origin of the aquifer usually determines how much natural denitrification is likely (Korom, 1992). The rock surface in Triassic Sandstone, for example, may hold enough carbon substrate for measurable denitrification to occur, but rocks formed under hotter conditions will have less substrate available. The process is used in industrial plants for removing nitrate from water. Methanol is added as a substrate – subject to controls that ensure that the microbes exhaust the methanol before the nitrate. There are also reports that glucose and methanol are being introduced into aquifers for this purpose.

Surface Waters

Because water is also drawn from the larger rivers for public consumption, nitrate in surface waters, like that in groundwater, would be perceived to be a health problem were it not for the medical evidence (Chapters 9 and 10). If nitrate is a problem in surface waters it is part of the wider problem of nutrient enrichment which often manifests itself in the form of algal blooms.

Algal blooms

The blue-green algae are very small single-celled plants of the *Cyanobacteria* class of species that grow on the surface of practically anything, including water. Some of them are toxic, and others are a problem because of buoyancy conferred by the gas vesicles they contain, which enables them to rise to the surface of the water during calm conditions. The resulting 'bloom' or 'scum' is often blown by even gentle breezes to the edge of the lake or river, where it is unsightly and offensive. Algal blooms took on a greater significance a decade or so ago when it was discovered that some species of *Cyanobacteria* were toxic to humans and dogs swimming in or drinking affected water (Bell and Codd, 1996). This toxicity may seem to be a recent phenomenon, but Bell and Codd noted that the first case of its kind was in Australia in 1853. It is also a widespread problem, having occurred in more than 30 countries. Another important concern is that when algae die, the bacteria that consume them use dissolved oxygen to do so and this lessens the supply to fish and other desirable organisms, which may die as a result.

Nutrient enrichment

Plant nutrients are usually beneficial in the soil, but if they get into waters they become a nutrient enrichment problem because they trigger unwanted growth of plants, both small and large. Three terms are widely used to indicate the degree of enrichment – oligotrophic, mesotrophic and eutrophic. These imply in essence low, medium and high degrees of enrichment, and the idea has been extended to the concept of 'mean trophic rank' as a means of assessing and comparing rivers (Dawson *et al.*, 1999). The word 'eutrophication' is, of course, widely used.

The concept of the limiting nutrient

The idea of nutrient enrichment leads to the question, 'Which nutrient is the problem?' Here we need to look back to Box 1.3 to Justus von Liebig, the originator of the Law of the Minimum, the idea that plant growth is limited by the element least available in the soil. The limiting nutrient concept was extended by K. Brandt to nutrients in water in which photosynthesis may

Table 7.1. The Redfield ratios and the Law of the Minimum.

Fixed ratios of atoms found by Redfield in aquatic algae and macrophytes:

106 carbon atoms : 16 nitrogen atoms : 1 phosphorus atom

Multiplying the number of atoms of each element by its atomic weight gives the ratio on a mass basis:

40 g carbon : 7 g nitrogen : 1 g phosphorus

The law of the minimum suggests:

- When the ratio mass of available N : mass of available P exceeds 7 : 1, production of algae or macrophytes will be limited by the additions of P.
- When it is less than 7:1, production will be limited by added N.

(Carbon is freely available in carbon dioxide and unlikely to limit growth.)

occur. He correlated the abundance of plankton in lakes in Germany with nutrient concentrations. A.C. Redfield later suggested that the Law of the Minimum should be interpreted in terms of the ratios of nutrients found in living algae (Table 7.1).

The principle of the limiting nutrient is central to any discussion of nutrient enrichment of fresh water or sea water, because it provides a means of understanding the system and because it offers the possibility of controlling algae or macrophytes by managing inputs of a single nutrient. The evidence that phosphate is the limiting nutrient for algal growth in fresh water is summarized below, and the complex problem of estuarine and coastal waters and the oceans is discussed in the following chapter.

Algal growth in fresh water: limited by phosphate?

The algae that cause the blooms are *Cyanobacteria*, which can fix nitrogen from the atmosphere. This suggests that nitrogen is unlikely to be the nutrient that limits their growth. Research reviewed by Sharpley *et al.* (1994) and Ferguson *et al.* (1996) showed that algal blooms were more sensitive in freshwater lakes to the concentration of phosphate rather than to that of nitrate. Regressions relating chlorophyll concentrations (as a proxy for the quantities of algae in the water) to the phosphate concentration were valid over five orders of magnitude of the phosphate concentration (Ferguson *et al.*, 1996). The corresponding regressions for nitrate concentration were valid over only two orders of magnitude and gave a poorer fit.

One experiment in particular was decisive in establishing that phosphate was the limiting nutrient in lakes. This was a 'whole lake' study made by an international team in Canada (Schindler, 1974) in which an oligotrophic lake with two similar basins was divided by a plastic curtain. One basin was treated with carbon and nitrogen, the other with carbon, nitrogen and phosphate. The basin that was treated with just carbon and nitrogen remained in apparently pristine condition. The other, which received phosphate as well, developed massive algal blooms.

Large water plants

Algal blooms are not the only problem caused by nutrient enrichment in surface waters. Much larger plants also grow to excess, usually with unwelcome effects. Reeds grow to excess, narrowing waterways and potentially overloading and damaging banks. The proliferation of underwater plants fouls the propellers of boats, entangles the tackle of fishermen and blocks water supply conduits, thereby damaging machinery. Although these plants need nitrogen, there is usually enough in the water, and their growth is limited by phosphate.

Nitrous Oxide in the Atmosphere

Nitrous oxide is produced in the soil as a result of incomplete denitrification or during nitrification (Chapter 3). Both arable and grassland agriculture contribute to its production, as do other sources including industry (Table 7.2). Nitrous oxide is formed during the manufacture of adipic acid, which is used in the production of nylon. Human activities contribute to atmospheric nitrous oxide slightly less than natural emissions. On a global basis, the amount of nitrous oxide-N added annually to the atmosphere is about 4.5 million tonnes greater than in pre-industrial times (Granli and Bøckman, 1994). Sources cited by these authors suggest that before 1700 the concentration of nitrous oxide in air was about 285 mg/m^3 but that this had risen to 310 mg/m^3 by 1990, with much of the increase occurring after 1945.

Lægreid *et al.* (1999) calculated that on average about 1.25% of the nitrogen added to soil as fertilizer or manures or by biological fixation becomes nitrous oxide. This loss is trivial in agronomic or economic terms but far from

Table 7.2. Emissions of nitric oxide from human activities and natural sources worldwide. (From Lægreid *et al.*, 1999.)

Source	Nitrous oxide emitted (Mt N/year)
From human activities	
Agricultural soils	3.3
Associated with cattle	2.1
Burning of wood and other biomass	0.5
Industry	1.3
Total from human activities	7.2
Natural emissions	9.0

trivial in environmental terms. These authors also cited the Intergovernmental Panel on Climate Change (IPCC, 1997) as estimating that 2.5% of leached nitrate ended up as nitrous oxide. Fowler *et al.* (1996) described the rather complex relation between nitric and nitrous oxides near the soil surface and estimated the annual emission of nitrous oxide from soils in the UK at 27 kt N, with grassland responsible for about 60% of the emissions. They noted that soils will become the main producers when the manufacture of adipic acid ceases here.

The release of nitrous oxide to the atmosphere is an example of a global pollution problem. Nitrate lost to surface waters may pollute a lake, a river system or even a large area of coastal water such as the Gulf of Mexico, but nitrous oxide affects the atmosphere of the whole planet. It is the global nature of the problem that has encouraged the measurements of large-scale nitrous oxide emissions discussed in Chapters 5 and 6.

The layers of the atmosphere

The atmosphere is divided into two layers:

- The troposphere is the lower part of the atmosphere extending from the earth's surface to a height of about 8 km at the poles and 16 km at the equator. In the troposphere the temperature usually decreases with height.
- The stratosphere is the upper part of the atmosphere, above the troposphere and extending to a height of about 80 km. The temperature increases with height in the stratosphere.

Nitrous oxide contributes to two major atmospheric problems – the 'greenhouse effect' which is at least partly responsible for global warming at the earth's surface, and the destruction of ozone in the stratosphere.

Global warming

Global warming (Box 7.1) is regarded by many as the most serious threat to the survival of the planet and has stimulated a debate of almost religious intensity between 'believers' and 'sceptics'. This is entirely proper, because we are dealing with the ultimate in reciprocal risk problems (Chapter 12). If the believers are right, and global warming is occurring as a result of greenhouse gases released by human activities, our failure to act now to restrain the release of these gases could be disastrous for future generations. But if the sceptics are right, and the changes turn out to reflect natural cycles in the climate that are beyond human control, science, which is already suffering a slump in public trust, will be totally discredited. Worse still, the almost unimaginably large sums of money due to be spent on implementation of the Kyoto Treaty are sorely needed for development in the Third World, particularly in Africa, where they could make the difference between life and death for the poor. A recent report stated that the money to be spent on Kyoto could provide the Third World with fresh water in perpetuity. If global warming turns out to have been no more real than the 'new ice age' about which climatologists warned us in the mid-1970s, its cost may have to be counted in millions of lives as well as billions of dollars.

Box 7.1. Global warming: a brief introduction and some questions.

Weather and climate

There is a distinction between *weather* and *climate*. This is a scale issue similar to those discussed in Chapter 5, and we are concerned with temporal rather than spatial scale.

Weather occurs in a specific area at a particular time. For example, on 4 August 2003 (when this was written), the weather was warm and sunny over the south-east of England.

Climate could be said to be weather summarized over time. The key variates in the summary are the average rainfall and average temperature. A more detailed summary would include the average run of wind, sunshine hours, number of frosts and similar data. Yearly averages are usually calculated for each month of the year and for the whole year, and long-term averages for the 30 years preceding the current year. Averages are not the whole story, and the summary needs to include deviations from the average and extreme weather events. The example could be that the south-east of England enjoys a temperate climate with about 750 mm of rain annually and a temperature that rarely exceeds 30°C or goes below 0°C.

Weather is always in the singular, but climate can be in the plural. The climates of East and West Africa, for example, differ appreciably.

The greenhouse effect

Global warming, or more specifically the 'greenhouse effect' which causes it in part, has a long and distinguished intellectual pedigree. Houghton (1992) records that it was first recognized in 1827 by the French mathematician and scientist Fourier, perhaps remembered more for his series. He pointed out the similarity between:

- The effect of the inner surface of the glass in reflecting and thereby retaining the heat energy of a greenhouse.
- The effect of the atmosphere in re-absorbing radiation reflected from the earth's surface and thereby warming the planet.

Fourier could be described as the 'father of the greenhouse effect'.

The Irish physicist Tyndall, best known for the 'Tyndall effect' observed when sunshine passes through a dusty atmosphere, measured the absorption of infra-red radiation by carbon dioxide and water vapour in about 1860. He suggested that decreases in carbon dioxide and its greenhouse effect could have caused the ice ages. This idea was pursued further in 1896 by Arrhenius, the Swedish 'founder of physical chemistry', who seems to have foreseen the possibility of global warming in the current sense. He first calculated (without a computer!) the effect of the water vapour and carbon dioxide in the atmosphere, showing that between them they increased the temperature of the earth's surface by about 33°C, from –18 to 15°C. He then calculated the effect of increasing the concentration of greenhouse gases on the warmth of the planet, and his estimate, that doubling

continued

the concentration of carbon dioxide would increase the global average temperature by 5–6°C, is not far out of line with assessments from the computer models used today.

The greenhouse gases

Water vapour and carbon dioxide are the gases mainly responsible for the greenhouse effect but methane, nitrous oxide, chlorofluorocarbons and ozone (in the troposphere rather than the stratosphere) also contribute to the warming effect. That order of importance is probably correct, but assigning an order is not simple. The effectiveness of a gas depends not just on how much of it is in the atmosphere but also on its residence time there and the wavelengths of radiation it absorbs. Nitrous oxide is significant as a greenhouse gas because it absorbs radiation of particular wavelengths, but it is also important environmentally as a destroyer of the ozone layer. There is considerable argument about the extent to which greenhouse gases can be controlled.

Water vapour varies greatly over time at a single location. The barometer in my living-room has during the last few years registered pressures ranging from 960 to about 1025 mbar. Nitrogen and oxygen together make up a constant 99% of the gases other than water in the atmosphere, so this substantial variation in pressure with time is due almost solely to large changes in the amount of water vapour in the air. The barometer goes down when the atmosphere is moist simply because the molecular weight of water, 18, is appreciably less than those of nitrogen and oxygen, 28 and 32 respectively. Atmospheric water vapour also varies between locations – think of the Sahara Desert and the rainforests not far to the south of it – and it is totally beyond human control.

Carbon dioxide emissions increased by more than 20% in the last century as a result of human activity. To restrict these emissions we need to cut energy consumption, but the costs of just stabilizing them would be huge (see main text). Some, such as Stott (P. Stott, London, 2003, personal communication), doubt whether any useful control can be exerted by trying to manipulate just one variable (carbon dioxide) of the many that influence the climate. Others (e.g. Houghton, 1992) argue that we have no alternative but to try.

The concentration of methane in the atmosphere has doubled since 1880 (Houghton, 1992), but to cut emissions we would need, amongst other things, to restrict the activities of the coal, natural gas and petroleum industries, cut back on the production of rice in paddies, decrease worldwide the number of domestic animals and constrain the amount of domestic sewage treated and the amount of rubbish put into landfill. Before embarking on this ambitious project, we perhaps need to be sure that we understand the natural production of methane in wetlands and its natural removal by reactions in the atmosphere. We also need to consider the crucial alternative view of the role of methane set out in the Gaia hypothesis (Lovelock, 1995a), which is summarized in Box 7.2.

Nitrous oxide is the fourth most important greenhouse gas, but it is also important from an environmental standpoint as a destroyer of the ozone layer. There is no argument about the need to minimize losses, but we need to note the alternative view of nitrous oxide in the Gaia hypothesis (Box 7.2).

The greenhouse controversy

There is general agreement about the nature of the greenhouse effect and the gases involved but beyond that there is controversy. The international body charged with responsibility for research on climate change and the implementation of measures to counter it is the Intergovernmental Panel on Climate Change (IPCC). This is a body with a political as well as a scientific standpoint. The view of the IPCC, that climate change is happening and that it is driven mainly by carbon dioxide emissions resulting from human activities, is now largely accepted as the orthodox view. But not all scientists agree, and some such as Stott (see above) argue strongly against it.

Some questions about the science

There is no doubt that the average global temperature is rising at present and that this temperature and atmospheric carbon dioxide were significantly correlated during the last 100 years (e.g. Kuo *et al.*, 1990). But two variables increasing simultaneously does not imply that one causes the other, even if they are correlated. During the latter half of the last century the birth rate in Europe declined, as did the population of storks – obviously implying that granny was right after all in telling you that babies were brought by storks. Most of the arguments for global warming are therefore made on the basis of computer models. These models raise a few questions.

1. All our heat comes ultimately from the sun, so the fluctuations that occur in the sun's energy output must have some influence, however small, on the temperature of the earth. Should the models take into account these fluctuations? The IPCC models do not seem to do so. There seems to a comprehensive disagreement about the reliability and significance of the measurements of the sun's energy output. The general confusion is exemplified by a recent paper (Laut and Gundermann, 1998, cited by Lomborg, 2001). This purported to support the IPCC but stated that the increase in direct solar radiation during the last 30 years accounted for about 40% of the global warming observed, something the IPCC does not accept.

2. Carbon dioxide not only causes global warming but also is itself caused by warming. Atmospheric warming will be accompanied by soil warming, which will accelerate mineralization of soil carbon (Chapter 3), releasing more carbon dioxide into the atmosphere. We need to know which came first, the warming or the carbon dioxide. Kuo *et al.* (1990) reported that, although temperature and atmospheric carbon dioxide had been significantly correlated for the previous 30 years, changes in carbon dioxide lagged those in temperature by 5 months. Each year the increase in carbon dioxide came *after* the warming. Do the climate models simulate this lag?

3. If you plot the carbon dioxide concentration and the variation in global temperature since 1900, say, against time on the same graph, both variates show an overall upward trend, but the patterns within the trend differ considerably (Fig. B7.1.1). Apart from an annual fluctuation, the carbon dioxide concentration follows a smooth, gradually accelerating upward curve. The temperature curve, by contrast, is highly irregular and shows that a large proportion of the warming occurred before 1940, by which date only about 20% of the release of carbon dioxide from the human activities had occurred. Can the models explain this, perhaps by taking extra factors into consideration?

4. Prior to their detailed statistical study (mentioned above) on the then most recent 30 years, Kuo *et al.* (1990) divided the temperature curve from 1880 onwards into four 30-year segments. The temperature increased in only three of these segments, and in the fourth, from about 1940 to about 1970, the temperature decreased while the carbon dioxide concentration continued to rise. It was that 30-year decline in temperature that led to the last alarm about the climate. Were we entering a new ice age? We then proceeded almost seamlessly from worrying about the new ice age to worrying about global warming. How do the climate models explain that 30-year decline, which occurred in spite of rising carbon dioxide concentrations?

continued

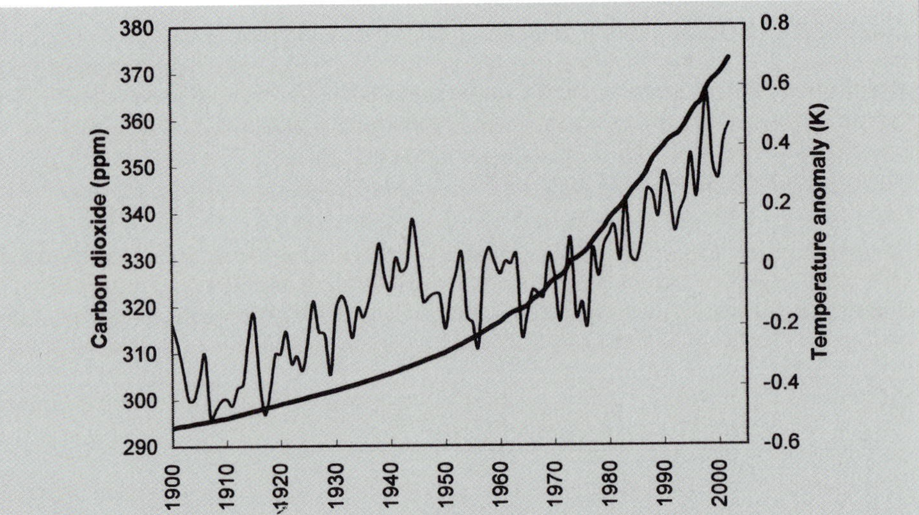

Fig. B7.1.1. The increase in the atmospheric concentration of carbon dioxide (heavy line) plotted with the global temperature anomaly (light line). (Diagram courtesy of Dr Jack Barrett.)

5. The climate models predict that the temperatures of both the land surface and the lower troposphere should rise as the carbon dioxide concentration increases. As noted in the last section, the surface temperature increased for much of the last century but decreased for about 30 years in the middle of it. The temperature in the troposphere, measured by instruments carried on balloons, did not increase between 1957 and 1976. It increased briefly in 1976 and 1977, but since 1979, when temperature measurements by satellite were started, neither type of measurement showed an increase in temperature. Thus the warming predicted by the climate models at the land surface during the last century was in reality a decrease in temperature for 30 of the last 100 years and the warming predicted in the lower troposphere hardly occurred at all. Can the climate model really be said to have been validated?

6. Another interesting question is whether the modellers should be taking more account of chaos theory. The first instance of chaos discussed by Gleick (1987) was in a weather model. One of the characteristics of a chaotic system is that it 'flips' from one form of behaviour to another, which appears to be just what we see in Fig. B7.1.1.

The climate modellers may well be able to answer these questions in due time, perhaps quite soon. But until they can, I prefer to remain 'agnostic' about global warming.

Feelings run high in the debate. The sceptics were recently pilloried as 'toxic sceptics' in a national magazine (*New Internationalist*, No. 357, June 2003, p. 13), and some scientists seem to agree with this extreme view. This bodes ill for rational debate on the topic. More worryingly, more than one of the sceptics has told me that it is difficult to get a paper published that does not agree with the orthodox view. I have given a brief introduction to the science of global warming and listed some questions about it in Box 7.1 but, as Houghton (1992) and others much more knowledgeable than myself have written books on global warming, and only a relatively

Box 7.2. The Gaia hypothesis and the greenhouse gases.

The essence of the Gaia hypothesis developed by Lovelock (1995a) is that the earth is a complex, living, self-regulating organism which he named Gaia after the Greek goddess of the Earth. In the terminology of Complexity Theory, Gaia is an emergent system. The earth as we know it is thermodynamically highly improbable and it is living matter that maintains it in this improbable state. This state is essential for the life that maintains it. The composition of the atmosphere provides one measure of this improbability. Table B7.2.1, based on Table 2 of Lovelock (1995a), compares the composition of the atmospheres of Venus and Mars with that of Earth as it is and would be without life. The most striking difference is that without life Earth would have an atmosphere made up of almost entirely carbon dioxide and would therefore be very hot.

Table B7.2.1. Comparing the composition of the atmospheres of Venus and Mars with that of Earth as it is and would be without life. (After Table 2 of Lovelock, 1995a.)

	Atmosphere of			
Gas/state	Venus	Mars	Earth with life	Earth without life
Carbon dioxide (%)	98	95	0.03	98
Nitrogen (%)	1.9	2.7	78	1.9
Oxygen (%)	Trace	0.13	21	Trace
Argon (%)	0.1	2	1	0.1
Surface temperature (°C)	477	–53	13	290 ± 50
Atmospheric pressure (bar)	90	0.064	1	60

The reason for the inclusion of this box is that three of the greenhouse gases have implications in the Gaia paradigm which differ greatly from those carried in Box 7.1.

Carbon dioxide

Lovelock (1995b) believes that there is cause for concern about the carbon dioxide emissions in terms of the Gaia paradigm, but his concern is not primarily about global warming. He feels that Gaia's regulatory system for carbon dioxide may be nearing the end of its capacity because, until the additions of the last century, the carbon dioxide concentration of the atmosphere had become rather small. This makes it a very bad time for humans to add carbon dioxide to the atmosphere, for the same reason that it is unwise to put a hypothermic animal into a warm bath. The shock of sudden change could be too much for the system at its limit. The hypothermic animal will recover if left to warm slowly, and Gaia will recover only with a slow increase in carbon dioxide. This conclusion obviously depends on the assumptions of the Gaia paradigm.

Methane

The role of methane in the Gaia paradigm is more heroic than its conventional role as the third worst greenhouse gas. Under Gaia, methane saves us, perhaps not from the flames of hell but certainly from an increased risk of combustion. Lovelock (1995a) pointed out that oxygen is a hazard because of its chemical potential. The present oxygen concentration in the atmosphere, 21%, represents a delicate balance between risk and benefit.

continued

Forest fires do occur, but not often enough to interfere with the level of human and other activity that 21% oxygen permits. Increasing the concentration from 21% to just 22% would greatly increase the risk of fire. The probability of a forest fire being started by a lightning flash increases by 70% for each 1% increase in oxygen concentration above the present level, and at 25% little land vegetation at any latitude would escape the inferno.

The maintenance of the oxygen concentration at 21% is clearly vital, and Lovelock (1995a) suggests that methane acts as a regulator for oxygen in two ways. Some methane reaches the stratosphere, where it is oxidized to carbon dioxide and water. The water eventually dissociates to hydrogen and oxygen, and the hydrogen escapes to space while the oxygen descends to the troposphere. In the troposphere, methane takes oxygen out of the system when it reacts with it in a series of complex reactions. Methane production is therefore important in the Gaia paradigm. It is produced by anaerobic bacteria in marshes and wetland, and also in estuaries and the seabed. Lovelock sees these zones as a key powerhouse for the maintenance of Gaia.

Nitrous oxide

Lovelock (1995a) considered nitrous oxide 'a puzzling gas' within the Gaia paradigm but he was able to suggest two roles for it. One was to complement the role of methane in oxygen regulation. Methane removes oxygen from the atmosphere but nitrous oxide brings it into the atmosphere. I think Lovelock saw the two gases as together having a 'fine tuning' function in maintaining the oxygen concentration at the appropriate level.

The other role Lovelock saw for the gas was a control on ozone in the stratosphere by the mechanisms shown in equations (7.1)–(7.3) in the main text of the chapter. He argued that too much ozone might be as undesirable as too little, and he pointed out that vitamin D was formed in the skin as a result of exposure to ultraviolet radiation. 'Too much ultraviolet may mean skin cancer,' he said, 'too little almost certainly means rickets.'

small proportion of the warming is caused by nitrous oxide and other nitrogen oxides, Box 7.1 is the limit of the comment on the problem in this book.

I have to declare myself as 'agnostic' about global warming. I have an energy-efficient flat with cavity-wall insulation, double glazing all round, water-heating only 2 hours per day and low-energy light bulbs. I also run a small economical Skoda car, which I aim to use on only 4 days a week. But I regard these measures as stewardship of increasingly scarce energy rather than precautions against global warming. And I am not convinced that the measures undertaken at the Kyoto convention will make much difference to the mainly natural processes of climate change, particularly as the USA declines to be involved.

In the long run the problem may well sort itself out. Dr Roger Bentley, speaking at the Scientific Alliance conference '2020 Vision – Powering the UK's Future' in May 2003, pointed out that the resource-limited peak in oil production was likely to occur in 10 years' time and the peak in all hydrocarbons about 5 years later. Dr Bentley foresaw that the world would soon experience oil and gas shortages, and this seems to be supported by recent events at Shell Oil, where directors have been removed as estimates of the company's oil reserves have been down-graded for the third time. An oil and gas shortage could save a lot of argument.

Nitrous oxide and ozone destruction

Nitrous oxide is very stable in air but, by an unfortunate mischance, the only processes that remove it from the atmosphere occur in the stratosphere and some involve the destruction of ozone. Ozone is an atmospheric contaminant at ground level, where it can be dangerous to people with respiratory problems and toxic to crops. Nitric oxide and nitrogen dioxide (known collectively as NO_x) catalyse its production there from volatile organic compounds (Fowler *et al.*, 1998). However, 90% of atmospheric ozone is in the stratosphere, where it forms a vital layer that protects life on the surface of the earth from ultraviolet radiation. The fluctuations in the ozone layer, particularly the hole in the layer above the South Pole, have understandably attracted widespread public concern. Nitrous oxide is not the only threat to the ozone layer. Like nitrous oxide, the chlorofluorocarbons used in refrigeration are largely inert at ground level but reactions they undergo in the stratosphere make them a threat to the ozone there.

The chemistry of nitrogen oxides in the stratosphere is somewhat complex, but the main reactions of nitrous oxide have been summarized by Granli and Bøckman (1994). Nitrous oxide reacts with excited singlet oxygen atoms (O*) formed by the photolysis of ozone to give nitric oxide. The nitric oxide reacts with ozone to give nitrogen dioxide and oxygen:

$$O_3 + hv \rightarrow O_2 + O^* \quad (7.1)$$

$$N_2O + O^* \rightarrow 2NO \quad (7.2)$$

$$NO + O_3 \leftrightarrow NO_2 + O_2 \quad (7.3)$$

Although it is the nitric oxide that destroys the ozone, the role of nitrous oxide is crucial. Nitric oxide, although fairly stable to decomposition, is a free radical and never reaches the stratosphere on its own because it reacts readily with other free radicals, of which there are plenty in the troposphere, before it gets to the stratosphere. Nitrous oxide travels to the stratosphere largely without reaction and therefore mainly intact. In this way it acts as a kind of 'Trojan horse' enabling nitric oxide to get into the stratosphere.

Two other reactions also remove nitrous oxide from the stratosphere but do not destroy ozone in the process (Granli and Bøckman, 1994). In one, nitrous oxide undergoes photolysis to nitrogen and a singlet oxygen atom, and in the other it reacts with an excited singlet oxygen atom to form normal molecular nitrogen (dinitrogen) and oxygen:

$$N_2O + hv \rightarrow N_2 + O \quad (7.4)$$

$$N_2O + O^* \rightarrow N_2 + O_2 \quad (7.5)$$

Table 7.3. Nitrogen species that may be present in the atmosphere and deposited on land.

Name	Chemical formula
Nitric oxide	NO
Nitrogen dioxide	NO_2
Dinitrogen pentoxide	N_2O_5
Nitrous acid	HNO_2
Nitric acid	HNO_3
Peroxyacetyl nitrate	$CH_3COO_2NO_2$
Organic nitrates	RNO_3
Ammonia	NH_3

Additional species that may be present include aerosols of ammonium and nitrate and nitrate radicals formed when nitrogen dioxide reacts with ozone.

The chemistry of atmospheric nitrogen is very complex, and the main chemical species likely to be found are listed in Table 7.3. These interact with each other under the influence of photons (from sunlight), ozone and hydroxyl and peroxyl radicals. Because of the influence of photons in some reactions, the chemistry in daytime differs appreciably from that at night-time. These species may be deposited on land in droplets of cloud or rain, by dry deposition, which means that they are transferred directly from the gaseous phase on to a surface, or as aerosols. Deposition on the soil surface and on vegetation both can cause problems. These are not as sensational as those caused by nitrous oxide in the stratosphere, but they merit attention.

Deposition from the Atmosphere

Deposition on farmland

Rainfall at Rothamsted was first analysed for nitrate and ammonium 127 years ago but not much was found. Between 1877 and 1915 the total deposition of nitrogen recorded was 227 kg/ha, about 6 kg/ha per year on average, but these measurements did not include all forms of deposition. More recent studies, such as those of Goulding (1990) at four sites in south-east England, included nitrate deposited on dry matter and dry deposition of nitrogen oxides, nitric acid and ammonia. These more comprehensive estimates showed an annual deposition of 35–40 kg/ha. More recent measurements at Rothamsted (United Kingdom Review Group on Impacts of Atmospheric Nitrogen, 1994) suggested that about 37 kg/ha was deposited annually on bare soil and about 48 kg/ha on soil with a well-established crop of winter wheat. The difference is due to the larger surface area for deposition offered by the crop.

The nitrogen deposited when the wheat was grown corresponded to one-quarter of the average application of nitrogen fertilizer at the time of the measurement and could be more than a quarter now. It came from industry and motor traffic, each of which generates about half the nitrogen oxides, and from the farm and neighbouring farms, which are mainly responsible for the ammonia. The deposition occurs whether or not there are crops taking up nitrogen and can readily add to the untimely nitrate (Chapter 2) that is leached. One estimate from a model (Goulding *et al.*, 1998) suggested that about half was taken up by the crop and about 30% was leached. Some currently unpublished runs of another model (N.A. Mirza, Harpenden, 1995, personal communication) suggested that a fairly similar proportion, 25%, was leached and that it contributed about 15% of the nitrate leached overall.

This calculation implies that nitrate deposited from the atmosphere may contribute 10–15 kg/ha to annual nitrate losses from the soil. Farmers can be encouraged to view this deposition as a resource and to make allowance for it, but they can hardly be held responsible for its consequences. Assuming that the through drainage is about 250 mm, as in many parts of eastern England, its contribution to the nitrate concentration leaving the soil is about 20 g/m^3, or 40% of the EC nitrate limit. This leaves farmers little room for error.

Deposition on natural vegetation

Nitrate or ammonium deposited on farmland need not be a problem, as we saw in the previous paragraph. Deposition is definitely a problem for natural vegetation and the soils on which it grows and its effects can include (United Kingdom Review Group on Impacts of Atmospheric Nitrogen, 1994):

- Loss of mosses and lichens.
- Loss of heather.
- Damage to trees from nitrate and ammonium in fog and cloud.
- Changes in species composition and loss of biodiversity.
- Decreases in tolerance of plants to stresses such as frost, drought and feeding by animals and insects.
- Changes in soil solution chemistry, including acidification and loss of cations.

Critical load and nitrogen saturation

Two concepts, critical load and nitrogen saturation, are important for understanding the problems caused by deposition of nitrate, ammonia and nitrogen oxides from the atmosphere.

A *critical load* is defined in roughly the same way as a 'no effect' level (Chapter 12) as 'an exposure below which harmful effects do not occur'. The main difference is that 'no effect' levels commonly relate to a single species or to microcosms whereas critical loads usually relate to entire ecosystems at the landscape scale. The formal definition of a critical load, according to the United Nations Economic Commission for Europe, is 'a quantitative estimate of exposure to one or more pollutants below which significant harmful effects on sensitive elements of the environment do not occur according to present knowledge'.

The critical load for nitrogen may relate either to its acidifying effect or to its nutrient effect. The assessment can be made by calculating the critical load for each of these and using the lower value. Critical loads have been estimated for a wide variety of ecosystems including forests (Grennfelt and Thornelof, 1992) with, as the authors admit, various degrees of reliability. Their values range from 5–15 kg/ha of nitrogen per year for arctic and alpine heaths to 20–30 kg/ha for neutral or calcareous species-rich grassland. A mesotrophic, rather than an oligotrophic, fen has a critical load of 20–35 kg/ha. All these load estimates are less than the amounts of nitrogen measured by Goulding (1990) to have been deposited from the atmosphere. Goulding's measurements were in an agricultural rather than an environmentally sensitive setting, but nitrogen oxides can travel some distance in the atmosphere.

Nitrogen saturation

Nitrogen limits growth in most natural ecosystems so nitrogen deposited from the atmosphere initially increases the growth of vegetation. But ecosystems, like crops, have a limit to the amount of nitrogen they can use. Giving too much nitrogen fertilizer to a crop not only causes problems such as lodging (when the stem becomes too weak to support the weight of the crop) but also results in nitrate losses. Similarly, if more nitrogen is deposited on an ecosystem than it can use in its current state, the excess will cause some kind of change, which may well be deleterious, and nitrate will be lost from the system. Various measures of nitrogen saturation have been suggested (e.g. Grennfelt and Thornelof, 1992), but nitrate loss may be the most reliable. A small child tells you when it has had enough food by throwing the rest on the floor, and both natural and agricultural ecosystems do so by losing nitrate.

Perakis and Hedin (2002) provide excellent examples of ecosystems unsaturated and saturated with nitrogen. They showed that substantial amounts of dissolved organic nitrogen, but no nitrate, were lost in stream water flowing from pristine forests in Chile. (The nature of this organic nitrogen was discussed in Chapter 3.) These authors found that a forest in the vicinity of New York lost mainly nitrate in stream water. The second forest received a substantial load of inorganic nitrogen from the atmosphere generated probably by industry and road and air traffic.

8 Nitrate in Coastal Waters

Contributed by Candace A. Oviatt and
Arthur J. Gold, University of Rhode Island

Much of the nitrogen that gets into streams and rivers ends up in the sea, where it remains a potent source of controversy, particularly in the estuaries and bays in which rivers meet the ocean. Chapter 1 detailed the enormous increase in the use of nitrogen fertilizer since 1900. This has been accompanied by large increases in the deposition of nitrogen oxides generated by traffic and industry and, as a result of this human activity, the quantity of nitrogen circulating in the global nitrogen cycle has doubled since pre-industrial times. Rivers now contribute twice as much nitrogen to marine waters as they did 100 years ago (Vitousek *et al.*, 1997). Eutrophication is therefore a problem in marine as well as in fresh waters. It has caused startling changes in the growth of aquatic plants and algae in coastal waters (Nixon, 1995). The excessive algal growth in particular has destroyed critical aquatic habitats and lowered the concentration of dissolved oxygen, thereby killing fish and shifting the balance between marine organisms (Howarth *et al.*, 2000). This eutrophication threatens the long-term sustainability of fisheries and the use of coastal waters for recreation.

This chapter introduces the environmental consequences of the enrichment of marine waters by plant nutrients, particularly nitrogen (Table 8.1), together with the principal concepts needed to understand it and the methods used to study it. It reviews the evidence implicating nitrogen in the fouling of marine waters and presents several case studies of large marine ecosystems. These studies, of the Baltic Sea, the Gulf of Mexico and the open waters of the Pacific, provide a warning against assuming that any single factor can explain all the complexities and intricacies of marine waters.

Table 8.1. Responses of aquatic ecosystems to eutrophication.

Increased biomass of phytoplankton and suspended and attached algae.
Decreased transparency of water column.
Shift in phytoplankton composition to bloom-forming species, some of which may be toxic.
Accumulation of carbon within the system.
Changes in vascular plant production and species composition.
Decrease in living aquatic habitats, including seagrasses and coral reefs.
Depletion in deep-water oxygen concentration, resulting in hypoxia.
Changes in species and production of fish.
Decline in aesthetic values.

Consequences of Eutrophication of Coastal Waters

Stratification and oxygen depletion

Two levels of oxygen depletion are defined for marine studies. Either can arise from enrichment of coastal waters with plant nutrients:

- Hypoxia, defined as when the supply of oxygen is low.
- Anoxia, which implies there is no oxygen supply at all.

Hypoxia resulting from eutrophication has caused widespread damage to fisheries ranging from Chesapeake Bay and the Gulf of Mexico in North America to the Baltic and Black Seas in Europe (Howarth *et al.*, 2000). Oxygen depletion occurs, as it does in fresh water, as an after-effect of the stimulation of algal growth by a supply of nutrients in the water. Large algal blooms fix carbon dioxide from the atmosphere and increase the organic biomass in the water. (Biomass is simply the mass of living material.) These algae increase the oxygen concentration in the water while they are photosynthesizing during active growth, but the microorganisms that decompose them when they die consume oxygen. The resulting oxygen demand can be a problem for the whole ecosystem.

This is a particular problem in stratified aquatic systems that are isolated from the oxygen in the atmosphere. Summer stratification occurs in freshwater lakes deeper than 3–5 m when the upper water layers absorb heat from the sun and become warmer and therefore less dense than the water at the bottom of the lake. When this stratification occurs, a layer with a steep temperature gradient known as the *thermocline* effectively isolates the water at the bottom and prevents oxygen from diffusing from the atmosphere to the water beneath the thermocline (Wetzel, 2001). Estuarine systems have a related problem. When fresh water from the river meets salt water coming in from the ocean, a layer of low-salinity water floats above a layer of denser saline water. This, combined with the temperature gradient, gives rise to a steep density gradient known as the *pycnocline,* which also prevents oxygen from diffusing from the atmosphere to the water at the bottom. The combination of the pycnocline and decomposing algal biomass settling to the bottom can cause severe oxygen depletion in the water at the bottom.

Changes in species composition

Enrichment with plant nutrients may favour the metabolic processes of one set of organisms at the expense of another, allowing the favoured organisms to grow more rapidly and dominate the ecosystem. It can therefore alter the species composition and the biodiversity, ultimately affecting the structure and food web of the whole ecosystem. In estuaries and bays, eutrophication has been linked to harmful algal blooms described as 'red tides' that can kill many fish and other marine organisms (Vitousek *et al.*, 1997; Howarth *et al.*, 2000).

Seagrass destruction

Patches of seagrass form important spawning and nursery habitats for fish and other marine organisms in shallow estuaries. The rates of seagrass destruction rival those of tropical forests and its loss causes serious problems in the Baltic Sea, the Gulf of Mexico and in estuaries along the east coast of the USA. The survival and growth of seagrass depend on the amount of light that can penetrate through to it (Howarth *et al.*, 2000). Increasing the supply of nitrogen in the water decreases the penetration of light to the surface of the seagrass leaf by stimulating the growth of two undesirable types of species: phytoplankton (floating algae) in the water above the seagrass and epiphytes (attached algae) growing directly on seagrass leaves (Fig. 8.1).

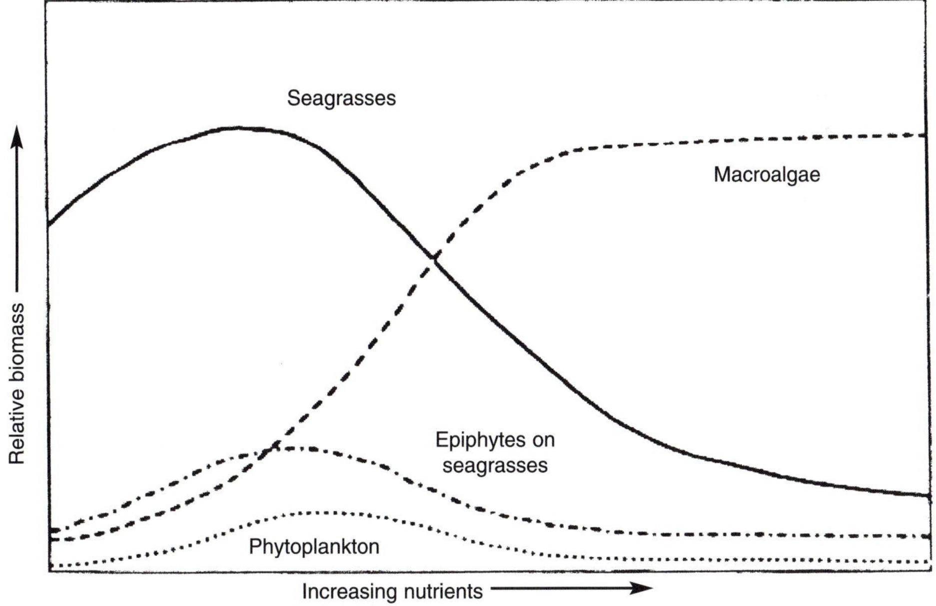

Fig. 8.1. Effect of nitrate with other nutrients in shallow marine systems. Decline in seagrass beds with increase in other species. (Redrawn from Harlin, 1995.)

Seagrass is important not just as a spawning habitat, but also because it stabilizes sediments on the bottom. Its loss therefore increases suspended sediment in the water, further decreasing water clarity and hastening the loss of submerged aquatic vegetation. To restore the habitat of the seagrass, we need to control the supply of nitrogen in the water.

Eutrophication can also stimulate troublesome blooms of *benthic macroalgae*, or 'green seaweeds' that require less light than seagrass (Fig. 8.1). The macroalgae form thick extensive mats over seagrass beds and the sediment surface, thereby decreasing spawning habitats, especially for fish that need mineral substrates to spawn. The macroalgae die off periodically, and their decay depletes oxygen and kills some fish. They also leave an unpleasant accumulation of rotting remains along the shoreline.

Degradation of corals

Coral reefs are among the most productive ecosystems on the planet, but they thrive best in sea water that is entirely clear and has very small concentrations of plant nutrients. Nutrient enrichment is therefore a threat to them, and both nitrogen and phosphate are involved. Only slight enrichment of coral reefs is needed to stimulate the growth of macroalgae, which attach themselves to the reef and inhibit the propagation of the coral. Periodic blooms of the macroalgae also cause hypoxia in the reef environment. Degradation of coral reefs by plant nutrients is a global problem, and examples are reported from the Great Barrier Reef, the Caribbean and Hawaii.

Principles of Nutrient Enrichment in Coastal Waters

The effects of nutrient enrichment in fresh water were discussed in Chapter 7. The work of the early researchers Liebig, Brandt and Redfield on plant nutrients in fresh water is also relevant to coastal and marine waters (e.g. Smith, 1998).

Which nutrient limits the growth of unwanted biomass in coastal waters?

The concept of the limiting nutrient – that is, that one nutrient limits the growth of biomass in an ecosystem – suggests that we should be able to manage aquatic systems by controlling a single limiting nutrient (Smith, 1998). We saw in Chapter 7 that there is plenty of evidence that in fresh water this nutrient is phosphate, but in many coastal marine systems the limiting nutrient is usually nitrogen (Howarth *et al.*, 2000). We need to consider carbon too, but this is rarely limiting, being readily available from the atmosphere. Aquatic ecosystems, particularly estuarine systems, are notoriously complex, and these insights came only after extensive experimentation and vigorous debate. The literature shows that some issues remain unresolved. To understand this important difference between freshwater and marine systems, we need to understand the cycles of the two nutrients.

Differences between the phosphorus and nitrogen cycles

Nitrogen and phosphorus are both brought into aquatic systems in both soluble and particulate forms, but the two nutrients subsequently undergo markedly different processes. Biological processes have a large effect on the supply of nitrogen available for biomass production in aquatic systems. Blue-green algae (cyanobacteria) fix nitrogen from the atmosphere, so adding it to the water. Nitrogen as nitrate is lost from the water when denitrifying bacteria reduce the nitrate to gaseous nitrous oxide and dinitrogen. Nitrogen can accumulate in bottom sediments, but remains vulnerable to loss in gaseous form, which reduces the potential for mineral nitrogen (ammonium and nitrate) to be released from the sediments.

The behaviour of phosphorus contrasts with that of nitrogen in that biological processes neither introduce phosphorus nor remove it from an aquatic ecosystem. The quantity of phosphorus in a water body is controlled solely by chemical and physical factors. Rivers are the primary source of phosphorus for marine ecosystems, and the amount of phosphorus in a river depends on its geochemistry and oxygen concentration and on physical factors such as residence time and dilution.

Influence of nutrient cycles on eutrophication in fresh and marine waters

We saw in Chapter 7 that a principal factor determining whether nitrogen or phosphorus is the limiting nutrient in an aquatic ecosystem is the ratio of nitrogen to phosphorus. Redfield suggested that if the N:P atomic ratio was smaller than 16:1 nitrogen might be the limiting nutrient, and if it was greater phosphorus might be. Why does the ratio tend to be more than 16:1 in lakes and less than 16:1 in coastal waters? One reason is simply that the sources of water for lakes and coastal systems differ. Both receive inputs of water from the land and from the atmosphere, but coastal systems also receive inputs from the ocean, and the water from the ocean tends to have low N:P ratios because of denitrification on the continental shelf. Another reason is that blue-green algae seem to fix less nitrogen in estuarine waters than they do in lakes (where the blooms they form are often a considerable problem).

The final reason lies in differences in geochemistry between the two water systems. In freshwater systems, phosphate tends to react with iron and settle into sediments in which it becomes unavailable to plants or microorganisms. In sea water, anions such as hydroxyl, sulphate and borate compete for sites on particles, and their concentrations are orders of magnitude greater than that of phosphate. As a result, particles carried in water from land tend to release the phosphate they are holding when the water transporting them mixes with water in a marine system (Froelich, 1988). Sulphate concentrations in particular are very large in marine waters and iron seems to sorb sulphate in preference to phosphate. Fresh water usually contains small concentrations of sulphate, so that iron sorbs more phosphate in fresh than in sea water.

The conclusion that phosphorus is the limiting nutrient in fresh water but nitrogen in sea water seems generally applicable but should not be regarded as invariant.

Experimental Methods for Marine Systems

Assessing the limiting nutrient – another scale problem

The problem of scale, which we encountered on land in Chapter 5, emerges again when we take to the sea (or at least to its fringes). Nutrient enrichment and the responses to it involve processes and responses occurring within individual components of the ecosystem. We can study these individual processes and responses, often perhaps in a relatively small volume. We might, for example, use laboratory assays to study the responses of specific algae to particular nutrients. But many of the processes interact with each other, and serious errors will result from simply aggregating the individual processes. As in the soil, so in the marine ecosystem, the whole is much more than the sum of the parts.

For example, those laboratory bioassays for the response of specific algae to particular nutrients give useful information on which nutrients limit growth, but simple experiments such as these may fail to take account of important processes such as the regeneration of nutrients from sediments or factors such as locations and substrates for nitrogen fixation. To address the complexity of the marine ecosystem, we need to use a *'whole system'* approach in our experiments. In practice, however, questions of logistics, cost, replicability and time usually restrict the scope of the experiment to the simplest 'whole system' capable of answering the question posed.

Whole system studies are not necessarily on a large scale. What is essential is to have an experimental unit that incorporates all the critical processes and factors that characterize the natural system under study. According to the nature of this system, the study may be made in a *microcosm* (commonly a test tube), a *mesocosm* (a large tank or bag) or a homogeneous patch of open ocean. Given the complexity of the ecosystem and the concern about interactions between processes, would it to be better to study entire bays and estuaries? The problem with this idea is that experiments of this nature tend to run into problems with uncontrolled boundaries and unknown interactions. Also, data may have to be collected in rough seas and high winds or with a recalcitrant crew – problems unfamiliar to scientists working on land.

Whole system experiments in aquatic ecosystems

Soil scientists will be interested to learn that aquatic scientists trace the origin of their whole system studies back to the long-term experiments begun by Lawes and Gilbert at Rothamsted that were described in Chapter 1. These whole system studies aim to make experiments on bodies of water that can be isolated from dilution and contamination during the course of the study, but are subjected to the critical environmental conditions of the ecosystem under study.

Some of the earliest aquatic studies were concerned with phytoplankton metabolism and were made in bottles. In an experiment in Massachusetts in the 1920s, water from a reservoir was placed in either dark or clear bottles and incubated in the water so that respiration and photosynthesis could be measured simultaneously. This study established the basis for methods still in use today and which are described in detail below. Working from the Plymouth Laboratory in the UK in 1922, Aikens hung water bottles in the English Channel and measured the decrease in phosphate concentrations, from which he estimated the amount of biomass produced in the water. He also employed a new colorimetric technique for measuring phosphate that allowed important new insights into nutrient depletion during short periods of time (Mills, 1989).

Large floating bags and spheres became available in the 1960s and 1970s. These enabled large volumes of water, 10 m^3 or more, to be isolated and subjected to the natural variations of climate and temperature, thereby giving further insights into nutrient and population dynamics. Kiel towers, which were widely used in the western part of the Baltic Sea, modified the floating bag concept by attaching the bag to undisturbed natural sediment, enabling interactions between plankton and sediment to be studied.

There is no magic formula for designing whole system experiments, nor can any design, scale or statistical test be said to be optimal. But successful whole system experiments all meet three conditions:

- The experiment is not dominated by any artefact.
- There is enough replication to detect change.
- All the natural components and processes that control the system have been incorporated.

These conditions are not easily satisfied, and failure to meet them has compromised many studies.

Microcosms: Trade-offs between Size and Replicability

Where there is a need for replication, perhaps due to variability in a natural process, experimental volumes often have to be small and they are accommodated in microcosms. Bottles provide small microcosms that often work well, better perhaps than their reputation suggests. They have been used to determine which nutrient limits plankton metabolism by measuring oxygen use in dark and clear (or light) bottles and tracing the fate of ^{14}C-labelled carbon dioxide. The oxygen measurement in the dark bottle estimates respiration from the decrease in oxygen, while that in the light bottle estimates the net production of phytoplankton from the increase in oxygen. The sum of the two is the gross production.

The experimental procedure is essentially to add each nutrient to its set of replicated microcosms, and find which one gives the largest changes in oxygen. Until recently, the method was restricted to fairly productive systems because oxygen could not be measured sufficiently precisely to detect the small changes in its concentration that were found in systems with low productivity such as pristine lakes and the open ocean. The carbon-14 technique has frequently been used in these low-productivity systems because of its sensitivity, but it does not measure respiration and, according to the length of the incubation, it gives a measurement that lies between net and gross production of phytoplankton.

Bottles have been popular but they have attracted a cautionary large literature about their use. The concerns raised include:

- Artefacts from bacterial growth on their walls during longer incubations.
- Leaching of nutrients or toxic material from the glass or the plastic of which the bottle is made.
- Exclusion of ultraviolet radiation.
- Size effects.
- Grazing effects.

Perhaps the most important concern is that the size of the bottle may exclude processes that are important to the system (Hecky and Kilham, 1988). But bottles have played an important part in research because they are cheap, readily replicated and have simple logistics.

Mesocosms: Optimizing Size and Replicability

Mesocosms may comprise large tanks or bags or even exclusion towers extending from the sea floor to the surface. They provide a more comprehensive description of the complexity of the system at the expense of increased cost and some loss of replicability, but they do not avoid all problems. Artefacts, particularly the wall effects noted with bottles, can confound the experimenter's intentions more than lack of replicability. Growth on walls out-competes growth in the water unless it is controlled.

Sediments have to be considered for some nutrients but the way they are included can introduce artefacts. For the Kiel towers, for example, the bags were attached to the otherwise undisturbed sediment. But this apparently reasonable approach failed because differences in pressure between the inside and the outside of the bags forced water through the sediments and into bags, thereby altering the waterborne concentrations of nutrients and other ions (Smetacek *et al.*, 1982). Researchers sometimes seem to be so concerned about artefacts that they pay more attention to the artefact than to the ecological problem that was the reason for the research facility.

Another serious problem is that the financial resources and engineering skills used to establish large structures in coastal areas can occasionally be wasted because the structures have a short lifetime in the stormy conditions that prevail in such areas. Despite all these difficulties, research facilities at Loch Ewe, Kiel and CEPEX, and in Maryland, Rhode Island, Florida, New Hampshire, Norway, The Netherlands and other centres, have improved the understanding of nutrient dynamics in coastal waters.

The MERL Facility

The Marine Ecosystems Research Laboratory (MERL) at Narragansett, Rhode Island, has mesocosms that were designed to sustain natural processes at a size at which repeated sampling would not have a detrimental impact. There is a set of 14 tanks, each more than 5 m deep, and each tank is filled with 13 m^3 of natural sea water and a 37 cm depth of undisturbed sediment (Fig. 8.2). The tanks have precise controls that enable them to mimic many of the natural processes that control primary production in an estuary. These processes include light penetration, the residence time of the water, temperature, tidal mixing and inputs and exports of water. The mesocosms have proved very useful for investigating the following:

- The dynamics of nutrients in estuaries and the rates of their uptake by phytoplankton.
- Grazing.
- Sedimentation and sediment diagenesis.
- Nitrogen fixation, denitrification and nitrogen cycling under different regimes.

The mesocosms enable processes in both the water and in the sediment to be studied, and dozens of characteristics of both can be sampled repeatedly for more than a year. Seasonal cycles can thus be investigated thoroughly without compromising the integrity of the mesocosm as an ecosystem.

The MERL experimenters expended considerable effort to make sure that artefacts such as growth on the walls were minimized and natural processes dominated the system. Walls were cleaned on a regular schedule to avoid excessive growth. Not all problems could be eliminated. For example, rates of vertical mixing exceeded those in natural systems. But the tanks provided simplified replicated systems that could be sampled on a seasonal or annual basis.

A simple whole system experiment with added nutrients in the mesocosms at MERL established that nitrogen could be the nutrient that limited productivity in estuaries (Oviatt *et al.*, 1995). The treatments were:

- Control – no nutrients added.
- Nitrogen.

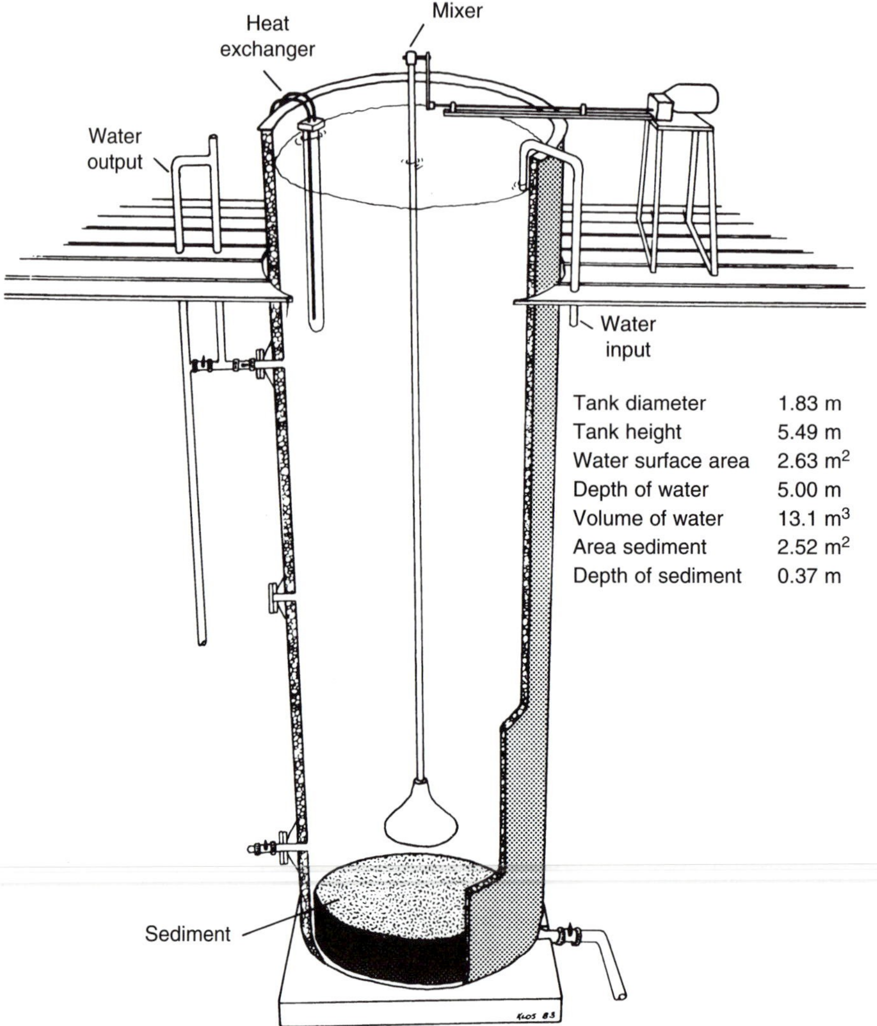

Fig. 8.2. A Marine Ecosystems Research Laboratory (MERL) mesocosm and its input and output pipes, mixer, sediment container and heat exchanger. The mesocosm is made of fibreglass-reinforced resin, and the walls are white to maximize the reflection of sunlight. Sea water is fed in a pulsed flow of 10 l/min for 12 min every 6 h and its temperature is controlled by the heat exchangers. It is mixed by the plunger, 50 cm in diameter, which moves up and down over 60 cm at 5 cycles/min for 2 in every 6 h. (Courtesy of the University of Rhode Island.)

- Phosphorus.
- Nitrogen+phosphorus.

Phytoplankton biomass and metabolism were about five times greater in the nitrogen and nitrogen+phosphorus treatments than in the control and phosphorus treatments. The phosphorus treatment differed little from the control. Furthermore, rates of denitrification exceeded those of nitrogen fixation in the control and phosphorus treatments, emphasizing the nitrogen limitation in those treatments (Oviatt *et al.*, 1995).

A more elaborate experiment at MERL examined nutrient limitation and the processes controlling it along a gradient in salinity extending from fresh water (0 g/l of dissolved ions) to saline water (25 g/l of dissolved ions) (Doering *et al.*, 1995). Bioassays, nutrient concentrations and N:P ratios all showed that phosphorus was the limiting nutrient when the salinity was low but that nitrogen was limiting at 25 g/l. Examining nutrient effects over the whole range of salinity suggested that solubility of phosphorus dictated the switch in limiting nutrient.

Whole ecosystem studies and long-term field monitoring

Why not test hypotheses about limiting nutrients at the ecosystem scale – on whole lakes, bays or harbours? This would provide the ultimate whole system study, with all the processes, interactions and feedbacks of natural ecosystems. But, even if society and particularly environmental pressure groups were to accept the risk to such ecosystems, there remains the problem that studies of this kind cannot be subjected to the normal scientific regime of controls and replication. There would be no way of separating treatment effects from those of variation. It is probably worth accepting this limitation if the effect found is large enough to leave little doubt, as was the case with the lake studied by Schindler (1974). This study, discussed in the previous chapter, had a decisive effect in persuading decision makers that phosphate was the nutrient that limited algal blooms in fresh water (Smith, 1998).

Whether such a study is needed in a coastal water ecosystem is questionable, given the evidence already discussed and the case studies discussed below.

Case Studies: Eutrophication of Large Coastal Estuarine and Marine Systems

Intensive studies on a number of large coastal ecosystems have inevitably highlighted both similarities and differences in the responses of the systems to enrichment with nutrients. In this section we summarize the results – and key questions – that have emerged from the dedicated, long-term efforts of the many scientific teams who have worked in the Gulf of Mexico, Chesapeake Bay, San Francisco Bay and the Baltic Sea. Each of these ecosystems has unique attributes that illustrate the dynamic and complex nature of such systems. Each also shows how we need to assess the current condition of the system and the options for managing it in the light of its natural patterns and historic behaviour.

Gulf of Mexico

The detection of a large and expanding hypoxic zone at the outlet of the Mississippi River to the Gulf of Mexico has stimulated great interest in the effect on marine ecosystems of nitrogen lost from the land. The hypoxia has been linked to nutrients carried in the Mississippi from farms in the mid-western states of the USA all the way to the Gulf.

The zone of hypoxia, in which the concentration of oxygen in the water is 2 mg/l or less, covers more than 20,000 km^2 of water at the bottom of the Gulf. The seasonal development of the zone, its size and duration, and its causes and impacts, have been studied by Rabelais and her co-workers since 1985 (Rabelais and Turner, 2001). The worsening hypoxia has been attributed to increases in farming activity in the Mississippi catchment, and this has led to considerable ecological and economic debate, because that catchment covers 41% of the main part of the USA (excluding Alaska and Hawaii).

The hypoxia in the water develops in late spring and persists until late summer (Rabelais and Turner, 2001). First the water stratifies as described above, isolating the water at the bottom. This happens because the surface water warms in spring and the wind and its mixing effect decline. This stratification is enhanced by the greater flow of less dense water from the Mississippi in spring. The flow peaks between March and May and is low during summer and autumn. The spring flow in particular carries nutrients that stimulate blooms of phytoplankton. The organic matter from these blooms sinks, either directly or as faecal pellets from grazers, to the water beneath the pycnocline where it adds considerably to the oxygen demand.

Hypoxia in the Gulf of Mexico is not a new phenomenon. It has been increasing since about 1900, but the concern has arisen because of the rapid expansion of the problem in recent decades. Palaeo-oceanographers have found that organisms characteristic of well-oxygenated water were common between 1700 and 1900 but disappeared completely from the sediment record in the early 1900s (Sen Gupta *et al.*, 1996). The area in which the bottom water was hypoxic in midsummer was around 8000–9000 km^2 between 1986 and 1992. By the period 1992–2000, the extent of the hypoxia had increased to 16,000–20,000 km^2 (Rabelais and Turner, 2001). Hypoxic water is found typically between 5 and 30 m beneath the surface along near-shore to along deep-shelf locations, west of where the Mississippi flows into the Gulf. During early investigations the hypoxia seemed to develop in spots associated with the deltas, but it is now more uniformly distributed to the west of the deltas.

The extent of the hypoxic zone increases in years when the river is in flood and carries a large load of nutrients and decreases when the water flow and nutrient flux are lessened by drought. Indicators of annual phytoplankton production show that it is of the order of 300 g/m^2, and during the past 50 years this production has been correlated with the amount of dissolved inorganic nitrogen in the river.

During recent years, the Mississippi has discharged 1.6 Mt of nitrogen into the Gulf of Mexico each year, of which about two-thirds was nitrate and one-third organic nitrogen (Goolsby *et al.*, 1999). The mean annual concentration of nitrate was relatively constant between 1905 and 1950 but had doubled by the 1990s (Rabelais and Turner, 2001). There was an equally dramatic change in the silicate concentration, which halved between the 1950s and the 1990s. This could be a significant change because silicate is a key nutrient for diatoms, a common type of algae in the Gulf. If the supply of silicate limits the growth of diatoms, non-siliceous algae could become more plentiful. This could alter the foundations of the food chain and have a serious impact on the ecosystem of the Gulf. There does not seem to be a problem yet. In the 1990s, the atomic ratios for N:Si, N:P and Si:P were 1.1:1, 15:1 and 14:1, close to the values established by Redfield (Justic *et al.*, 1995). The ratios for silicon remain appropriate for the production of diatoms at present, while the N:P ratio shows a tendency towards nitrogen limitation.

The impacts of the nutrient enrichment and the hypoxia on fish and crustaceans in the Gulf have not yet been well quantified, but there are signs of some negative impacts. Analyses of trawl data suggest that bottom-dwelling species avoid not only specifically hypoxic bottom water but also areas in which the concentration of dissolved oxygen is between 2 and 5 mg/l (Craig *et al.*, 2001). Brown shrimp also seem to avoid hypoxic water. Zimmerman and Nance (2001) found that the catch of brown shrimp was significantly and negatively related to the area of hypoxia during July and August. There is little indication that mobile species are killed by the hypoxia – they just change their distribution to avoid it.

These changes from the traditional distribution of fish stocks can affect commercial fisheries. Poor catches of menhaden in 1995, for example, may have resulted from hypoxia in near-shore waters off Louisiana during late July and August (Smith, 2001). Species such as clams that cannot move fare less well and have declined in numbers in response to the hypoxia. The recovery in the communities of such organisms may take up to 2 years following a hypoxic event (Rabelais *et al.*, 2001; Rosenberg *et al.*, 2002).

Chesapeake Bay

Chesapeake Bay, the largest and most productive estuary on the east coast of the USA, has also suffered increased hypoxia in bottom waters during recent decades. Several states have cooperated in efforts to curtail the input to the bay of nitrogen in waste water, and oyster reefs have been promoted as a means of cleansing the overlying water by natural filtration.

The pattern of development of the hypoxia is similar to that in the Gulf of Mexico. Spring runoff in the Susquehanna River creates a pycnocline and stratifies the main stem of the bay, isolating the water at the bottom (Cerco, 2000). A phytoplankton bloom usually begins in February and ends in May. Its size depends on the quantity and nutrient concentration of the water in the river. Much of this bloom sinks beneath the pycnocline, increasing oxygen demand in the bottom water and leading to anoxia.

Hypoxia in the bottom water begins in late May, with large rates of respiration in the organic matter under the pycnocline. Later, the autumn winds break the stratification down and mix the hypoxic with aerobic surface water. Flemer *et al.* (1983) concluded that the volume of anoxic water in Chesapeake Bay had increased by an order of magnitude between 1959 and 1980, but this conclusion proved controversial and there were doubts as to whether the evidence available was sufficient to support it (Cerco, 2000). Further studies, including stratigraphy of the sediments and long-term monitoring, have led to the general conclusion that the volume of hypoxic water depends on the amounts of water and nutrients flowing into the bay each year.

The nutrient that limits the phytoplankton bloom may switch between seasons, from phosphorus in early spring, when the river flow is high, to nitrogen in summer, when the river flow is much smaller (D'Elia *et al.*, 1986). All studies, however much they may disagree otherwise, suggest that the river is the main source of nutrients, both nitrogen and phosphorus, into the bay.

During the 1960s and 1970s, the extent of seagrass meadows in Chesapeake Bay decreased dramatically, stimulating investigations (Orth and Moore, 1983) into possible causes such as turbidity, nutrients and herbicides in the flow from the river. The eventual consensus was that increases in nutrients of agricultural origin in the river had stimulated phytoplankton blooms which shaded out the eelgrass. But the issue is not as simple as this, because of other changes. A few decades ago, the number of oysters in the bay was sufficient to filter the entire bay every few days, but by the 1990s the oysters had become so scarce that they could barely filter the bay in a year (Newell, 1988). The absence of the oysters would also have allowed the blooms to proliferate and prevent light from reaching the seagrass meadows.

San Francisco Bay

The studies in San Francisco Bay illustrate further the importance of filter feeders in controlling eutrophication (Officer *et al.*, 1982). Benthic bivalves, such as mussels and cockles, restrict phytoplankton to small amounts during summer and autumn by filtering them out of the water as fast as they grow despite large nutrient concentrations in the water and abundant light. But during spring, stratification isolates the surface water from the bottom-dwelling bivalves allowing a bloom to develop. Over shoals the phytoplankton biomass may peak in June giving concentrations of chlorophyll of more than 35 μg/l, but during late summer and autumn, these concentrations fall to less than 10 μg/l. Averaging nutrient concentrations over several years shows a low N:P ratio, suggesting that nitrogen is the limiting nutrient in the bay (Hammond *et al.*, 1985).

Large nutrient concentrations are supported in the bay by inputs of sewage effluent, and these should support a high level of primary productivity, but rates of net production measured in the South Bay are relatively small (Cloern, 1999). This is because the potentially large production of phytoplankton is controlled by a large population of benthic filter feeders. The size of this population is a result of the inadvertent introduction to the South Bay of several large, exotic and fecund species, including the Japanese cockle (*Japes japonica*) and the Japanese mussel (*Muscular senhonsia*) (Nichols, 1979).

The Baltic Sea

The Baltic Sea has since the 1960s experienced increasing inputs of sewage and nutrients from agriculture that have led to worsening hypoxia or even anoxia in the bottom waters (Elmgren, 2001). Concentrations of nutrients during winter increased from the 1960s to the 1980s, since when they have more or less levelled off. A fourfold increase in the concentration of nitrogen and an eightfold increase in that of phosphorus have been correlated with increasing oxygen depletion of the deep Baltic. But the oxygen depletion has also corresponded with changes in the period of stagnation of the bottom waters, and the involvement of stagnation suggests that episodes of oxygen depletion could on a historic scale have been related to natural climate variability. As in other eutrophic areas, increased organic matter production in surface waters has led to an increase in oxygen demand in bottom waters.

The Baltic is not very saline compared with other large seas, with 6–8 g/l salinity in surface waters, depending on ice-melt and rainfall. There is a strong density discontinuity between the surface water and the deep water in which the salinity ranges from 13 to 20 g/l. The bottom water derives from irregular inflows coming from the Kattegat through the Danish Straights. These erratic inflows can result in periods of stagnation lasting from 2 to 4 years and occasionally up to 16 years (Elmgren, 2001). During the longer periods the deep waters may become hypoxic.

The management of nutrient inputs to the Baltic Sea has led to the question of whether nitrogen or phosphorus is the more important nutrient to remove from sewage effluent. Experience with lakes and a Redfield N:P ratio of 44:1 initially suggested phosphorus was the limiting nutrient (Elmgren and Larsson, 2001) but growth experiments in the 1970s suggested that nitrogen limited phytoplankton production. The latter experiments were not entirely conclusive (Graneli *et al.*, 1990). The spring bloom appeared to be limited by nitrogen, but the brackish Baltic waters experience summer blooms of nitrogen-fixing cyanobacteria. These summer blooms may be limited by phosphorus or even iron. The current management solution is to remove both nitrogen and phosphorus.

The Open Ocean

Until recently, nobody knew which nutrients limited production in the deep oceans, but nitrogen and phosphorus were assumed to do so. Two recent experiments have, however, revealed a more complex situation. In the first, Chisholm and Morel (1991) showed that iron was the nutrient that limited production in large areas of the Pacific Ocean. And the second set, the US Joint Global Ocean Flux Study (Buesseler, 2001), comprised a decade-long time series of measurements in the north subtropical gyres (large-scale counter-clockwise current circulation patterns) of the Atlantic and Pacific Oceans. These showed that iron played a critical role in production in the Pacific but that production was limited mainly by phosphate in the Atlantic.

Ocean patch studies

The study of nutrient dynamics in the open ocean has been accelerated recently by the use of ocean patches for experiments with added nutrients. An ocean patch comprises a large area of homogeneous water that is defined by the addition of conservative tracers. The patches are often as large as 8 km^2 in area but mix minimally with the surrounding water for weeks at a time, even while moving distances of up to 1000 km. A research vessel distributes tonnes of nutrients on to the patch and then follows it and measures the growth of phytoplankton. Such studies require great skill and dedication, particularly when made in the rigours of the Southern Ocean surrounding Antarctica. But the experimenters nevertheless relate the concept of whole system studies like these back to Broadbalk and the other experiments begun by Lawes and Gilbert in the peace of the English countryside.

Ocean patch studies – conclusive results

Some regions of the Pacific and Southern Oceans are characterized by large nutrient concentrations but small concentrations of chlorophyll (from phytoplankton, etc.). Experiments in replicated bottles suggested that iron was the limiting nutrient in these regions, but the bottles were suspected of having excluded critical processes, and the diatoms they contained were thought to be an artefact because they were so rare in nature. The bottles could have excluded grazing organisms that maintain small populations of diatoms in nature.

Martin and Fitzwalter (1988) were the first to develop clean techniques that enabled trace metals to be determined accurately in sea water. The concentrations of iron they found where nutrients were plentiful but chlorophyll scarce were of the order of a picomole (10^{-12} M), leading them to suspect that iron was the limiting factor for primary production. Martin and his co-workers made incubations in bottles in the Pacific sub-Arctic, the tropical Pacific and the Southern Ocean which showed in all three regions that adding iron in nanomolar concentrations stimulated diatom blooms. Then, faced with continuing scepticism about bottle experiments, they made iron fertilization experiments in patches of open ocean which showed that the diatoms grew as long as the iron was present. Grazing and similar processes did not control the growth of phytoplankton. It is worth noting that the simple bottle experiments had correctly predicted the results of the patch experiments, and seemed therefore to have incorporated all the essential natural processes that controlled the diatoms.

The first two fertilized patch experiments were made in the tropical Pacific and a third was made by other workers in the Southern Ocean (Martin *et al.*, 1994; Coale *et al.*, 1996; Boyd *et al.*, 2000). All showed iron to stimulate primary production, causing originally scarce diatoms to bloom. The first experimental patch was dragged beneath the surface by currents after 4 days and the iron disappeared rapidly, resulting in only a minor bloom. During the second experiment, iron infusions were added for several days and resulted in a major bloom (Frost, 1996). The experiment in the Southern Ocean also led to a major bloom, but an attempt to measure sinking organic matter was not successful because of the mixing processes occurring at the time (Boyd *et al.*, 2000).

The ocean patch experiments were a turning point in the debate about the limiting nutrient for primary production in the open ocean. Their decisive impact in establishing the critical role of iron in the Pacific and Southern Oceans compares with the impact of the whole-lake nutrient addition experiments in establishing that phosphorus was the limiting nutrient for eutrophication in fresh water.

Differences between the Atlantic and Pacific Oceans

The ocean patch experiments provided insights into crucial differences between the Atlantic and Pacific Oceans in the role of phosphorus in primary productivity. Wu *et al.* (2000) recently showed in a time-series experiment that phosphorus concentrations were 100 to 200 times greater at a Pacific station near Hawaii than they were at an Atlantic station near Bermuda. Iron is probably the key to this difference. The Atlantic receives a generous supply of iron-containing dust in winds blowing from the Sahara but the Pacific receives very little. If iron limited diatom blooms in the Pacific, it probably also limited bacterial growth and processes, including nitrogen fixation, so phosphorus concentrations were greater in the Pacific than in the Atlantic. Rates of nitrogen fixation in the iron-rich Atlantic Ocean (~72 mmol N/m^2 per year) are about double those in the Pacific (31–51 mmol N/m^2 per year) and it is phosphorus that becomes the limiting nutrient.

Iron limitation in the ocean and global warming

Iron limitation of primary production in a major part of the world's ocean has implications for the regulation of carbon dioxide in the atmosphere (Martin and Fitzwalter, 1988). These authors suggested that, during the dry windy climate of the last glaciation, dust containing iron was deposited in the ocean, where it stimulated production of phytoplankton. This decreased the concentration of carbon dioxide in the atmosphere, a process summarized as the 'biological pump'. The quantities of dust and iron in ice cores from the Antarctic support this hypothesis, as do proxy concentrations for silicate from diatoms in sediment from the Southern Ocean (Falkowski *et al.*, 1998). But the ocean patch experiments have not yet been able to quantify the efficiency of the 'biological pump' in sequestering carbon from the atmosphere into primary production which, when it dies, falls to the bottom of the ocean.

The conclusion from this chapter has to be that, as with so many natural systems, it is unwise to generalize about the waters of the sea without adequate information. In the context of the book we need to note that nitrogen is the nutrient that limits primary production in coastal and estuarine waters off the Atlantic and Pacific coasts of the USA. But there seems to be no evidence that nitrogen limits production in the open waters of the Pacific Ocean, which are probably iron-limited, or the subtropical waters of the Atlantic, which may be limited by phosphorus. The Baltic Sea is the most enclosed of the systems discussed and suffers severe eutrophication for which nitrogen and phosphorus seem to share the responsibility.

9 Nitrate and Health

BASED ON MATERIAL CONTRIBUTED BY NIGEL BENJAMIN, ST BARTHOLOMEW'S HOSPITAL

The general chemistry of nitrate (Chapter 2) is neither complicated nor particularly exciting – unless you do something foolish with ammonium nitrate. Virtually all nitrate in the natural world is dissolved in water and, in this state, nitrate is chemically stable except that it is able to accept electrons and thereby act as an oxidizing agent (Box 2.1), being itself reduced in the process. But during the last 12 years or so, this otherwise rather boring ion has become very interesting to medical researchers. This is because nitric oxide, the free radical that can be formed from nitrate and is responsible for the destruction of ozone in the stratosphere (Chapter 7), also has vital functions in human metabolism. Its function as a signalling molecule in the cardiovascular system led to the award of the 1998 Nobel Prize in Physiology to the discoverers of this role, and nitric oxide itself was declared 'molecule of the year' by the journal *Science* in 1992 (Box 9.1). The role of nitrate in human physiology is treated in greater detail by L'hirondel and L'hirondel (2001).

This chapter deals with the physiological contribution made by nitrate to human health and well-being. The defence of nitrate against the accusations that it was the cause of methaemoglobinaemia (blue-baby syndrome) in infants and stomach cancer in adults is presented in the chapter on politics and economics (Chapter 10).

The Peculiar Physiology of Nitrate in the Human Body

Humans were first shown to produce nitrate in their bodies as long ago as 1916 (Mitchell *et al.*, 1916), but this process was not quantified rigorously

Box 9.1. Nitric oxide. Molecule of the year 1992.

The citation for nitric oxide as 'molecule of the year' said of nitric oxide that, 'a startlingly simple molecule unites neuroscience, physiology, immunology and revises scientists' understanding of how cells communicate' (Culotta and Koshland, 1992). Its many functions within the body include the following:

- Communication between cells as a 'messenger molecule'.
- Regulation of the immune system.
- The relaxation of smooth muscle in blood vessels.
- Inhibition of platelet aggregation (important in the control of heart attacks and strokes).
- Killing of pathogenic bacteria in the stomach and mouth.

until the 1980s when it was shown to derive at least in part from processes through which nitric oxide is produced (Green *et al.*, 1981). Nitrate has been known to be managed in a peculiar way in the human body since the 1970s, and the most notable oddities (Benjamin, 2000) are the following.

- When we swallow nitrate it is rapidly absorbed in the stomach and upper small intestine, and at least 25% is concentrated from the blood into the saliva by the salivary glands via a mechanism that has not yet been characterized. The nitrate concentration of saliva is therefore at least ten times that of the blood plasma.
- The nitrate (NO_3^-) in the saliva is rapidly changed (reduced) to nitrite (NO_2^-) in the mouth by bacteria described below, which live on the surface of the tongue. Each day we produce about 1 l of saliva, which we swallow. This saliva contains large amounts of nitrite which becomes nitrous acid when it reaches a normal human stomach.
- Nitrate can be produced in the human body (Mitchell *et al.*, 1916; Benjamin, 2000). The enzyme nitric oxide synthase acts on an amino-acid, L-arginine, to produce nitric oxide, NO, and this nitric oxide is converted to nitrate when it encounters superoxide (O_2^-) or oxidized haemoglobin. Consequently, even someone on a nitrate-free diet has an appreciable nitrate concentration in their blood plasma and a much greater concentration in their urine.
- The human body is so reluctant to lose nitrate that there is an active transport mechanism in the kidneys that pumps nitrate back from the urine into the blood (Kahn *et al.*, 1975). About 80% of the nitrate in urine is recovered in this way.

The peculiar behaviour of the body with respect to nitrate led Benjamin *et al.* (1994, 1997) to suspect that the oddities described above were all part of an underlying purpose, probably to protect against disease-causing organisms, and to make further investigations. They first investigated the reduction of nitrate to nitrite on the tongue.

Conversion of Nitrate to Nitrite on the Tongue

The bacteria that reduce nitrate to nitrite in the soil are facultative anaerobes, a class of bacteria introduced in Chapter 3. Sasaki and Matano (1979) showed that in humans the reduction (see Box 2.1) of nitrate occurred almost entirely on the surface of the tongue. This reduction must have been effected by one of the *nitrate reductase* enzymes which are found in animals, plants and soils. Sasaki and Matano suggested that the enzyme involved was a mammalian enzyme. When Duncan *et al.* (1995) studied the surface of the tongue of a rat, they too found substantial nitrate reductase activity, which was confined to the rear two-thirds of the tongue. Further examination of the surface of the tongue revealed a dense population of bacteria, 80% of which showed considerable nitrate-reducing activity *in vitro*.

The experimenters suspected that nitrate was being reduced on the rat's tongue by the bacteria, and this was confirmed by the observation that rats bred in a germ-free environment which prevented colonization by bacteria had no nitrate-reducing activity on their tongues. Further evidence came from experiments in which healthy human volunteers were treated with the broad-spectrum antibiotic amoxycillin. The antibiotic decreased the nitrite concentrations in their saliva (Dougall *et al.*, 1995). The research has not been able to characterize the organisms on normal human tongues because this would require too much of the tongue to be removed for biopsy. Most of the bacteria are at the bottom of deep pouches (the papillary clefts) on the surface of the tongue. The species of bacteria on the rat's tongue have been reported.

Benjamin (2000) and his colleagues are convinced from these and other studies that the bacteria are truly in a symbiotic relationship with the mammalian host and that the latter actively encourages the growth of nitrite-forming organisms on the surface of the tongue. The bacteria are facultative anaerobes (Chapter 3) which, in the absence of oxygen, use nitrate instead of oxygen as an electron acceptor to oxidize carbon compounds to manufacture ATP and derive energy. In doing so they reduce the nitrate to nitrite. For the bacteria, nitrite is a waste product but, for the mammalian host, the provider of the tongue, the nitrite is an important antimicrobial agent.

Acidified Nitrite and the Production of Nitric Oxide

When nitrite formed on the surface of the tongue is acidified it becomes an effective antimicrobial agent. This can happen in two ways. If it is swallowed, it is acidified rapidly in the stomach. If it stays in the mouth it encounters the acid environment around the teeth provided by the acidifying bacteria thought to be a major cause of dental caries. We look first at the fate of the nitrite that is swallowed.

Nitrite and nitric oxide in the stomach

Nitrite, NO_2^-, is in equilibrium with nitrous acid, HNO_2:

$$H^+ + NO_2^- \leftrightarrow HNO_2 \quad (9.1)$$

Nitrous acid has an acid dissociation constant of 3.2, so in the very acid environment of the normal fasting stomach (pH 1–2) the equilibrium is displaced strongly towards nitrous acid. But nitrous acid is unstable and decomposes spontaneously to nitric oxide, NO, and nitrogen dioxide, NO_2. If there is a reducing agent present, more NO will be formed than NO_2. Lundberg *et al.* (1994) showed that there was a very large concentration of nitric oxide in gas expelled from the stomach in healthy volunteers. This increased with greater nitrate intake and decreased when the acidity of the stomach was decreased using the drug omeprazole to restrict the activity of the 'pumps' that transfer hydrogen ions into the stomach.

McKnight *et al.* (1997) made detailed studies that measured the amount of nitric oxide produced when healthy human volunteers ingested nitrate through tubes that went up their noses and down into their stomachs. Sixty milligrams of nitrate – the amount contained naturally in a large helping of lettuce – gave a large increase in the nitric oxide concentration in the head-space of the stomach. This peaked after about an hour but remained greater than in a control for at least 6 h. (See also Box 9.2.) The concentrations of nitric oxide measured would be lethal after about 20 min if breathed continuously.

Enhancement of nitric oxide activity in the stomach

The concentration of nitric oxide in the stomach is much greater than would be expected from the concentration of nitrite in saliva and the pH measured

Box 9.2. Nitrate, nitrite, nitrous oxide and lunch.

Professor Benjamin told me of a volunteer, fitted with a tube going down into the head-space of the stomach, who ate a good helping of lettuce. This resulted in a spectacular increase in the nitric oxide concentration in the head-space. This is not surprising, because lettuce is recognized as a notable accumulator of nitrate.

We were at the time travelling on a Eurostar train to Paris for a meeting. When we arrived, we were intrigued to note during lunch that before the main meat course we were served with a generous portion of salad, including lettuce. This is a frequent feature of a French meal, so perhaps experience has taught the French that it is a good idea to embark on the meat course with plenty of nitric oxide in the head-space of the stomach. The nitric oxide in the head-space is probably not as necessary as in former years, but could it be that the acidified nitrite/nitrous oxide mechanism has influenced the gastronomy of the nation that gave the world the word 'cuisine'?

in the solution in the stomach. *In vitro* studies suggest that the concentrations of nitrite observed would generate about one-tenth of the nitric oxide that is actually measured at the pH of the stomach. The enhanced production is probably due to a reducing agent such as ascorbic acid (vitamin C), which is actively secreted into the stomach, or thiols, organic sulphur compounds that are found in high concentrations in the mucous lining of the stomach.

Nitrite and nitric oxide in the mouth

Saliva is generally neutral or slightly alkaline, so the researchers were surprised to find that nitric oxide is also generated from salivary nitrite in the mouth. The most likely mechanism for this production is acidification at the gingival margins, where the teeth emerge from the gums. The acidity needed to generate nitric oxide from nitrite in saliva might be produced by organisms such as *Lactobacillus* or *Streptococcus mutans*, which are implicated in dental caries. This idea is not yet clearly established because the nitric oxide formed by the acidity would inhibit the acidity-producing organisms. This apparent anomaly needs to be clarified as soon as possible because nitric oxide produced in the mouth may explain why saliva seems to be important in protecting against dental caries (Mendez *et al.*, 1999).

As happens in the stomach, acidifying saliva produces larger amounts of nitric oxide than would be expected from the concentration of nitrite it contains, suggesting the presence of a reducing agent. Saliva contains ascorbic acid, and using *ascorbic acid oxidase* to minimize its concentration decreases the amount of nitric oxide produced to that expected from the concentration of nitrite. There are probably other reducing agents in saliva that augment nitric oxide production.

The idea that saliva has antibacterial properties may explain certain behavioural habits. When we injure ourselves, cutting a finger, for example, we have a strong instinct to lick the injured part (Benjamin *et al.*, 1997). Dogs have the same instinct, and sometimes have to be fitted with rigid plastic bonnets to stop them from excessive licking, in which they can remove all of the coat and even some skin from a sore or injured foot. This instinct to use the antibacterial properties of saliva also explains why an otherwise fastidious cat will lick its bottom.

Nitrite and nitric oxide on the skin

Weller *et al.* (1996) used a nitric oxide detector, which was based on chemiluminescence and able to detect as little as 1 ppb of nitric oxide in air, to show that nitric oxide was generated from normal human skin. They also established that this nitric oxide was not coming from under or within the skin and that the flux was enhanced by acidity but lessened when the subject was given an antibiotic. Applying nitrite to the skin surface substantially increased the flux of nitric oxide, suggesting that nitric oxide was again being generated by the reduction of nitrate to nitrite which was then acidified by acids on the skin. This suggestion was strengthened by the fact that normal human sweat contains nitrite at precisely the concentration needed generate the flux of nitric oxide observed from the skin.

The system inferred above has yet to be confirmed, but bacteria living on the skin are known to contain the nitrate reductase enzyme that can reduce the nitrate in sweat to nitrite. The researchers suggest that the production of nitric oxide on the skin provides a defence against skin infections, especially those caused by fungi. This idea is supported by the results of Weller *et al.* (1998) who showed that nitrite acidified with an organic acid is an effective treatment for *tinea pedis* (athlete's foot).

What Exactly Kills the Bacteria?

Nitric oxide (NO) is not the only oxide of nitrogen formed when nitrite is acidified and may not be the only molecular species responsible for killing bacteria (Fang, 1997). Individual pathogens may differ in their susceptibility to particular nitrogen oxides or other reactive nitrogen species. The additional stress provided by the acidity should also make the organisms more vulnerable. Nitrous acid (HNO_2), dinitrogen trioxide and nitrogen dioxide are all effective nitrosating agents (NO donors) and this could add to their ability to kill bacteria.

Nitrous oxide and acidity

Many bacteria that cause gastroenteritis in humans are remarkably resistant to acid on its own. Incubating *Candida albicans* in a very acid solution (pH 1) for an hour has no detectable effect on its subsequent growth. Adding nitrite to this solution at the concentrations found in saliva results in an almost complete elimination of the organism (Benjamin *et al.*, 1994). *Escherichia coli* survives at a pH of 3 but adding as little as 0.03 mg/l of nitrite slows its subsequent growth. The concentration of nitrite in saliva ranges between 3 and 30 mg/l depending on the amount of nitrate ingested.

The common causes of stomach and bowel problems, *Salmonella typhimurium*, *Yersinia enterocolitica*, *Shigella sonnei* and *E. coli* O157 are largely sensitive to the combination of acidity and nitrite (Dykhuizen *et al.*, 1996). Most of these organisms would survive exposure to moderate acidity (pH 3) for 1 h, but would succumb to the nitric oxide generated from nitrite at the concentrations normally found in saliva (Table 9.1). Although the fasting stomach has a pH of 1–2, the pH rises – that is, the stomach becomes less acid – after a main meal.

One final factor in bacterial killing is nitrosation. Reactions of this kind, described in Box 9.3, almost certainly add to the effectiveness of nitric oxide in killing bacteria.

Table 9.1. Interaction between nitrite concentration and pH in the control of various organisms. Minimum nitrite concentration needed to ensure that no growth of the organism occurred when it was transferred from acidified nitrite into a recovery medium after 30 min. (Adapted from Table 1 of Dykhuisen *et al.*, 1966.)

	Minimum nitrite concentration (mg/l) at pH					
Organism	2.1	3.0	3.7	4.2	4.8	5.4
Salmonella enteridis	0[a]	6	61	460	> 460	> 460
Salmonella typhimurium	0	38	307	> 460	> 460	> 460
Escherichia coli O157	46	307	> 460	> 460	> 460	> 460
Yersinia entercolitica	0	1	61	307	> 460	> 460
Shigella sonnei	9	77	460	> 460	> 460	> 460

[a]The zero implies that the pH alone controlled the organism.

Box 9.3. Nitrosation.

Nitric oxide (NO) is described as a free radical because it has an unpaired electron in its electronic configuration. Like other free radicals, it is very reactive. The addition of the NO group to another compound is 'nitrosation' and produces a nitroso compound. Two particular groups of compounds are of interest here, *N*-nitrosamines and *S*-nitrosothiols. They are produced when nitrite or nitrous oxide reacts with an NH or SH group. The NH group would probably be in a secondary amine and the SH group in a thiol (an alcohol with the OH group replaced with an SH group). Both *N*-nitrosamines and *S*-nitrosothiols are of considerable medical interest.

The *N*-nitrosamines are carcinogenic, and their mode of formation was at the root of the supposed link between nitrate and stomach cancer. The *S*-nitrosothiols are interesting as 'nitric oxide-donor drugs' (Al-Sa'doni and Ferro, 2000) because of the therapeutic properties of nitric oxide (Box 9.1), particularly in cardiovascular disorders. They also probably play a part in bacterial killing.

Nitric oxide itself may attack bacteria by nitrosating the cell surface. But the *S*-nitrosothiols are more than nitric oxide donors, they are effective nitrosating agents. Fang is the leading authority on the killing of bacteria by nitric oxide and was cited by de Groote *et al.* (1996) as suggesting that the generation of nitrosothiols could be the main mechanism by which pathogenic bacteria such as *Salmonella* are killed. Thiols are plentiful in the stomach lining and *S*-nitrosothiols will inevitably be formed when nitrite and acid are present as well. Thiocyanate, which is plentiful in saliva, and chloride ions, which are at high concentrations in the stomach, both catalyse nitrosation by the formation of more reactive intermediates and may thereby add to the toxicity of acidified nitrite.

Nitric oxide can react with superoxide (O_2^-) to form peroxynitrite ($ONOO^-$) (e.g. Al-Sa'doni and Ferro, 2000). This is a potent oxidizing agent and nitrosating agent and has been shown to nitrosate proteins and nucleic acids as well as cause lipid peroxidation. It could well play a part in the killing of bacteria.

There is, however, one organism, *Helicobacter pylori*, that seems considerably more resistant to acidified nitrite than others (Dykhuizen *et al.*, 1998). It may be able to withstand the effect of nitrogen oxides by using the urease enzyme to generate ammonia from urea, or it may have developed other biochemical protection mechanisms. This is an important disease-causing organism and its protection mechanisms present an interesting challenge to medical research.

Nitrate – an Old Friend?

Despite its current pariah status among the politically correct, nitrate has a long and honourable history in which it has served mankind as a food preservative. Our ancestors realized its value several centuries ago, probably because they discovered empirically that it was about the only thing that stood between them and a particularly unpleasant form of death that we now recognize as the outcome of ingesting the botulinum toxin which is produced in stored meat by a bacterium, *Clostridium botulinum*. Nitrate and/or nitrite are still in today's meat products and with good reason – nitrite is acidified when added to meat and the resulting nitric oxide is one of the few effective killers of botulinum spores.

The evidence for the protective effects of nitrate collected by medical researchers is convincing, but the researchers feel they do not yet have sufficient evidence that this mechanism fully protects humans exposed to a contaminated environment. Research now in progress aims to find out whether our bodies' capacity to protect themselves from gastroenteritis using acidified nitrite can be enhanced by simply increasing the amount of nitrate we ingest. Millions of people die from bacterial gastroenteritis each year, most of them in the developing world. If the simple acidified nitrite system can be developed into a cheap readily applicable therapy for the disease it will be an achievement of the greatest importance. It will also be a severe indictment of those who still seek to stigmatize as a poison one of mankind's oldest friends, the nitrate ion.

10 The Politics and Economics of Nitrate

Listing the most unpopular professions has become something of a sport in the UK – even among journalists, who are frequently at the less desirable end of the list. It would be interesting to know where, relative to each other, politicians, economists and farmers would appear. Another interesting comparison that can be made between these professions is in the amount of responsibility each bears for the 'nitrate problem'. The farmer is the obvious scapegoat for the nitrate problem, but is he the right one? There is no objective measure by which blame can be assigned but, if such a measure could be devised, politicians and economists might fare at least as badly or even worse than farmers, and we definitely need to examine the consequences of political and economic decisions as well as the activities of farmers. To keep the chapter within reasonable bounds, I have concentrated largely on what happened in the UK and did not stray beyond Europe.

The Common Agricultural Policy

The Treaty of Rome, signed in 1957, committed the six founder members of the European Community (EC) to the Common Agricultural Policy (CAP), which has undoubtedly been the largest political influence on nitrate in the European environment. This policy has been criticized frequently, and it is helpful to look back to the circumstances that led to its origin.

Europe was in a bad state in the late 1940s. The combatant nations were financially and mentally exhausted after 6 years of war. The 1946/47 winter was an exceptionally cold one and fuel was in short supply. Food shortages were so bad that the whole of Europe was officially designated a famine area in 1947. The rationing of food that came in with the war was to continue well into the 1950s. (After 50 years, I can still remember the first sweets I bought when they were de-rationed.) Food security must have been as high a priority in Europe in the early 1950s as it is in parts of Africa now.

The founders of the CAP were no doubt well aware that there were many small or part-time farmers in Europe, all of whom had votes. Because of the way land is inherited in some EC countries, the proportion of small and part-time farmers is larger in those countries than in the UK. The founders may also have had in the backs of their minds that Hitler's rise to power depended in no small way on support from disgruntled farmers. Any or all of these considerations could have led to the CAP, which had five founding aims:

- Increased productivity.
- A fair standard of living for farmers.
- Stable markets.
- Regular food supplies.
- Reasonable prices for consumers.

The CAP was based on three principles:

- A single market in farm products, with common prices and free movement of agricultural goods within the community.
- Preference for community members.
- Shared costs.

In many ways the CAP was a success. I remember being driven through France and Belgium in 1957 and 1958 on the way to family holidays in Switzerland. Many of the towns and villages had not fully recovered from the war and looked scruffy and rather despondent. By the time I did similar drives in 1970 and 1972 there had been a marked change, with everything looking much smarter and more prosperous. The economic benefits from the EC and the CAP were clear to see and famine was a distant memory.

The problem with the CAP can arguably be summed up in two words, 'intervention buying'. To ensure security for Europe's many farmers, agricultural produce was brought into intervention whenever there was a glut and prices fell. This gave farmers the incentive to produce whether or not there was a market for their produce, and the result was over-production leading to the notorious 'butter mountains' and 'wine lakes'. Virtually all productive activities involve pollution of some sort, and agriculture is no exception. One of the main pollutants from agriculture is nitrate, so over-production of agricultural commodities meant an unnecessary increase in loss of nitrate from the soil. Thus a direct line can be drawn from the political decision to implement the CAP to the increased loss of nitrate. This is a somewhat simplified argument and, although I have no doubt it is essentially true, I expand it in the context of the UK below.

The effects of the CAP on nitrogen use by UK farmers

The UK became a member of the EC in 1973 and came under the CAP at the same time. This had direct and indirect effects on the use of nitrogen fertilizer.

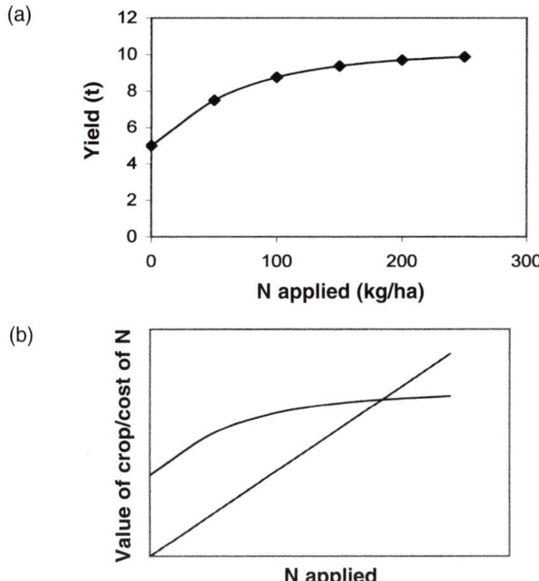

Fig. 10.1. (a) Diminishing returns in yield as nitrogen fertilizer is applied. (b) The optimum nitrogen application, N_{opt}, occurs where the extra cost of the nitrogen (straight line) is no longer met by the extra value of the crop (curved line).

Direct effects

The main direct effect was that assured prices for crops made it economic to use more nitrogen on each crop. This effect is not straightforward, because the relationship between nitrogen application and yield is not a straight line but a curve, which shows a diminishing return in yield with each increment of nitrogen (Fig. 10.1). The amount of nitrogen that gives the optimum response in economic terms, N_{opt}, is obtained as the amount at which the extra grain (in the case of wheat) just pays for the extra fertilizer. This optimum obviously depends on the prices of both grain and fertilizer, and increases when the grain price goes up. N_{opt} has economic rather than biological significance.

There are several problems with this optimum. One is that it varies with the type of fertilizer used and the state of the grain market. Some fertilizers are cheaper than others and grain prices are better at some times than others. Another is that, as more nitrogen is given, the crop becomes more prone to disease and the farmer needs to spend more on chemicals. The cost of the chemicals ought to be included in the computation of N_{opt} but often is not. There is also the intrinsic problem that N_{opt} is known only when the experiment is harvested. Farmers planning to apply nitrogen depend on information from previous experiments.

How does the response curve that gives us N_{opt} relate to the 'surplus nitrate curve' in Fig. 5.12, which gives us the 'upturn point' at which the nitrate loss increases sharply from the constant amount occurring at smaller

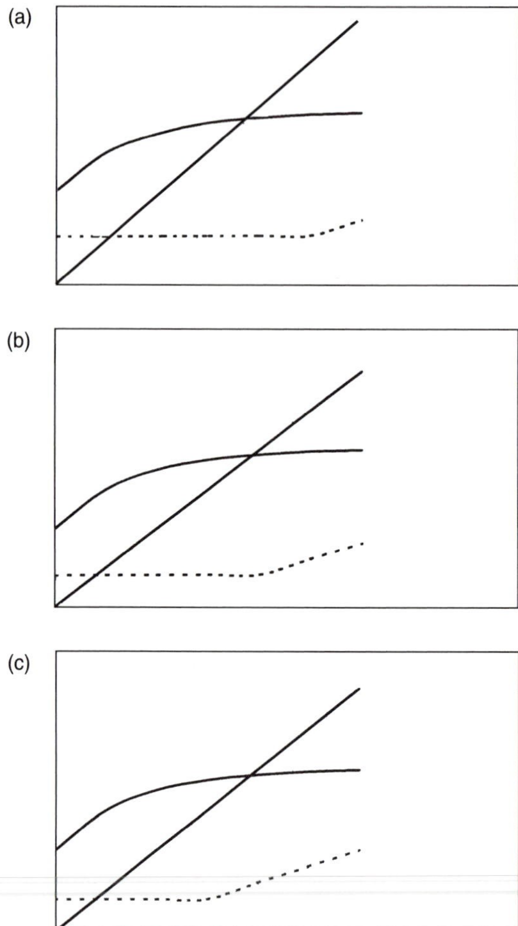

Fig. 10.2. The relationship between the N response curve (solid curved line) and the surplus nitrate curve (broken line). (a) N_{opt}, identified as in Fig. 10.1b, before upturn in surplus nitrate curve, (b) N_{opt} and the upturn coincide, (c) N_{opt} after the upturn.

nitrogen applications? If N_{opt} was determined biologically, it could reasonably be expected to coincide with the upturn point, but it is determined on an economic basis and the two may not coincide. Figure 10.2 shows three possibilities: that N_{opt} occurs before the upturn in nitrate remaining in the soil, that it coincides with it, or that it occurs beyond it. Obviously the first possibility is the safest from an environmental viewpoint. The second, that N_{opt} and the upturn point coincide, is also acceptable. The two do coincide approximately in some experiments, but this cannot necessarily be expected to be the norm, because N_{opt} is determined by economics and the upturn point by biology. The third possibility, in which N_{opt} is beyond the upturn point, is clearly undesirable.

The precise shape of the curve in Fig. 10.1 that gives N_{opt} is important.

Back in the 1980s, a colleague in the Statistics Department at Rothamsted, Alastair Murray, was computing the curves describing the yield response of winter wheat to nitrogen fertilizer in various field experiments and computing N_{opt} values. I asked him to compute how much of the yield would have been obtained with half the N_{opt} value in each case. He found that half the N_{opt} would have given about 90% of the yield in the experiments he analysed. This implies that at the optimum the first half of the N_{opt} gave 90% of the yield while the second half added only another 10%. It is obviously this second half that is at greatest risk of pushing N_{opt} beyond the upturn point and causing a sharp increase in nitrate losses.

The key point is that the second half of the N_{opt} still paid for itself at the prices then prevailing. Farmers are human, and given an assured market and a guaranteed price, they would have applied the whole of the N_{opt} without hesitation and probably added a bit extra as 'insurance'. That the CAP price structure caused over-production is beyond doubt. This was because the prices of wheat and other crops were kept higher than they would otherwise have been. This implied that N_{opt} was kept higher than it would otherwise have been and therefore more likely to go beyond the upturn point. It did not necessarily do so in all cases, but if you take the whole of the UK or the whole of the EC, the CAP increased the proportion of cases in which N_{opt} went beyond the upturn point and led to increased losses of nitrate. It is worth recalling that in 1976, 3 years after the UK came under the CAP, UK farmers switched from applying less nitrogen to winter wheat than the crop took up to applying more (Chapter 5). These direct effects provide more detailed evidence of the link between the political decision to implement the CAP and the increase in nitrate losses to the environment. They were supplemented by some indirect effects.

Indirect effects

The CAP had some economic effects that the founders of the EC may not have envisaged. One was that, as the CAP made farming profitable, land became more attractive as an investment, notably to large institutions such as pension funds. Land prices therefore rose about fivefold between 1970 and 1980 (Fig. 10.3) (but by rather less when corrected for inflation). The rise had started by 1973 but accelerated at that time. Interest rates also rose during this period, making it important that investments gave a good return. Farmland could best be made to give a good return by using more nitrogen fertilizer, which increased crop yields and enabled stocking rates (the number of cows (etc.) per acre) to be increased. The average application of fertilizer nitrogen for all arable crops and grassland together rose by 75% between 1970 and 1984, from 80 to 140 kg/ha. This change may not have had a large direct effect on nitrate losses, but it increased the amount of nitrogen in the soil nitrogen cycle and must have increased losses of nitrate in the long term.

Other influences

The CAP was not the only influence on nitrogen use in Britain. Until 1966, there was a government subsidy on fertilizers, a relic of the thin years

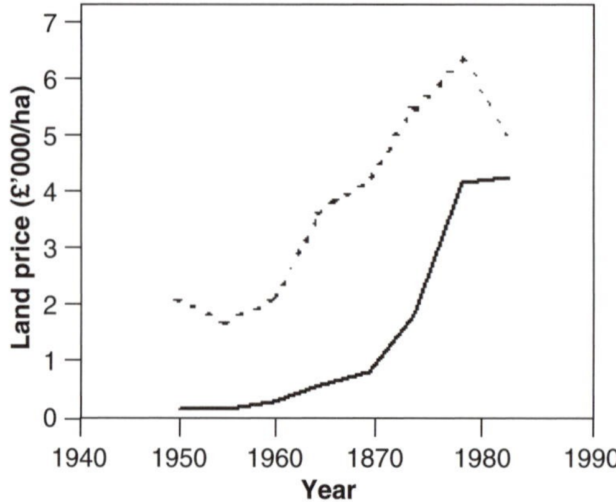

Fig. 10.3. Increase in land prices between 1950 and 1985. Current values at the time (solid line) and values adjusted to 1986 prices (dashed line). (From Marks, 1989.)

during and after the war. I was working for the Lawes Chemical Company in 1966, and the removal the subsidy was bad news for us. The British Government still seemed enthusiastic about agriculture in 1975, when it stated in a white paper that 'the continuing expansion of food production will be in the national interest', but its enthusiasm waned subsequently. And another influence not to be forgotten was the demand of the public for cheap food which could only be met through intensification of agriculture.

What Really Went Wrong?

The direct and indirect effects of the CAP cannot altogether be separated from each other, but both point to the direct link between the CAP and increased nitrate losses. To draw attention to this link is not to condemn the principle of the CAP. I saw for myself the benefits it brought to post-war France and Belgium. Even the notorious 'mountains' were not entirely bad. Back in 1988, I looked up the amount of cereal grain in intervention stocks in the UK for a popular article about the nitrate problem (Addiscott, 1988b). It was 1.7 million tonnes. This was before the 1988 harvest had been stored. But what if that harvest had failed? That 1.7 million tonnes of grain would have lasted about 1 month. Was it a monstrous grain mountain or a pathetically meagre reserve? Our balance of payments was bad enough at the time without massive grain imports.

Economics: solution or problem

With the CAP we need to ask how something which started with such good intentions ended up by causing pollution. Part of the answer is the inappropriate involvement of economics. Economics has its uses, but it is far too blunt an instrument to be let loose in anything as delicately balanced as a biological system. The optimum amount of fertilizer, N_{opt}, depended on the prices of grain and fertilizer and took no account of the biological requirements of the crop. Fertilizer applications were also driven by land prices and interest rates, neither of which had any biological relevance. We hear a lot of glib talk from politicians and economists about 'market solutions', how this omniscient and all-powerful being 'the market' will provide answers to all our problems. In this case, the market solution was pollution.

The reaction to the nitrate problem

The impact of the CAP on farming practice was soon clear. Agriculture was intensified and nitrogen fertilizer was a key part of the intensification package. Crop yields and stocking rates for farm animals increased considerably as more nitrogen was used. In some places the landscape altered appreciably as hedges were removed, and this change was exacerbated in the UK by the loss of virtually all the elm trees to Dutch Elm Disease.

It is difficult to be certain when the public became aware that there was a 'nitrate problem', but this was probably in or shortly after 1980. One key event was the publication by a British newspaper of an article on 'The nitrate time bomb' in the early 1980s. This 'time bomb' was the downward movement of nitrate released by the ploughing-up of old permanent grassland (discussed below). This was a groundwater problem, but the article also stirred up concerns about increasing nitrate concentrations in rivers and questions about nitrate in drinking water. Nitrate in drinking water was discussed as a cause of stomach cancer and concerns were raised about the role of nitrate in methaemoglobinaemia (blue-baby syndrome).

It is clear with the benefit of hindsight that 1980 or thereabouts would have been a good time for the EU to review the side-effects of the CAP as well as its very definite benefits. Some fine-tuning of the provisions of the CAP at this point might have saved a lot of trouble later. Taxation of nitrogen fertilizer might also have been considered, but the shape of the curve relating yield to application of nitrogen fertilizer would probably have meant an unacceptably large tax to achieve the desired effect. Instead, the EU chose to legislate, and imposed the 50 mg/l on nitrate in drinking water in a series of directives starting in 1980. The imposition of that limit marked the beginning of the real nitrate problem in Europe.

The EC 50 mg/l nitrate limit

It was once said of a British TV personality that 'he rose without trace'. Much the same could be said of the 50 mg/l limit for nitrate. There was no clear rationale for it. No systematic toxicological tests of the type described in Chapter 12 seem to have been made. The limit seems rather to have sprung from limits for nitrate used by the World Health Organization

(WHO) that were probably based on the rather questionable data from the USA discussed below. The WHO set European standards for nitrate in drinking water in 1970, which stated that less than 50 mg/l was 'satisfactory', 50–100 mg/l was 'acceptable' and more than 100 mg/l was 'not recommended'. The EC adopted 50 mg/l as its upper limit, a seemingly arbitrary decision given that there was little medical evidence of any risk at 100 mg/l. It also applied the limit without taking note that nitrate concentrations in water vary and show peaks. The peaks must never exceed 50 mg/l, and this is far more difficult to ensure than to keep the monthly or yearly average within bounds. The EC has never published any explanation or justification of the limit (L'hirondel and L'hirondel, 2001), and a letter I sent to the 'Administrateur des Eaux' in Brussels, stating that I was about to write this book and inviting comment on the basis of the limit, received no reply.

Medical evidence

Nitrate in drinking water caused two main health concerns, methaemoglobinaemia and stomach cancer.

Methaemoglobinaemia is a condition suffered by infants up to about 1 year old, and is caused by *nitric oxide* rather than *nitrate*. This free radical converts the haemoglobin in the blood to an abnormal oxidized form known as methaemoglobin that cannot bind oxygen properly. The nitric oxide could in principle come from the chemical or bacterial reduction of nitrate in drinking water or vegetables. But Hegesh and Shiloah (1982) found that when methaemoglobinaemia occurred, gastroenteritis was almost exclusively the reason for the admission of the infant to hospital. They found no correlation between methaemoglobinaemia and the ingestion by the infant of nitrate in food or water. Gastroenteritis stimulates a defensive reaction in the body that produces nitric oxide and thereby causes methaemoglobinaemia. Contact with oxidized haemoglobin or superoxide (O_2^-) converts the nitric oxide to nitrate and can cause a large concentration of nitrate in the blood. The gastroenteritis may be caused by a bacterium or a virus, but the bacterial form is more likely in the context of water supply, particularly from wells.

Only one case of methaemoglobinaemia has ever been recorded in the UK when the concentration of nitrate in the water was less than 100 mg/l. In that case the concentration was 95 mg/l and the water was from a well polluted with bacteria (Ewing and Mayon-White, 1951). The last fatal case in the UK, reported by the same authors, occurred 53 years ago and was associated with well-water with a concentration of 200 mg/l. The problem occurred typically when well-water was used to make up bottle-feeds for infants. There has never been a case involving mains tap water.

In the USA, the condition is also associated with water from wells. It was first identified by Comly (1945), who called it 'well-water methaemoglobinaemia'. His comments on the state of the wells from which the water

Box 10.1. Problems with wells.

Two descriptions of the state of wells associated with cases of well-water methaemoglobinaemia. Comly (1945) was the first to identify the condition. He wrote of the wells involved:

> The water came from very undesirable wells. In many cases the wells were old, dug rather than drilled, had inadequate casings or none at all, and were poorly covered so that surface water, animal excreta and other objectionable material could enter freely. In every one of the instances in which cyanosis developed in infants the wells were situated near barnyards and pit privies.

Romania is one of the few places in which well-water methaemoglobinaemia still occurs. Ayebo *et al.* (1997), cited by L'hirondel and L'hirondel (2001) in their Box 5.3, gave the following description of the type of well that led to problems:

> The typical well has a dugout less than 8 metres deep (with some as shallow as 3 to 4 metres), without a casing and with no protective cover over the wellhead. The growth of algae on the inside walls was common, and most wells were located close to human traffic areas. Nitrate contamination from the agricultural use of fertilizers and animal manure may occur, but the most important source of nitrate in the wells observed in this study was local human and animal waste. Well construction methods, well placement and general hygiene were found to be the primary causes of poor water-quality and high nitrate levels. Household laundry and washing of utensils are done around the open shallow wellhead. The nitrate concentrations of the water in most of the wells used for infant feeding at the homes visited exceeded 250 mg l^{-1}, as estimated in the field using colorimetric strips.

responsible came (Box 10.1) make interesting reading for those with strong stomachs. Walton (1951), cited by Avery (1999), gathered together data from more than 278 cases in the USA, all of which were associated with wells, and 214 of which were accompanied by data. Critically, Walton's information did not include whether or not the water was polluted by bacteria, and Box 10.1 shows how important this information was likely to have been. These cases resulted in 39 deaths, 31 of which were in the mid-western states Minnesota, Iowa and Illinois. Table 10.1 summarizes the nitrate concentrations associated with these cases. Of these, 98% involved concentrations of more than 89 mg/l and 81% more than 226 mg/l. Only 2% of the cases were at nitrate concentrations less than 89 mg/l, and we do not know whether or not these involved bacterial pollution. L'hirondel and L'hirondel (2001) warned that this collection of cases may be biased. No cases are shown in the 0–44 mg/l range (Table 10.1) because many of the data originated from two earlier studies from which wells with these smaller nitrate concentrations were excluded by intent.

Table 10.2 shows some of the nitrate concentrations in wells associated with particular cases. Some of these were very large, and one well had a concentration of 0.02 M nitrate! Even so, the fatalities were not particularly associated with the largest concentrations, perhaps reflecting the doubts of those who collected the early American data (Box 10.2). More recent surveys have shown no correlation between methaemoglobinaemia and nitrate in water until the concentration of nitrate exceeds 100 mg/l. Concentrations

Table 10.1. Summary of cases of methaemoglobinaemia in the USA and the concentrations of nitrate in the well-water involved. (Based on data of Walton, 1951.)

Concentration of nitrate (mg/l)	Number of cases	% of total cases
0–44	0	0.0
44–89	5	2.3
89–222	36	16.8
222–443	81	37.8
443+	92	43.1

It is not known in how many of these cases the water was polluted by bacteria. The nitrate concentration was unknown in 64 cases.

this great are often associated with organic pollution which may imply bacterial pollution.

The nearest equivalent of present-day toxicological tests (Chapter 12) of nitrate in infants were made by Cornblath and Hartman (1948). These tests, described as 'ethically questionable' by L'hirondel and L'hirondel (2001), administered oral doses of 175–700 mg of nitrate per day to infants and also to older people. None of the doses to infants caused the proportion of haemoglobin converted to methaemoglobin to exceed 7.5%, strongly suggesting once again that nitrate alone, without bacterial pollution, does not cause methaemoglobinaemia.

L'hirondel and L'hirondel (2001), who were doctors, concluded from their very detailed study that there was little evidence that nitrate in drinking water is the prime cause of methaemoglobinaemia, and Avery (1999), who made a study of the early reports of methaemoglobinaemia, doubted whether nitrate, rather than bacterial pollution, was responsible for the condition. Other evidence seems to support these authors completely.

Stomach cancer is a painful and unpleasant way to die, so we cannot ignore the link suggested between this cancer and nitrate in water. There are

Table 10.2. The concentrations of nitrate reported in association with some specific cases of methaemoglobinaemia. (P) indicates that the well was bacterially polluted, but its absence does not imply that the well was clean. (F) denotes a known fatal case.

Source	Country	Cases	Nitrate concentration (mg/l)
Comly (1945)	USA	2	388, 619
		Anecdotal	283–620
Choquette (1980)	USA	1	1200
Busch and Meyer (1982)	USA	1	545
Johnson *et al.* (1987)	USA	1	665 (F)
Ewing and Mayon-White (1951)	UK	2	200 (F), 95 (P)
Hye-Knudsen (1984)	Denmark	1	200 (P)

Box 10.2. The standing of the early American water data. Adapted from Box 6.2 of L'hirondel and L'hirondel (2001).

Some comments from those involved with the collection of water samples connected with some of the early cases of methaemoglobinaemia. Bosch *et al.* (1950), who reported on 139 cases, stated:

> The samples from many of the 25 (cases) which contained 21–50 ppm (93–221.5 mg NO_3^- l^{-1}) were collected a year or more after a methaemoglobinaemia case had occurred, and sometimes after the well had been abandoned subsequent to the illness.

The American Public Health Association Committee on Water Supply (APHA, 1949–1950) investigated the connection between methaemoglobinaemia and the nitrate content of well waters, and stated:

> The committee does not have at its disposal detailed epidemiological and technical data connected with cases associated with water found to contain less than 50 ppm nitrate nitrogen (i.e. ca. 220 mg NO_3^- l^{-1}) when the samples were collected. It is evident, however, that many uncertainties prevail, such as that of samples of water collected after cases were reported may have contained a lower concentration of nitrates than when the water from the same well was consumed by specific infants.

The committee concluded:

> Therefore, it is impossible at this time to select any precise concentration of nitrates in potable waters fed to infants which definitely will distinguish between waters which are safe or unsafe for this purpose.

sound theoretical reasons for proposing such a link. Nitrite produced from nitrate could react in the stomach with a secondary amine formed when digesting meat or some other protein to produce an *N*-nitroso compound. Some such compounds are carcinogenic (Tannenbaum, 1987), so the reaction could lead to stomach cancer. But when this hypothesis was tested by epidemiological methods, four studies suggested that it did not stand.

1. A test at the Radcliffe Infirmary at Oxford (Forman *et al.*, 1985) identified two areas in which the incidence of stomach cancer was exceptionally high and two in which it was exceptionally low. These differences were so marked that the mortality in the high-risk areas was about double that in the low-risk areas. People coming to hospitals in these areas as visitors rather than patients were asked to give saliva samples. The hypothesis suggested that the samples from the high-risk areas should contain more nitrite and nitrate than those from the low-risk areas. The reality was that the samples from the low-risk areas had nitrate concentrations 50% greater than in those from the high-risk areas (Table 10.3).
2. Another study (Beresford, 1985) examined nitrate concentrations in 229 urban areas of the UK between 1969 and 1973 and related them to stomach cancer in the same areas at the same time. The hypothesis suggested that there should have been a positive relationship between them. There was in fact a negative relationship.

Table 10.3. Nitrate concentrations in saliva samples from people living in areas of low or high risk with respect to stomach cancer. (From Forman *et al.*, 1985.)

	Nitrate concentrations (mg/l) for					
	Low risk		High risk		Both low	Both high
	Area I	Area II	Area III	Area IV	Areas I + II	Areas III + IV
All samples	208	157	107	108	–	–
Refined samples[a]	172	150	97	117	162[b]	106[b]

[a]Refined sample: the donor had not eaten or drunk in the 2 h before the sample was given.
[b]There is a less than 1 in 10,000 probability that this difference occurred by chance.

3. Van Loon *et al.* (1998) looked for an association between stomach cancer and nitrate intake in a group of 120,000 people and failed to find one.

4. Every time I have been in a fertilizer factory I tasted fertilizer in my mouth within minutes of arriving, as the traces of fine dust in the air dissolved in my saliva. If anyone is going to suffer stomach cancer from exposure to nitrate, it is workers in fertilizer factories. But a study by Al-Dabbagh *et al.* (1986) on fertilizer workers showed no difference between their mortality from stomach cancer and that of comparable workers in other jobs. Twelve fertilizer workers died from stomach cancer during the 35 years of the study, while 12.06 would have been expected to die given the mortality rate among the other workers (Table 10.4). Not only were the fertilizer workers

Table 10.4. Mortality from cancer of the stomach among 1327 male workers in an ammonium nitrate fertilizer plant, 1 January 1946 to 28 February 1981. Number of deaths observed (obs) among those exposed to nitrate compared with the number of deaths that would have been expected (exp) from statistics from workers in the locality with comparable jobs. Data for heavily and less heavily exposed workers combined. (From Al-Dabbagh *et al.*, 1986.)

	Number of deaths	
Disease	Obs	Exp
Stomach cancer	12	12.06
All cancers	91	86.83
Respiratory diseases	36	51.04
Heart disease	92	113.72
All causes	**304**	**368.11**

The differences for cancer were not statistically significant. For respiratory and heart disease they were significant at $P < 0.05$, i.e. there was only a 1 in 20 chance that the difference was accidental. For all causes the difference was significant at $P < 0.001$, i.e. there was only a 1 in 1000 chance that the difference was accidental. The differences for respiratory and heart disease probably arose because smoking would not have been allowed in the plant because of the explosion risk. (See Box 4.2.)

no more prone to stomach cancer than the others, they were actually healthier, in that fewer of them died from heart and respiratory disease. This was almost certainly partly because they were not allowed to smoke at work because of the explosion risk, but recent research suggests a more positive role for nitrate in cardiac health (N. Benjamin, London, 2003, personal communication). (See also Box 9.1.) L'hirondel and L'hirondel (2001) record several other studies that show no increase in stomach cancer among workers in fertilizer factories.

These epidemiological studies are supported by the observation that the incidence of stomach cancer has been declining during the past 35 years, while nitrate concentrations in water have been increasing. The absence of a link was formally accepted by the British Chief Medical Officer more than 15 years ago (Acheson, 1985).

What justification for the limit is left?

Now the link between nitrate and stomach cancer is intellectually and administratively dead, what is left to justify the EC 50 mg/l limit? It would appear to stand or fall on the methaemoglobinaemia issue. But we saw above that methaemoglobinaemia is caused by nitric oxide produced in a defensive reaction stimulated by gastroenteritis and not by nitrate. The propinquity of wells and pit privies described by Comly (1945) seems to have led to a very expensive misunderstanding.

Does nitrate pose any other threats to human health? A statistically based study made by a group in Leeds (McKinney *et al.*, 1999) suggested that nitrate in drinking water is a factor in insulin-dependent diabetes in young people. Their findings show that the threshold concentration for the effect is 15 mg/l, which is both worrying and puzzling. It is worrying because this threshold is less than one-third of the EC limit and it will be exceeded in some areas of the country just as a result of deposition of nitrogen from the atmosphere, without any contribution from farming. It is puzzling because this threshold, based on a study in part of Yorkshire, is already comprehensively exceeded in other parts of the UK, seemingly without causing increased incidence in insulin-dependent diabetes in young people. These results must be regarded as open to question at the moment because of the seeming absence of effect outside Yorkshire and because another study (van Maanen *et al.*, 1999) showed no association between insulin-dependent diabetes and nitrate. Could they have had more to do with diet, given the increased consumption of junk food?

Could the EC limit itself be a threat to health?

It seems virtually certain that methaemoglobinaemia is not caused by nitrate in water, but by nitric oxide produced in a defensive reaction against bacterial gastroenteritis. The nitrate scare and the nitrate limit have led us in completely the wrong direction. It is not just infants who are at risk from gastroenteritis. The number of notifications of food poisoning in the UK rose from 35,000 in 1987 to 106,000 in 1997. The cause is not specified in the tables, but bacteria will have been at the root of many of these cases. Nitrate plays a key role in protecting against bacterial gastroenteritis, but many of those suffering from this condition will have had nitrate removed from their water in response to the EC nitrate limit. The removal of nitrate from water may not have had anything to do with the increase in gastroenteritis, but the coincidence is unfortunate.

The greatest overall threat posed by the nitrate limit is probably its enormous expense, its capacity to devour financial resources that could have been used for other more beneficial purposes. This is discussed further in Chapter 13.

Ploughing up Old Grassland During and After the War

The CAP sprang from the miserable aftermath of the Second World War, and also perhaps from memories of the rise of Adolf Hitler in the 1930s. Hitler and the war also played a part in another aspect of the nitrate saga, the ploughing up of old grassland to make way for arable crops and the resulting release of mineral nitrogen. This was a change that had begun in the 1930s but it accelerated when Hitler started the war by attacking Poland in 1939 and became crucial when in the early 1940s his U-boats were sinking large tonnages of allied shipping bringing food to the UK. Needless to say, no one gave much thought to the consequences of the ploughing when Hitler was trying to starve the UK into submission.

The consequences remained unforeseen as the ploughing continued after the war into the 1960s and to a lesser extent into the 1980s. By 1982 the area of permanent grass was one-third of what it had been in 1942. Enlarging the area of arable land must have increased the use of nitrogen fertilizer, but that is unimportant compared with the release from soil organic matter of mineral nitrogen, nearly all of which was either nitrate or soon nitrified to nitrate. How much did it contribute to the overall nitrate problem? And how long did its effects last?

Temporary and permanent grassland

Grassland is categorized as 'temporary', which usually means that it is left in place for about 1–5 years, or 'permanent', which has been there as long as anyone can remember. Ploughing up either category of grassland releases nitrogen from the organic matter that has accumulated under the grass canopy. Temporary grassland is often described as a 'ley'. Ploughing it up releases 100–200 kg/ha of mineral nitrogen, mainly as nitrate, with the amount released depending broadly on the length of the ley.

Ploughing up permanent grassland has more dramatic effects. Some

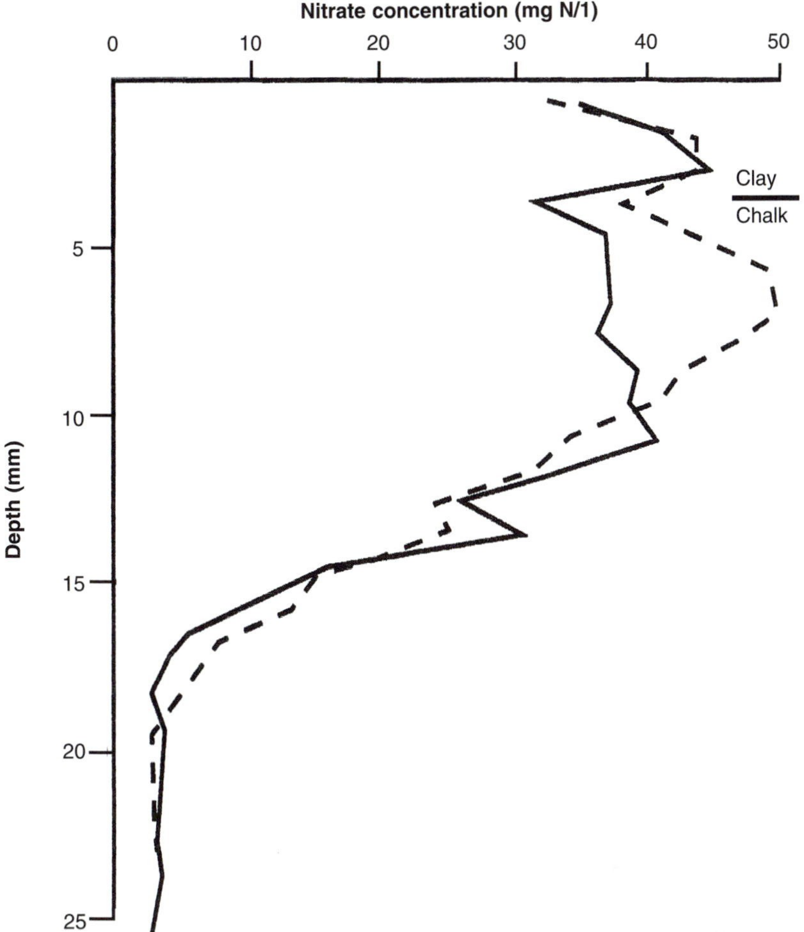

Fig. 10.4. The 'bulge' of nitrate in the interstitial water in a chalk core under Highfield at Rothamsted. Old grassland ploughed in December 1959, cores taken 18 years later. Measurements (solid line) by Water Research Centre, Medmenham; simulation (dashed line) by the author assuming that all nitrate came from the ploughing.

old permanent grass ploughed up on Highfield at Rothamsted in 1959 had released about 4 t/ha of mineral nitrogen in the first 18 years. Much of this nitrogen was then found as a bulge of nitrate in the interstitial water in the chalk below when a team from the Water Research Centre took cores from the chalk. This bulge could be simulated quite well with a computer model (Addiscott *et al.*, 1998) on the assumption that nitrogen lost from organic matter in the topsoil was leached into the chalk as nitrate (Fig. 10.4).

Like the Drain Gauge experiment (Chapter 5), this study emphasizes that large amounts of nitrate can be released into the environment from organic nitrogen. The difference between the two experiments is that the Highfield site was in permanent grass, was ploughed in 1959 and grew arable crops from then onwards, whereas the soil of the Drain Gauge was in arable use for many years before the Gauge was built but was never cultivated afterwards.

Computing the effects of the ploughing

Whitmore *et al.* (1992) used the data from Highfield with data from grassland ploughed up in Lincolnshire to develop a simple model for estimating nitrate concentrations in drainage leaving the soil as a result of the ploughing. This enabled them to produce maps (e.g. Fig. 10.5) showing where the greatest concentrations of nitrate in drainage were to be found in the UK in 1945. This was in the area to the west and south-west of the Wash. These maps also showed that if the EC nitrate limit had been in force at that time it would been seriously disobeyed over a wide area. The offending nitrate would have come from organic matter in the soil, as was the case when the Drain Gauges were 'in breach' of the limit in the 1870s.

The nitrate from the ploughing was the nitrate of the 'nitrate time bomb' mentioned earlier, and we need to check what its implications are now. We can do this using a relationship from the model of Whitmore *et al.* (1992). This exponential relationship describes how the amount of organic nitrogen in the soil N_{org} decreases with the time t from the ploughing up of the permanent grassland:

$$N_{org} = 6733 + 3954 \exp(-0.13t) \tag{10.1}$$

This relationship was used to show what contribution the ploughing of old grassland in 1943 and at subsequent 10-year intervals could be making to nitrate-N in the soil in 2003 and, if it were leached, it then would make to the nitrate concentration in drainage, assuming 250 mm of rainfall through the year (Table 10.5). The calculation assumed that all nitrogen lost from organic matter became nitrate, but the same assumption made in Fig. 10.4 gave a satisfactory simulation of the nitrate concentration in similar circumstances.

The calculations in Table 10.5 should be regarded as well-informed guesswork rather than fact, but the suggestion that the war-time ploughing is now a spent force in respect of nitrate in and leached from the topsoil is probably broadly correct. The same appears to be true for any ploughing done 10 years or so after the war. The ploughing was beginning to come to an end by 1980, but any that was done then could still now be contributing enough nitrate on its own to take the current nitrate concentration in drainage over the EC limit.

What about that 'time bomb'?

So far, we have considered only the production of nitrate in the topsoil and its concentration in drainage from it. The story that led to the 'time bomb'

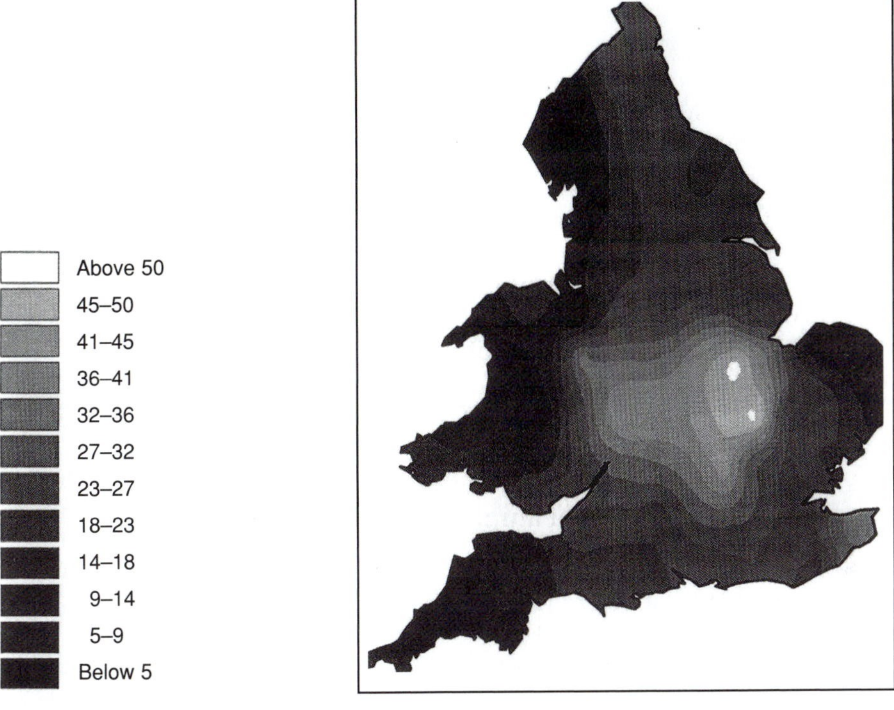

Fig. 10.5. The estimated contribution of ploughing permanent grassland to nitrate-N concentrations (mg/l) in drainage in England and Wales between 1939 and 1945, assuming that the only source of nitrate was ploughed grass. (From Whitmore *et al.*, 1992.)

scare was that a large concentration of nitrate produced by the ploughing of old grassland was moving slowly but inexorably downwards towards the surface of the groundwater, where it would be added to the water from which a substantial proportion of our domestic supplies are drawn. Young *et al.* (1976), who played an important part in the study of this process, showed that water and the nitrate it carried moved very slowly in some of the rock formations which provide our aquifers. They found, for example, that nitrate moves down at 0.8 m per year in unsaturated chalk. This means that the war-time ploughing may not be a spent force in some deep aquifers.

We can do another simple calculation which combines equation (10.1) with the rate of downward movement of 0.8 m/year found by Young *et al.* (1976) to show what concentrations of nitrate arrive at the surface of the groundwater at particular times as a result of war-time ploughing over aquifers of various depths. This was done for nitrate arriving at the water surface in 1980, when the time bomb scare occurred, and in 2003 when this chapter was written (Table 10.6). These calculations make the simplifying assumption that the nitrate concentration in a band of water did not change as it moved down. This is not entirely correct because there will have been some dispersion (spreading out and dilution of the concentration band) but

Table 10.5. Ploughing up old grassland at various dates. Estimated contribution in 2003 to nitrate nitrogen in soil and, if leached, to the nitrate concentration in drainage from the soil, assumed to be 250 mm/year. All nitrogen lost from organic matter assumed to become nitrate.

	Contribution in 2003 to	
Year of ploughing	Nitrate-N in soil (kg/ha)	Nitrate concentration in drainage (g/m)
1943	< 1	< 1
1953	< 1	1
1963	3	5
1973	11	19
1983	40	71
1993	149	264

Young *et al.* (1976) found that the bands of nitrate maintained their integrity fairly well initially. The calculation should again be regarded as informed guesswork rather than fact.

Some of the larger concentrations in Table 10.6 are almost certainly greater than would have been found. This is because dispersion depends on time and, after passing through the chalk for 30–60 years, the bands of nitrate would have become increasingly spread out. An even simpler form of dilution will also have applied. If, of the land above an aquifer, only one-quarter (say) was ploughed out of old grass, the concentration resulting from the ploughing will have been diluted by a factor of four when the

Table 10.6. Concentrations of nitrate from the war-time ploughing (assumed for convenience to have occurred in 1943) calculated to have arrived in 1980 or 2003 at the groundwater surface in aquifers of various depths, assuming a rate of downward movement for nitrate of 0.8 m/year.

		Concentration (mg/l) of nitrate arriving in	
Depth to water surface (m)	Travel time (years)	1980	2003
5	6.25	17	1
10	12.5	39	2
15	18.75	89	4
20	25	203	10
25	31.25	463	22
30	37.5	0	51
35	43.75	0	116
40	50	0	264
45	56.25	0	602
50	62.5	0	0

A zero in a column implies that the nitrate had not yet arrived. The largest concentrations in particular will have been subject to the various attenuating factors discussed in the text.

whole aquifer is considered. Denitrification has also played a part in removing nitrate from some aquifers, but the process needs a carbon substrate and the availability of such a substrate depends considerably on the conditions under which the rock of the aquifer was formed (Korom, 1992). Even with these allowances, Table 10.6 shows very well why people were concerned about a nitrate time bomb in 1980. It also suggests that by now the worst effects of the time bomb have probably passed in all but the deepest aquifers and, even in those, the attenuating factors have had a long time to work on the largest nitrate concentrations.

Although the ploughing of old grassland seemed to have ended, there are reports that it has restarted in the UK in response to the economic crisis afflicting British farming during the last year or so. The economic return on sheep farming in particular has become impossibly low and some farmers are reported to have ploughed some of the old pasture on which the sheep grazed. We can only note the polluting effect of an economic phenomenon and hope that this is the last instance.

11 Nitrate in Africa: the 'Western Hegemony'

WITH A CONTRIBUTION FROM KEN GILLER,
WAGENINGEN UNIVERSITY

The word hegemony comes from the Greek word *hegemonia*, which implies the dominance of one power in a group in which all are *supposedly equal*. It was originally applied to the relationship of Athens with the other city states. More recently, it was applied to the dominance of the USA and the USSR over the rest of the world. But with the collapse of the USSR it applies to the power held by 'the West' over the rest of the world, particularly over Africa and the developing world.

The Implications of the Hegemony

The 'Western hegemony' implies dominance by the West not only in trade but also in the field of ideas. This chapter is concerned with two aspects of this hegemony, political correctness and market dogma. Both are relatively recent arrivals on the Western scene, and both are probably alien to African thought. The chapter is included for two reasons. One is that the fallacious idea that nitrate is a threat to health has led to seriously mistaken ideas about nitrate in an African context. The other is that market dogma seems to have had highly undesirable effects on the supply and use of fertilizer in Africa. Political correctness and market dogma probably disapprove strongly of each other, but their effects are remarkably similar. Both, as we shall see, tend to discourage poor farmers from ameliorating poor soils with fertilizers.

Political correctness

Only people with full stomachs worry about nitrate in the environment. In most of sub-Saharan Africa, the problem with nitrate is not that it is a threat of any sort, it is just that it isn't there. There is not enough nitrate in the soil for staple food crops to grow to their full potential. The same is true of phosphate and frequently of potassium and trace elements. Very modest amounts of fertilizer would do a lot to keep hunger at bay, but there seems to be a problem of political correctness. Certain people in the comfortable Western world seem to feel that because of various scares about nitrate, usually unfounded (Chapter 10), Africa needs to be protected from fertilizers and left untempted by fruit of the branch of the tree of knowledge. What they forget is that, in the West, nitrate is one of the 'effluents of affluence', but Africa is very far from affluent and most people there would be delighted to receive even a small part of the Western world's nitrate problem.

Market dogma

The other aspect of the Western hegemony has been the imposition on Africa of market dogma. This has been a mixed blessing in Britain (Chapter 12) and disastrous for many people in Russia. There, 'Structural Adjustment' to the economy has been effected in response to the requirements of the International Monetary Fund (IMF). The aim was to achieve capital liberalization that would make Russia more accessible to investment and to the changes implicit in globalization, but the process has created massive differences between rich and poor. One of the newly rich in Russia has been able to buy the English football club Chelsea and pour money into buying players, but back in Russia the number of people living in poverty went from 2% in 1989 to 24% in 1998 (Stiglitz, 2002). Stiglitz, a former chief economist at the World Bank and the 2001 winner of the Nobel Prize for Economics, also reports that 50% of children now live in impoverished families.

Market-based reforms are being enforced on Africa, but will their effects be any better than in Russia? The omens are not good. Stiglitz records that at the 1986–1993 ('Uruguay') round (Box 11.1) of trade negotiations, the subject of trade in services was introduced, but markets were opened mainly for services provided by the advanced countries. As a result of the trade agreement, sub-Saharan Africa, according to a World Bank calculation, suffered a 2% decline in income. It was this agreement that led to the formation of the World Trade Organization (WTO) from the General Agreement on Tariffs and Trade. Box 11.1 explains the origins and roles of the IMF, the World Bank and the WTO.

Box 11.1. What exactly are the International Monetary Fund, the World Bank and the World Trade Organization?

This box, which may be unnecessary for many readers, is based on information provided by Stiglitz (2002). He is a former chief economist at the World Bank and a Nobel Laureate in Economics.

The International Monetary Fund

The International Monetary Fund (IMF) started life at a conference at Bretton Woods in New Hampshire in July 1944, as did the World Bank. Both organizations were intended to finance the rebuilding of Europe after the Second World War and to save the world from further economic depressions like that which had struck in the 1930s. The IMF was founded on the belief that there was a need for collective action for political stability at the global level. Its task was to prevent another global depression and it was to do this by putting international pressure on countries that were not doing enough to maintain global demand by allowing their economies to slump. The IMF is a public institution established with money from taxpayers around the world, but it does not report to either those who pay for it or those whose lives it affects (Stiglitz, 2002). It reports to the finance ministries and central banks of governments, and countries exert influence roughly in proportion to their economic power at the end of the war. This means that it is run by the main developed countries and that one country only has a veto. That country is the USA.

The IMF has changed greatly in the 60 years since 1944. The Keynesian ideals on which it was founded presumed that the IMF was a necessity because markets often worked badly, but the IMF has become a fervent believer in the dogma of the infallible market. Stiglitz (2002) records that the main change occurred in the 1980s when Margaret Thatcher and Ronald Reagan were preaching free market ideology in their respective countries. He comments that the IMF and the World Bank became the new missionary organizations through which market ideology was imposed on poor countries.

Stiglitz (2002) considers that the IMF has not done what it was supposed to do – that is, provide funds for countries facing economic downturns to help them restore employment and play a full part in international trade. Economic crises have been more frequent since its foundation than they were before. They have also been deeper than before with one exception – the Great Depression of the 1930s was the deepest of all. He also feels that some IMF policies, particularly the premature liberalization of some capital markets, have contributed to global instability.

The World Bank

The World Bank started life at the Bretton Woods Conference as the 'International Bank for Reconstruction and Development' and it has usually been closely associated with the IMF in its activities, often as the junior partner (Stiglitz, 2002). The distinction between the two organizations was defined at Bretton Woods as follows. The IMF would concern itself with *macroeconomics* in its dealings with a country. It would be involved in matters such as the country's monetary policy, budget deficit, inflation, overseas borrowing and the like. The World Bank was to deal with *structural issues*, such as its financial institutions, its trade policies and its labour markets – generally, the purposes for which it used its money.

It seems that the IMF began to take the view that almost everything fell within its remit and, as a result, the two organizations became more and more tied together. This was particularly so when the World Bank moved beyond lending for specific projects such as dams to providing structural adjustment loans (broad-based support to help the country adjust to and weather some kind of crisis). It did so only with IMF approval and this approval came with conditions attached. Because the developing world suffered fairly frequent crises, the IMF played a major role in its life.

The World Trade Organization

The agreement at Bretton Woods had advocated a third new economic organization, a World Trade Organization, whose aim was to manage international trade and put to an end the tariff 'wars' in which countries raised tariffs to help their own economies thereby damaging those of other countries, a practice that had been blamed for The Great Depression. This organization aimed to prevent another depression by encouraging the free flow of trade in goods and services. The General Agreement on Tariffs and Trade (GATT) was the first stage in the process. It succeeded in lowering tariffs appreciably in a series of 'rounds' of negotiations.

The eighth round of negotiations was highly significant. It is known as the 'Uruguay Round' simply because it began in 1986 in Uruguay. It ended in Marrakesh in 1993 when 117 countries joined in a trade liberalization agreement. Part of the agreement was that GATT became the World Trade Organization (WTO), and this came into effect at the beginning of 1995. It arrived 51 years after the Bretton Woods conference, in a world that had changed greatly in the meantime. Those who suggested it in 1944 could have had no idea of the present extent of world poverty or the dominance of market dogma. As Stiglitz (2002) points out, its belated establishment has not benefited the poor, in Africa or elsewhere, and the main beneficiaries seem to have been Western banks.

Assessing the impact of the hegemony

To assess the effects of the two aspects of the hegemony, we need to look first at the soils of Africa, which are to a significant extent the key to the problems of Africa. We then go on to examine the impacts of political correctness and market dogma on the use of fertilizer and other resources and consider whether organic farming is the answer to Africa's problems or part of the hegemony.

The Soils of Africa

The problems of African soils are many – too many to be described in one section of one chapter of one book, but an excellent account of these problems is given in a collection of papers edited by Buresh *et al.* (1997a). The underlying problem is that Africa is geologically an old continent, and its soils are inherently deficient of the nutrients needed by crops. This problem has been exacerbated by long-term nutrient removal in 'slash-and-burn' agriculture (Nye and Greenland, 1960). This system is sustainable as long as the land can be 'rested' for several decades between periods of cultivation, but in many parts of Africa the population is now too large for this to be possible. A recent study by Smaling *et al.* (1997) estimated rates of nutrient depletion to be 'very high' in Ethiopia, Kenya, Rwanda, Malawi and Lesotho, and 'high' for the other East African countries. They were also 'high' in Nigeria and two other West African countries. Erosion is a major factor, particularly in losses of phosphate. The 'very high' rates are found in countries that are mountainous and whose soils are inherently fertile, so that a lot of nutrients are at risk of erosional loss or leaching. Nutrient losses in countries such as Zimbabwe rate only 'high' because they have relatively infertile soils derived from granite that have already lost most of their stock of nutrients. The sandy soils to be found extensively in southern and western Africa generally have small stocks of nutrients. Overall, the losses from Africa's soils are now several times greater than the inputs of fertilizers.

Human activity is frequently at the root of erosion, but it can also lead to another, totally different form of nutrient loss: inter-continental transfer of nutrients. Trade in food and animal feedstuffs, though beneficial in some ways, contributes to nutrient losses from poor countries (Cooke, 1986; Bouwman and Booij, 1998). Large tonnages of nutrients have been exported in these commodities from Africa to Europe, where they add to environmental concerns about nutrient excesses.

Nutrient deficiency – getting better or worse?

I worked on soil phosphate in Tanzania as a United Nations Association volunteer in 1964 and 1965. Some of the phosphate concentrations I measured in Tanzanian soils 40 years ago were extraordinarily low even then (Addiscott, 1969). Research by Smaling *et al.* (1997) and Buresh *et al.* (1997b) suggests that fertility has declined widely since then and that phosphate is now desperately needed over large areas of Africa, mainly because of the nature and history of the soil. Forty years or more ago, field experiments very rarely showed that crops needed more potassium than was supplied by the weathering of minerals in the soil (Duthie, 1953; Scaife, 1968) but potassium additions are becoming increasingly important now (Smithson *et al.*, 1993; Bekunda *et al.*, 1997). Nitrogen was widely needed then and the need has not declined, because losses of nitrogen are inherently large in a hot climate. This is at least partly due to the Birch effect (Birch, 1960) in which the re-wetting of dry soil causes vigorous mineralization of organic carbon and nitrogen. When the first rain breaks the long Tanzanian dry season the air smells invigorating because of the large emission of carbon dioxide from the soil, but there is a correspondingly large release of ammonium that soon becomes nitrate and vulnerable to loss by leaching or denitrification before the crop roots are well established.

Problems of replenishing nutrients

Some of the nutrients needed in African soils can be supplied from organic sources, but certainly not all. In large areas there is a desperate shortage of plant-available phosphate in the soil. Not much can be released by mineralization of organic matter in the soil because there is rarely much organic matter. Most phosphate needs to be supplied in some mineral form, ideally fertilizer (Buresh *et al.*, 1997b). Potassium is not released at all by mineralization of organic matter and the weathering process is becoming increasingly inadequate on its own and will soon need to be supplemented by fertilizer, if not now. Of the three major nutrients, nitrogen can best be supplied from organic sources, mainly because of help from legume crops which, thanks to the rhizobial nodules on their roots (Chapter 3), not only supply themselves with nitrogen but also leave some in the soil for following crops. Giller *et al.* (1997) discussed the role of legumes in African agriculture in more depth.

The Impact of the Hegemony

Political correctness and 'low-external-input' approaches

Organic materials alone are clearly not enough to nourish crops adequately, and appropriately used fertilizers are essential in Africa (Sanchez *et al.*, 1997). Nevertheless, politically correct donor organizations in European countries often refuse to fund research or development projects that involve fertilizers, preferring to push for 'low-external-input' approaches. Such approaches concentrate on recycling crop nutrients rather than replenishing them, and they help to restrict losses of nitrate and phosphate to natural waters from rich-world soils rich in these nutrients. However they become the problem rather than the solution if applied to soils short of these nutrients. They just recycle nutrient deficiency and thereby recycle poverty (Dudal and Deckers, 1993).

This low-external-input approach is part of the pressure for 'sustainability', a generally rather poorly defined term that is discussed in more detail in Chapter 13. In the Western world, one definition involves making every effort to pass on the land and natural water in a viable condition from one generation to the next – an entirely laudable objective. But there isn't enough time for that kind of sustainability in much of Africa. The concern is with the present generation. Does agriculture produce yields that give them a reasonable life now? Does it even enable them to survive – now?

Market dogma and structural adjustment

In much of the Western world, 'the market' is seen as an all-wise arbiter of man's affairs, the provider of 'market solutions' to our problems. We shall see in Chapter 12 that the market is no better than the models that underpin it and that these models do not seem to have been tested adequately even in a Western context. Stiglitz (2002) points to another problem – there is no single concept of 'the market'. There are marked differences between the

Japanese concept, the American concept and the Swedish concept, to name but three.

What Africa needs is an African concept of the market, one that takes account of the real concerns of her people. This is probably a matter of scale. Africans can use appropriate markets to great effect (e.g. Box 11.2). But the market concept that is gaining influence in Africa is one that benefits rich countries such as the USA (Stiglitz, 2002) and it is not working to the benefit of poor people. Stiglitz describes the disastrous consequences of the IMF's attempt to enforce this market model in Ethiopia, which was done largely on false premises.

Subsidizing fertilizers is one possible response to a shortage of food, and the UK government subsidized them during and after the Second World War until 1966. Sanchez *et al.* (1997) note that, although many African countries used to subsidize fertilizers in the past, they have had to cease doing so under the terms of Structural Adjustment Programmes imposed by the IMF. As a result, the ratio of the fertilizer price to that obtained for the crop has tripled or quadrupled and fertilizer is now too expensive for many smallholder farmers who need to use it.

Subsidizing fertilizer is a far from perfect answer to food shortage among the poor. The subsidies tend to favour richer farmers, who use the most fertilizer, and the subsidized fertilizer reportedly 'leaks' across national boundaries. African farmers might be helped more by the stabilization of the prices they get for their produce. They are very vulnerable to changing commodity prices, whether these result from international forces or from local factors such as the amount of rain falling in the wet season. Price stabilization helps them to aim for larger crop yields, perhaps by using fertilizers, without exposing themselves to too much risk. Price guarantees are helpful when given by the country in which the crop is produced. But if the crop is exported, a price guarantee from the importing country is even more beneficial, helping both the farmer and the producing country. This is one good reason for buying tea, coffee, food or goods marked with the 'fair trade' symbol.

Another mechanism for stabilizing prices, one for which Koning (2002) argues, is for African countries to protect their farmers by imposing import duty on food. Koning considers that the internal market in most African countries is fairly competitive and that prices for food would rise as a result of import duty. He suggests that there would an additional beneficial effect, in the form of improvements to the infrastructure which would further improve the lot of farmers. The WTO will be strongly against import duties on food but, as Koning points out, 'there can be little justification for declaring import duty in Africa as taboo as long as the European Union and the United States continue the disguised dumping of agricultural products by means of direct payments'.

Guaranteed or stabilized prices might remind us of 'intervention buying', the explosion in the use of nitrogen fertilizer and the associated nitrate problem in Europe. Will this happen in Africa? The answer is almost certainly 'no'. The nitrogen fertilizer that caused problems in Europe was very cheap but nitrogen is very expensive in most of Africa. African farmers cannot afford to cause a nitrate problem. And the infrastructure (roads, etc.) would place great restraints on their doing so. Another key difference between Europe and Africa lies in the deposition of nitrogen from the atmosphere, less than 5 kg/ha in Africa but up to 50 kg/ha in Europe, where about one-third is likely to be lost in water draining from the soil.

Market dogma, like political correctness, has discouraged African farmers from using much-needed fertilizer. The essential problem was probably that the Western concept of 'the market' was inappropriate at the scale at which it was used in an African context. Box 11.2 shows how the market concept worked very well when applied by the FARM-Africa charity at the right scale and in a manner that met the needs of local people in Tanzania.

The effects of the market go wider than fertilizer use. Uganda underwent a Trade Policy Review in 2001, and I came across a speech made by the US Ambassador to the WTO following that event. After noting the country's economic potential that sprang from its natural resources and the growth in its GDP, the ambassador commented on Uganda as a leading economic reformer in Africa. It had achieved macroeconomic stability, and privatization and civil service reforms were proceeding. Later he commented on the great emphasis that Uganda had placed on private sector development as an

Box 11.2. Making the market work for local people at the local scale. FARM-Africa* in the Nou Forest.

The Nou Forest is on the escarpment of the Great Rift Valley near Dareda in central Tanzania. It was in poor shape when FARM-Africa became involved with it. Trees had been damaged and other plants grazed to excess by goats – one reason for keeping them in pens (Box 11.3) – and the loss of vegetation had upset the flows of water within the escarpment so much that a number of streams had ceased to flow. Staff from FARM-Africa discussed the problem with the local community and it was agreed that goats would have to be kept out of some areas of the forest and that the trees that had been lost would have to be replaced. But how?

FARM-Africa's solution was to make a market in tree seedlings. Some local people, both women and men, were encouraged to set up tree nurseries and were lent, not given, tree seeds. (It is a strict rule of the charity that there are no free hand-outs.) When the seedlings were ready for planting out, they were sold to the community for the modest sum of 40 Tanzanian shillings each (about 4p in UK money) and used to replace the trees lost from the forest.

The system was in operation when I visited the area in 1999 with a group of 'Friends of FARM-Africa' (on a trip for which we paid ourselves). The 'market' had already proved very successful in that many trees had been replaced and six streams that had ceased to flow had started running again. Scale is important in processes in the landscape (Chapter 5), and it seems equally important in the application of market concepts in Africa.

* Details from FREEPOST, FARM-Africa, LON 14108, London WC1A 2BR, UK.

engine for economic growth but noted that the privatization process had slowed. He urged Uganda to move forward with the privatization of the Uganda Electricity Board and other such entities (note Box 11.1). He also urged the Uganda Government to move forward on the deregulation programme, which had been launched to dismantle 'unnecessary' licensing and regulatory procedures.

This speech emphasizes the hypocrisy of the WTO and others who push for the opening up of markets in poor countries while rich countries keep their markets closed to agricultural and other products from those poor countries (Stiglitz, 2002). The suggestions it contained seemed as likely to benefit the citizens of the USA as those of Uganda. It also seemed rather detached from the realities of life as I have seen it in Africa. A few of these changes may have worked to the benefit of the Ugandan people, but I cannot but wonder what impact privatization and deregulation had on the average Ugandan smallholder farmer and his or her family. Would they have really have been better off if the Uganda Electricity Board had been bought by a large multinational company? Even if electricity had reached their village? What they need is development at an appropriate scale that is specifically for their benefit. Uganda may eventually benefit from the ministrations of the WTO but our smallholder farmer will profit more if FARM-Africa provides a goat-breeding station to their village (Box 11.3).

Box 11.3. The FARM-Africa goat schemes.

Goats have caused problems in many parts of Africa when allowed to graze unrestricted (e.g. Box 11.2). They graze vegetation too low, making the soil vulnerable to erosion, and they damage trees by chewing the bark. But they can also be a lifeline. The FARM-Africa goat schemes build on the positive aspects of the goat, its resilience and its adaptability.

The native African goat is not particularly productive of either milk or meat. FARM-Africa therefore brings in high-quality Toggenburg goats from a breeder in Wales to improve the quality and productivity of the local goat population. So far they have established goat schemes, usually run by women, in Ethiopia, Kenya, Tanzania and South Africa. Goat schemes also form the spearhead of the recent expansion of their activities into Uganda. The goats are used in various ways.

- Some Toggenburg does (females) are given to widows, of whom there are all too many in Africa.
- Some villages are given a Toggenburg buck (male) and the villagers appoint a 'buck-master' to look after the buck and pay her (or sometimes him) to do so. The buck will serve the does in the village, improving the stock as he does so.
- Other villages will be given a 'goat-breeding station', comprising a buck and three does, enabling them to establish a high-quality herd in the village.

There are no free hand-outs where FARM-Africa is concerned, and a widow who receives a doe is expected to give the first weaned doe in the offspring from the doe she has received to another widow or poor person. Similarly, a village that receives a breeding station has to pass on a buck and three does to another village as soon as they are weaned.

A Toggenburg goat needs to be looked after well in an African environment, and the charity has devoted a great deal of effort and thought to this issue. Goat owners are encouraged to provide proper housing for their animals. This stops them roaming and damaging the environment, it makes them safer from marauders, and it enables their intake of food to be monitored and wilted to minimize infection by parasites. The pens are arranged so that the goat gets exercise stretching up for its food as it often would in the wild. They also have slatted floors, through which the manure falls. This helps to prevent the goat from re-ingesting parasites by picking up excrement from the floor with food that has dropped there. It also facilitates the collection of manure.

One person I particularly remember meeting in Tanzania was Mr Joseph Nicodemus, who lived not far from the Nou Forest (Box 11.2). He had some steeply sloping land on which he had built soil banks across the slope to prevent erosion. These banks were stabilized with leguminous *Leucaena* shrubs, whose vegetation was cut to feed his goats. The goat manure was used to fertilize his food crops and I remember he was running an experiment to find out whether manure placement under the crop paid off. My former boss, the late G.W. Cooke, who had a great interest in fertilizer placement, would have been delighted. Mr Nicodemus had another experiment in which he was alley-cropping beans and another crop so that some of the nitrogen fixed by the beans' root-nodule bacteria also benefited the other crop. Never underestimate the African smallholder farmer! In fact on a trip to Ethiopia two years later, we visited a site where several research projects were being run collectively by a group of smallholder farmers (FARM-Africa, 2001).

Parasites are but one of the health problems that goats encounter, and the goat schemes also include the training of 'Community Animal Health Workers' from the villages. They gain a basic understanding of treatment techniques for the diseases, parasites and problems of goats and are supervised by 'Animal Health Assistants', who are in turn supervised by veterinary surgeons.

The goat schemes have proved enormously popular in the countries where they have been established. They do much to ensure food security, but the benefits go beyond that. Milk that is not needed by the family can be sold, and the goats can well be the factor that means the children are not only fed properly but also able to go to school. There is no free education in most of Africa, and I remember meeting two boys on a roadside in Kenya. 'Why aren't you at school?' we asked. 'We want to, but our parents cannot afford for us to go.' Truancy is a Western problem.

The nitrogen resources of African farmers

African farmers have limited options for supplying nitrogen to their crops other than in fertilizer. These include legumes, green manures, organic residues and animal manures (Giller *et al.*, 1997). Legumes are more effective than green manures simply because, although farmers want to improve their soil, they need to grow crops that their families or their livestock can eat. They now have access to some good varieties of cowpea, soybean and pigeonpea which yield well while fixing plenty of atmospheric nitrogen (K.E. Giller, Harare, 2000, personal communication).

Manures from cattle and other animals are as valuable in Africa as they are elsewhere, but their usefulness is subject to a key constraint. Because the supply of nutrients is finite, the area of land that can be cropped adequately depends on the area that can be grazed to provide the manure. Small amounts of nitrogen fertilizer can counter this constraint by supplementing cattle manure and were very effective in experiments at the University of Harare, with just 20 kg/ha giving a worthwhile increase in yield (Nyamangara, 2001). This amount is about one-tenth of the application made to a winter wheat crop in the Western world, and is less than the crop actually uses, so there is no environmental risk. Similarly small amounts of phosphate fertilizer, carefully used, can benefit yields appreciably also without risk. Fertilizers in the bags usually seen in European or North American fields are way beyond the means of most African farmers and much in excess of their needs, but local entrepreneurs seem to have found that it is profitable as well as socially useful to buy large bags, split them and re-pack the fertilizer in smaller bags.

An alternative to bringing nutrients into the system as fertilizer is to bring them in as cut fodder. Goats in particular eat a wide range of vegetation, including shrubs such as *Leucaena*, and excrete conveniently pelletted manure. The farmers in the goat schemes run by the FARM-Africa charity are encouraged to keep their goats in pens with slatted floors. Food is brought to them and the manure is readily collected (Box 11.3).

Should Africa go organic?

Some people in Europe think Africa should turn to organic farming. I remember the 'Food Programme' on BBC Radio 4 getting very excited over a scheme under which organic food producers in Zambia would make a greatly improved living by exporting to Europe. I don't think the 'Food Programme' mentioned the nutrient export problem and I doubt if the producers were aware of it. I also don't remember hearing any more of that particular scheme.

African producers of organic food need to replace the nutrients they send to Europe in the crop but, if they are organic growers, they can only use potassium sulphate and rock phosphate as fertilizers. Potassium sulphate is more expensive than potassium chloride and the difference will add to the burden of the cost of imports. Rock phosphate is available in Africa and its phosphate will become available to crops in the acid soils which predominate there, but the supply is finite. Should precious African rock phosphate be used to feed people in Europe, where there is a problem of obesity, when it is desperately needed back in Africa to feed the hungry? The general answer is surely 'no', although a case can be made for African farmers exploiting niche markets in which they can obtain fair trading terms and good organic premiums.

Nitrogen fertilizer can be replaced in an organic farming system by growing legumes, but the crops needed for export may include non-legumes. Other crops benefit from the nitrogen left by the legumes, but in the long run they are likely to run down the supply of nitrogen in the soil.

Perhaps the Soil Association and the International Federation of Organic Agriculture Movements should give a special dispensation to organic producers in Africa and let them use urea. After all, urea is an organic chemical in the original sense of the word 'organic' (Box 2.2), whereas potassium sulphate, which is allowed in organic farming, certainly is not.

Organic farming is compulsory for many of Africa's farmers, but there seems no reason to make it the ideal. Africans have nothing to fear from fertilizers, and no African crop plant can tell the difference between ammonium, nitrate and phosphate from a fertilizer bag and the same nutrients released by organic manures. And crops use fertilizers more efficiently than organic manures when they are applied correctly. No African farmer will waste precious resources by putting on too much fertilizer.

Africa: the Underlying Problem

Political correctness and market dogma are not the whole story. The main reason why farmers in Africa do not use fertilizers is not environmental awareness or the removal of fertilizer subsidies, but poverty. They will stay poor as long as they are held back by economic constraints. The main problem is unfair trading arrangements. World trade is discussed regularly at summits at exotic locations such as Cancun and Davos. Pious platitudes are exchanged about the needs of the Third World, but American and EU farmers continue to be subsidized to the extent that they can undercut African producers in Africa. Put another way, African farmers are frequently the victims of 'dumping'. A recent report by the charity Christian Aid argued that the WTO's decision-making process is biased against the poor, that many WTO agreements harm rather than help the poor, and that there are many areas in which there are no rules where rules are needed. The charity listed seven existing trade rules that it sees as detrimental to the interests of

Box 11.4. Seven trade rules identified by Christian Aid as detrimental to the interests of the poor.

1. Developing countries are restricted from raising adequate barriers against cheap food imports, although rich countries retain export subsidies.
2. Countries must open up their service sectors to foreign investors.
3. Regulation of foreign investment is limited.
4. Developing countries are limited in their use of agricultural subsidies.
5. Governments are limited in their use of industrial subsidies to promote domestic manufacture.
6. Rich countries can retain import barriers and restrictions against exports from developing countries.
7. Countries must introduce patenting laws that can give transnational corporations rights over knowledge and natural resources.

the poor (Box 11.4). Global trade rules, says Christian Aid, need to be changed radically.

Regimes within some individual African countries, such as Zimbabwe, add to the problems that come from unfair trading rules. Africa's farmers do not need Western political correctness as an additional constraint. The line between environmental activism and eco-imperialism is all too thin. Africa also needs economic models that meet the real needs of her people. The needs of her poor must take precedence over the Western world's phobia about nitrate and its obsession with the market.

12 Risk

A Benchmark for Risk

Captain Pip Gardner was serving in 1941 with the Royal Tank regiment at Tobruk, in Libya, when he was sent with two tanks to rescue a pair of armoured cars that were out of action and trapped by enemy fire. He ordered one tank to give covering fire, advanced in the other towards the armoured cars and, under heavy machine-gun fire, crept out of his tank to secure a tow-rope to one of them. He then spotted a man lying beside it with both legs shattered and stopped to lift him into the armoured car before returning to his tank. As he did so he was shot in the leg. The tank set off with the armoured car in tow but the rope broke. Despite his wound, Captain Gardner got out of the tank again and carried the wounded man from the armoured car to his tank, this time being shot in the arm. He was awarded the Victoria Cross for his bravery on that occasion and the Military Cross for another episode in which he risked his life trying to save the life of a comrade. (From the obituary column of *The Week*, 22 February 2003.)

Risk in war is a specialized category of risk, because the captain and his comrades were expected to risk life and limb, and a fair number of them returned permanently injured from the desert or failed to return at all. We should remember them with gratitude. Risk in war has no further part in this chapter, but it is essential that we remember the risks taken by Captain Gardner and his comrades and use them as a benchmark by which to put other risks in perspective.

What Do We Mean By 'Risk'?

Captain Gardner survived his wounds and lived until 2003, but I wonder what this gallant officer made of the society in which he spent his latter years, safer than at any time in history but paranoid about some risks which were trivial beyond words compared with those he had faced. He could have been excused for wondering if, with the war 50 years behind us, we had forgotten what real risks were. In fact, are we altogether clear what 'risk' means now? A 1950s *Concise Oxford Dictionary* I consulted seemed to associate 'risk' and the associated word 'hazard' with the gambling table and the golf course, while the (lengthier) 1990s *Shorter Oxford English Dictionary* seemed to regard 'hazard' as a definition of 'risk'. What does 'risk' actually mean?

The origins of 'risk'

Bernstein (1996) considered that the word 'risk' has its root in the early Italian word *risicare*, meaning 'to dare'. He added that, 'In this sense risk is a choice rather than a fate', a percipient remark to which we shall return. This definition fits Captain Gardner's exploits perfectly. He dared in the best military tradition and he chose to risk his life repeatedly. But Bernstein was concerned with risks of a different kind. His book is a fascinating account of the mathematics of risk and the personalities involved in its development. The ideas that evolved were applied first to gambling, which probably explains the *Concise Oxford Dictionary*'s definition of risk. And they were applied later to economics and then market economics. I have, unfortunately, recently experienced the financial risk associated with the market and therefore include a brief comment, but most of the chapter is concerned with risks to humans and the environment, including those involving nitrate.

Financial Risk

Money in whatever form is a human invention. Financial risk is therefore largely a consequence of human behaviour, but this is not completely true as shocks from the natural world can play their part in it too. North (2000) took the view that financial risk-taking was not only inevitable but also compulsory in a free market society. He also discussed whether or not it was virtuous and suggested that the obvious relation with greed might obscure the possibility that it is also a species of heroism. He decided that risk-taking is necessary and, in that sense, virtuous. He justified this in terms of the social value of risk-taking, which is considerable because it contributes greatly to both wealth and employment.

For many retired people, including me, financial risk is virtually compulsory, in North's parlance. There is nothing particularly greedy, heroic or virtuous about us. It's just that when we save for retirement we have to put the money somewhere, and we have been encouraged to follow North's concept and to commit at least part of it to the stock market. But in the early 2000s, the free market was an unqualified disaster for us. We knew that there was a risk but our financial gurus did not prepare us for what actually happened. Is the problem simply an inevitable consequence of a 'market' which is 'free' and therefore beyond human control, or has it more to do with human failings?

Bernstein's (1996) book carries the reassuring message that in economics time changes risk, and that the market will bounce back in due course. If you stay invested, you should be 'all right in the long run'. The market has bounced back to some extent but I am not sure that the resurgence has a sound basis. I would like to believe that my finances are safe in the hands of the almighty market but I am not convinced, and the reasons for my lack of faith spring from my experience of developing and using computer models.

Models

I use the term 'model' to describe a theoretical construct usually written either in the language of mathematics or in a computer language that is used to describe the behaviour of the system under study. Such models are used widely in both science and economics, and potential faults in models lie at the root of much that we regard as risk. The sort of questions that I have had to address with models for soil processes are:

- Is the model credible?
- Has it been properly evaluated before being used in decision making?
- Does the model behave in a non-linear way with respect to any combination of input and output? Non-linearity (Box 12.1) is an issue that frequently arises in modelling and can lead to the related phenomena of chaos (Box 12.2) and complexity.

I was intrigued to find that these were exactly the kinds of questions raised by economist Paul Ormerod about three models that are central to market theory. He raised them in a critique of current economics, which he entitled *The Death of Economics* (Ormerod, 1994):

- The 'Law' of Supply and Demand.
- The Competitive Equilibrium Model.
- The Money Supply Equation.

The same questions are raised later about models used in an environmental context.

Credibility of economic models

The 'Law' of Supply and Demand failed hopelessly when Ormerod confronted it with a simple example. Teachers earn far less than do people with equivalent qualifications working in investment banks. According to the 'law', this must be because far more people want to be teachers than want to work in investment banks. This was readily exposed as the nonsense it was.

The Competitive Equilibrium Model is based on the highly idealized concept of a notional Auctioneer who coordinates the responses of an infinite number of rational producers and consumers, each of whom pursues their own interest. It is the task of the Auctioneer to clear the market – instantaneously. Ormerod described it as 'travesty of reality', mainly because the market consists in the main of a few very large multinational companies, rather than the near-infinite large number of smaller producers and consumers that is implicit in the model. Questions could also be asked about how exactly this mysterious being manages to deal simultaneously with all those transactions.

The Money Supply Equation is concerned with the amount of money M in circulation and the velocity V with which it circulates. M and V are related to the average price level P and the quantity Q of goods and services produced in an economy, such that $MV = PQ$. Money supply is claimed by the proponents of monetarism to be a means of controlling inflation, but Ormerod concludes that, '... the claim that inflation is always and everywhere purely caused by increases in the money supply and that the rate of inflation bears a stable and predictable relationship to increases in the money supply is ridiculous'. One reason it is ridiculous is simply that there is no unique definition of money supply. It is defined by a series of indicators, each of which has the prefix M followed by a number from 1 to 5. If the concept of money supply is real, these indicators ought to move more or less together in an economic zone but in practice M1, M2, M3, M4 and M5 all go off in different directions, like the motorways whose names they share.

Thus, all the three models central to market economics fail the credibility test comprehensively. Note that the evidence has been provided by Ormerod, himself an economist.

Evaluation of financial models before use in decision making

It is recognized in the sciences that no model is perfect, no modeller infallible, and that models tend to propagate error. Error propagation is a trap for the unwary, particularly in my own discipline of soil science where so many model inputs are subject to error in the statistical sense of *variation*. Soil scientists who develop models are unlikely to get them published unless they show an evaluation of the model against measured data. Even Einstein felt the need to test his theories against observations, but it seems that such activity is disdained in economics. According to Ormerod, 'in economics, pure theory is held to describe how the world actually operates. There is no perceived need to examine this empirically.' There is a hint here of what Medawar (1969, p. 17) described, when discussing pure and applied science, as 'the dire equation *Useless* = *Good*'. The distinguished philosopher Sir Karl Popper (Box 13.1) provided an intellectual framework for evaluating theories by a process of falsification (Popper, 1959) and, if economists are spurning this, they are setting a great gulf between themselves and scientists.

Non-linearity, chaos and complexity in financial models

Many scientists and, it would seem, most economists would prefer to live in a linear world. Unfortunately for them, the real world shows a definite preference for curves, and any system involving human beings is most unlikely to behave in a linear way. The problems caused by this propensity, particularly when non-linearity comes into contact with parameter variability, are shown in Box 12.1. And the further possibility that non-linearity can lead to chaotic behaviour is considered in Box 12.2.

Box 12.1. Non-linearity.

Non-linearity *per se* is a very simple concept. If plotting a relation gives a straight line, the relation is a linear one; if it does not, the relation is non-linear. The equation $y = 2x$ gives a straight line, but $y = 2/x$ does not. The former is linear and the latter non-linear. The consequences of non-linearity are numerous and include chaos (Box 12.2). Many are not simple. Books have been written on the topic, but all that is done here is to illustrate very simply the kind of problem non-linearity can cause. This particular case is an averaging problem.

Consider the Freundlich Equation:

$$x/m = kc^{1/n} \quad \text{(B12.1.1)}$$

where x is the amount of solute held on mass m of the solid, k is the sorption constant, c the concentration of solute in solution, and n is another constant. We rearrange the equation to evaluate c:

$$c = (x/km)^n \quad \text{(B12.1.2)}$$

This relation is clearly non-linear (Fig. B12.1.1).

Suppose the values of x, k and m are such that $c = 10^n$, and that n varies, being 2.0, 2.5 or 3.0. To evaluate the function, you take the average of n, which is 2.5 and work out $10^{2.5} = 316$. But someone else might add up 10^2, $10^{2.5}$ and 10^3 and take the average, getting an answer of 472. The discrepancy is due to the non-linearity in the relation. Be careful to average at the right point in the calculation!

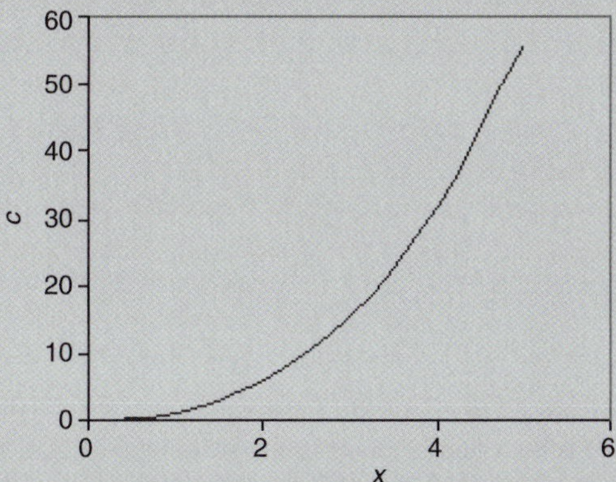

Fig. B12.1.1. c evaluated from equation (B12.1.2) for values of x, showing strong non-linearity. Other values: $k = 0.01$, $m = 10.0$, $n = 2.5$.

Box 12.2. Chaos.

The phenomenon of chaos is a consequence of mathematics itself (Rasband, 1990) and can appear in various physical, chemical, biological and economic systems. Chaotic behaviour is not to be confused with random or stochastic behaviour. A chaotic system evolves in a deterministic way, with the current state of the system always depending on the previous state. The results may appear to be random, and some systems which are regarded as random may actually be deterministic-chaotic.

Chaos always arises from non-linearity in the system, but this is not the only factor involved. The system has also to have more than one degree of freedom – that is, be non-autonomous. A pendulum follows a curved path and is a well recognized non-linear system. But it is not non-autonomous and therefore not chaotic. The 'kicked pendulum', one that is subjected to some force in addition to gravity, is non-autonomous and is often used as an example of a chaotic system. One of the commonest causes of chaotic behaviour is a feedback loop in the system.

Chaotic systems are inherently unpredictable and also never repeat their patterns of behaviour. One hallmark of a chaotic system is extreme sensitivity to small changes in initial conditions (Gleick, 1987). This suggests that examining the effects of small changes in the initial values of state variables should be a way of testing environmental or economic models for potential chaotic behaviour. Such behaviour may not be easy to find, because chaos only occurs within a certain range of the parameter space of the model. It is also necessary to watch for apparently chaotic behaviour that is caused by errors or inexactitudes in the model. I was recently excited, but also slightly worried, to find apparently chaotic behaviour in a computer model for phosphate sorption, but this vanished when some variates in the most non-linear part of the model were changed from single to double precision.

I am not aware that any economic models have been tested for chaotic behaviour, but it would not be surprising to find it because of the behaviour of cotton prices mentioned in the main text. One might envisage that the combined effects of supply and demand would lead to non-linear but not chaotic behaviour when the two forces were reasonably well balanced, but that an externally imposed shock, such as a drought or an outbreak of disease, might function like the kick to the pendulum and cause chaotic behaviour.

There is no doubt whatsoever that chaos, in the scientific sense, plays a part in economics. One of the classic examples of chaotic behaviour is found in cotton prices (Gleick, 1987). Introducing the concept of non-linearity into economics also brings the idea of 'increasing returns' rather than the standard economic concept of 'diminishing returns' (Waldrop, 1993). Ormerod (1993) seemed keen to introduce these new concepts into economics, but Bernstein (1996) was dismissive of them. He wrote:

> Students of chaos theory reject the symmetry of the (Gaussian) bell curve as a description of reality. They hold in contempt linear statistical systems in which, for example, the magnitude of an expected reward is assumed to be consistent with the magnitude of the risks taken to achieve it, or, in general, where results achieved bear a systematic relationship to efforts expended.

Bernstein seemed to show almost the exact response of a theoretical economist to real data that was cited from Ormerod in the previous section.

Gleick (1987, pp. 83–85) records a visit by the chaos theoretician Mandelbrot to the Harvard Professor of Economics Houthakker, who had been studying cotton prices. Gleick records that, 'No matter how he plotted them, Houthakker could not makes the changes in cotton price fit the bell-shaped model'. The students of chaos theory to whom Bernstein objected had excellent reasons for rejecting the symmetry of the bell curve. Data from the real world rejected it as well.

As for their 'contempt of linear statistical systems', the students of chaos theory could point to the chaos theoretician Brian Arthur who was also an economist. Arthur had plenty of evidence that it was time to reject linearized models. This evidence included the typewriter and keyboard layout whose top line starts with QWERTY. This was designed in 1873 to slow typists down and stop them jamming their typewriters. Why, in the brave new world of market economics, had this very inefficient keyboard not been replaced? Arthur's answer (Waldrop, 1993) was that everyone was using it and it had become 'locked into' the system. But isn't this an example of increasing returns, something that cannot happen with a linear system? Another example of increasing returns can be found in the fate of the Beta videotape format. The Beta format was technically better than its rival the VHS format, but the latter got a slightly larger market share initially and this small advantage rapidly turned into dominance of the market. This is a classic non-linear response that fully justified any 'contempt in which the students of chaos theory held linear statistical systems'.

Origins of financial risk

I am no expert in economics, but Paul Ormerod, whom I have cited extensively, is very much an expert. His book, together with my own interest in modelling, non-linearity and chaos, suggests to me that at least part the root of financial risk is to be found in the rather questionable models used by economists, their failure to subject these models to evaluation against measured data and their apparent dismissal of non-linear phenomena. Ormerod suggested that much current economic theory should be 'abandoned or at least suspended until it can find a sounder economic base'.

The necessary 'sounder base' can perhaps be found in research on the economics of information by Joseph Stiglitz, a former chief economist at the World Bank (Stiglitz, 2002). Like Ormerod, he does not believe that the simplistic models central to market theory work satisfactorily and referred (Stiglitz, 2002, p. xii) to 'the outworn assumption that markets by themselves lead to efficient outcomes'. Stiglitz's research on market imperfections, the problems arising in an imperfect real world from the economist's assumptions of perfect competition and perfect information, seems essential to the future credibility of 'The Market'.

Faulty economic theory is only one of the human failings involved in the risk. The dubious and apparently criminal accountancy practices revealed in the recent Enron, World-Com and Parmelat scandals have added to financial uncertainty and risk both directly and because of the suspicion that they are just the 'tip of the iceberg'. The behaviour of politicians can have a substantial impact on financial risk too. And the financial report on the radio often mentions 'sentiment' as a factor in market trends. Sentiment must surely be a non-linear phenomenon?

Risks to Humans and the Environment

Financial, human and environmental risk may be more related than might at first seem likely. One reason for saying this is that the environment, like money, is essentially a human concept. Another is that risks to humans and the environment, like financial risks, result to a large extent from human activities and often human greed. The two types of risk may be inter-related. Enterprises bearing financial risk may have this risk increased by environmental factors or environmental legislation. The other side of the coin is that these same enterprises may put habitats, species or humans at risk. Also, models are involved in the assessment of risk in both cases, but the definition of 'model' is broader for environmental risk. As with financial models, potential faults in models are an important consideration in risk, so inappropriate or unevaluated models are just as much of a problem in the environment as in the financial world.

Definition of 'risk' in a human or environmental context

The main difference between financial and other risks is that only the human species is likely to suffer financial risk, while many species, including humans, may be at environmental risk. Another difference is that while the definition of financial risk is essentially loss of money, the definition of risk in an environmental context is not so straightforward. Addiscott and Smith (1997) considered that both 'hazard' and 'risk' needed to be clarified in this context and suggested the following definitions.

> Hazard: Something exposure to which may cause undesirable effects.
> Risk: The probability that an undesirable effect will occur.

Probability has to be included in the definition of risk for two reasons. One is that, while there are some hazards to which exposure carries the certainty of ill effect or death, far more hazards have an uncertain outcome which can only be expressed in terms of their statistical probability. Another important reason for introducing probability is that it provides a means of comparing risks, something that is frequently essential.

Introducing probability, though essential, has its problems. One is that the public likes certainties and tends to be encouraged by politicians and the media to expect them. But the only certainties that can honestly be offered by scientists are those identified by Benjamin Franklin, death and taxes. Another problem is that the public is not numerate, and those presenting probability-based information do not always understand what they are presenting. Gigerenzer (2002) cited the story of a weather forecaster on an American television channel who stated that there was a 50% chance of rain on Saturday and a 50% chance of rain on Sunday – and therefore a 100% chance of rain during the weekend! Readers may like to check their understanding of probability by working out what he should have said!

Why risk assessment is needed

All chemicals are potentially dangerous, and the most widely used laboratory chemical is the most dangerous of all; people drown in it every year. But, used sensibly and in the right amounts, chemicals are a vital part of our daily lives. The public has been taught by the pressure groups to fear chemicals, and 'chemophobia' is now widespread. So many things are perceived as a risk that there is complete confusion about how to handle risk.

Politicians tend increasingly to follow public opinion rather than leading it, so that in effect the blind are leading the blind. We worry greatly, for example, about pesticides in water and have a stringent limit on their concentration, but we forget that practically all our pesticide intake comes from naturally produced pesticides in plants. In fact, 99.99% of all known pesticides are produced by plants wishing to repel insects. Most of us eat coleslaw quite happily, but some of us might be less enthusiastic if told that cabbage contains 49 natural pesticides, some of which cannot be used by farmers because they can cause cancer in laboratory rats (Ames and Gold, 1990). Coffee is particularly rich in rodent carcinogens (cancer-forming chemicals), so there is little point in worrying about minute concentrations of synthetic pesticides in the water used to brew it. But the legislation is concerned with the synthetic pesticides and no one stops to think about the potentially cancer-inducing brown liquid that most of us, myself included, pour regularly into our stomachs. Actually, nobody is going to die from the pesticides in the water or from drinking coffee, so we need to direct our worrying to better effect. Risk assessment should help.

We are not the only species on the planet, and most other species have no choice about the substances to which they are exposed. It is therefore proper for us to consider risks to them, but how we balance our own needs against theirs is a difficult decision. We need to meet our own requirements, but we need to consider the essentiality of these requirements against risks to other species. It is our environment that is diminished by the loss of these species, and it is that environment we leave to our children.

These more or less altruistic reasons for risk assessment are supported by economic realities. The resources available for keeping our environment wholesome are limited and must be used to combat the most serious risks first. Risks must be assessed and prioritized, and proper risk assessment is essential for this. We need to counter the assumption that new technologies are inherently more dangerous than old ones. The synthetic pyrethroid pesticides are surely preferable to the arsenic compounds used early in the 20th century. Arguably, the real problem technological advance has brought is that chemicals can now be measured, and therefore worried about, at far smaller concentrations than before. The European Community (EC) limit for any one pesticide in potable water is 0.01 ppm. This figure was chosen because it was the limit of detection at the time, so the legislators were effectively saying no pesticides should be found in water at all. If they were serious about protecting the public, they should perhaps have banned the sale of coffee as well.

The pesticide limit was determined by the limitations of analytical organic chemistry apparently without any reference to biology or any risk assessment. Decisions of this nature can become a risk themselves. If there are too many of them, we could find ourselves in an over-rigid regulatory system that inhibits useful or enjoyable activity without contributing to safety.

The reality is that there is no such thing as absolute safety, and if there were it probably would not be good for us. One reason suggested for the increase in asthma and allergy problems suffered by children is that they are now brought up in so antiseptic an environment that their immune systems are not challenged adequately by germs when they are young. Perhaps the reason so many people are so obsessed with environmental risks is that they have been so safe from other forms of risk for so long. What we need to do is to try to assess risks on as logical a basis as possible so that we optimize the balance between productive or enjoyable activity and the risk that it entails.

Reciprocal risk

The term 'reciprocal risk' is used here to define an important issue that has been addressed by others such as Gigerenzer (2002) but not, so far as I can see, given any particular name. When we consider the risk of doing something, it is essential that we compare it with the risk of not doing it. An obvious example is the inoculation of children against illness. People are naturally concerned that some inoculations carry a risk that something will go seriously wrong, possibly disabling the child. This risk is always extremely small, but it can discourage parents from going ahead with the inoculation. It is vital that they consider too the risk of not inoculating the child, which may have even more serious consequences, and not only to that particular child. The proportion of parents rejecting the MMR vaccine (for measles, mumps and rubella) in the UK has hugely increased the risk of an epidemic of measles. The data are usually there for parents to compare the probabilities associated with the two risks, but it is clear from media reports that the reciprocal risk is often ignored.

The great 'seat-belt debate' we had in the UK in the 1960s provides another example. A friend of mine used to argue that some people wearing seat-belts in their cars were killed or injured in accidents, so it was unwise to wear one. I could never persuade him that more were killed or injured when not wearing them.

Risks, familiar and unfamiliar, and choice

I suggested above that the very small concentrations of pesticide likely to be in water were insignificant if the water was made into coffee, because of the cancer-causing chemicals in the coffee. This information would probably have little impact on most coffee drinkers for two reasons, familiarity and choice.

People who drink coffee have usually done so for many years. It is completely familiar to them, and they feel that as it has not done them any harm in the past twenty years, it will not do them any harm in the next twenty. Pesticides are a different matter. They are alien where coffee is familiar, an unknown quantity that people have been taught to fear. More importantly, people who drink coffee do so by choice. They did not choose to have pesticides, even in minute traces, in the water supply. They feel that pesticides have been imposed on them. These questions of familiarity and choice make the idea of proper risk assessment harder to propound.

Assessing Environmental Risk

Before we embark on risk assessment we need to ask what we really want to know.

- What exactly is perceived to be at risk?
- Are we concerned with a particular species or an ecological community whose individual species may react to pollutants in different ways?
- Does the pollution need to be considered in terms of concentration or load? Some species will react to the intensity of the pollution as measured by the concentration and others to the total amount of the pollutant in the system.
- What is the standard, and how is it to be measured? It might, for example, be the concentration at which no effect is found in an ecotoxicological test.

The answers to these questions will determine what procedures are used, but these may include some of the following.

Tiering

Tiering is a useful way of making risk assessment more efficient (Committee on Risk Assessment of Hazardous Air Pollutants, 1994; OECD, 1995). The procedure involves a series of appraisals of increasing intensity and detail which continue until a definite decision can be reached. A tiered system might start at the level of a 'back of an envelope' calculation. If this shows no possibility of a risk, the assessor would stop there. If there was doubt about this, he or she might turn to look-up tables for help and then, if necessary, to larger-scale monitoring or modelling. The key point is that, if an appraisal at any level in the tiering below the highest level shows that the risk is negligible, higher-intensity appraisals can be omitted and the attention turned to the next risk.

One good reason for tiering lies simply in the ever-increasing number of substances entering the environment. There will never be enough resources of money and people with the appropriate skills to assess the risks posed by all substances to all species in all environments, so it is essential not to waste resources when even a quick check can show that the risk is negligible or non-existent. In particular, it is essential to establish as early as possible whether the substance is actually toxic. This point may seem self-evident, but we need look no further than the history of the nitrate problem to see why it is relevant.

Models

The OECD workshop cited above (OECD, 1995) included a working group which described models as 'essential tools' with several important roles in risk assessment. These included the following:

- Understanding the spread of the pollutant in the environment and its ultimate fate.
- Providing insights into the influence of external factors on its spread.
- Aiding decisions about exposure or risk, for example in screening and in planning research.
- Predicting concentrations in the environment. This can be important in the absence of data from monitoring, as may happen with new substances.

These four roles are all relevant, particularly the last one. Risk assessment will be very difficult without extrapolating from the behaviour of known to unknown substances and from monitored to unmonitored sites.

The group cited added that in certain circumstances 'predicted concentrations from models may be more cost-effective than implementation of monitoring programmes'. On the basis of 28 years' modelling experience, I suggest that this comment needs to be treated with caution. Models can certainly be used to give added value to measurements and to extrapolate from them, but we need to be very cautious about viewing them as an alternative to measurements.

Whenever a model is used in environmental risk assessment, it must be subject to the questions to which economic models were subjected.

- Is the model credible?
- Has it been properly evaluated before being used in decision making?
- Could its parameter values have been influenced in any way by deficiencies in the model or in any fitting procedure used?
- Does the model behave in a non-linear way with respect to any combination of input and output?

The third question was not included for the economic models, because I did not know how their parameters were obtained. Further details about these questions were given by Addiscott *et al.* (1995) and procedures for evaluating and parameterizing models were given by Whitmore (1991), Loague and Green (1991) and Smith *et al.* (1996b). These questions are very important because of the increasing use of models in decision making. As with economic models, a defect in an environmental model can result in very considerable financial losses.

Ecotoxicology

Models for physical processes can help us assess the spread of a pollutant within the environment. We need to know what problems it is likely to cause in various ecosystems, particularly water-based ones. This is the realm of ecotoxicology, which has its roots in human toxicology and pharmacology (Kooijman and Bedaux, 1996). Its primary concern is with the effects of concentrations of chemicals or other potentially toxic substances on organisms. Most procedures seem to involve subjecting the organism to increasing concentrations of the pollutant until it shows some adverse effect. The next concentration below that which gave the adverse effect is selected. In studies of human exposure to toxic substances, this was originally described as the 'threshold limit value' (TLV). As the tests became more systemized, the TLV became the 'no-observable-effect level' (NOEL). Subsequently 'observable' became 'observed', and later still the word 'adverse' was added to give 'no-observed-adverse-effect level' (NOAEL). The level that gave the adverse effect is noted as the 'lowest-observed-adverse-effect level' (LOAEL) (Committee on Risk Assessment of Hazardous Air Pollutants, 1994). The ultimate adverse effect is death, and this is recognized in the frequently used LD50 statistic, which is the concentration at which 50% of the organisms die. Toxicology was originally concerned mainly with risks to humans and an additional term, the 'acceptable daily intake' (ADI), was set by dividing the NOEL by 100.

There are several problems associated with these tests. NOAEL values are very variable between tests and can yield contradictory results, and Barnett (2004) warned that they often pose statistical problems because no dose–response model is specified. They assume implicitly that plotting the effect on the organism against the concentration of the pollutant gives a straight line, but this is often not the case and the reality may be more interesting. Hormesis (Trewavas and Srewart, 2003) is the action of a toxin that has a beneficial effect at very small concentrations but a toxic effect at larger ones. This action seems to be recognized in homeopathic medicine but a review by Trewavas and Stewart (2003) suggests that hormesis may occur with a range of perceived toxins way beyond those used in homeopathy. Some pesticides, for example, may decrease the incidence of cancer at very small concentrations, which raises further questions about the EC limit on pesticides. A more fundamental issue is that most toxicological tests made for the benefit of humans are performed on rats and other small mammals, and the physiology of the rat, for example, may not resemble human physiology in all relevant respects.

These issues arise with a single organism, but the simplest ecosystem comprises an assortment of organisms integrated in food webs. A pollutant acquired by one species may be transferred to another, and the decline in the population of one means less food for another. Food webs are not the only form of interaction. If the population of one organism increases too much, it may deprive others of light or oxygen, as well as food. This raises the question of whether we can determine NOAELs for ecosystems as well as individual organisms. There seems to be some interest in using micro- or mesocosms for this purpose, but at the moment risk assessments seem to depend to a large extent on representative or 'model' organisms. For freshwater ecosystems, algae, the water flea *Daphnia* and fish, particularly trout, seem to be the principal 'model' organisms. Algae are part of the food web and also central to the problem of eutrophication. *Daphnia* are also part of the food web, and fish are a vital part of freshwater ecosystems, with trout especially highly prized.

Kooijman and Bedaux (1996) discuss these problems in much greater detail and also show how models (as opposed to model organisms) can be used in ecotoxicological tests.

Risk and the Bayes Equation

We saw above that ecotoxicological tests are not always reliable. Neither are some of the tests used to screen humans for disease. Gigerenzer's (2002) excellent book showed how the Bayes Equation could be used to help those tested to work out from the information they are given what chance there is that they actually have the disease for which they were tested. This is very important because patients frequently misinterpret the results with dire consequences that can include suicide. The equation can also be used to cast light on the uncertainties attached to ecological tests and is therefore included here.

A key point about the equation is that it can be expressed in terms of either natural frequencies or conditional probabilities and, as Gigerenzer pointed out, it is very much easier to understand in terms of frequencies. He suggested this is simply because our minds (and even those of animals) are better adapted to thinking about frequencies than about probabilities. We need to allow for the fact that medical and toxicological tests are not infallible and can give a positive result when there is no disease (a false positive) or a negative result when there is disease (a false negative). The same is true of toxicological tests and pollution. Table 12.1 may make the point clearer. The proportion of true positives is a measure of the *sensitivity* of the test and that of true negatives measures its *specificity*. When we apply the Bayes Equation to a disease for which someone has been tested, we find that the probability of that person having the disease when they have

Table 12.1. Possible outcomes of ecotoxicological tests on a number of samples. (After Gigerenzer, 2002.)

	Pollution	
Test result	Yes	No
Positive	Proportion true positive (Sensitivity)	Proportion false positive
Negative	Proportion false negative	Proportion true negative (Specificity)

The sum of the proportions of true positives and false negatives is 1. The sum of the proportions of false positives and true negatives is 1. The *frequencies* of true positives and false positives are *a* and *b* respectively, in the Bayes Equation.

tested positive, $p(\text{disease}|\text{positive})$, is given in terms of natural frequencies by:

$$p(\text{disease}|\text{positive}) = a/(a + b) \tag{12.1}$$

where a is the number of people who test positive and have the disease and b is the number who test positive but do not have the disease.

When the equation is presented in terms of conditional probabilities it looks vastly more complicated:

$$p(\text{disease}|\text{positive}) = [p(\text{disease})p(\text{positive}|\text{disease})]/ [p(\text{disease})p(\text{positive}|\text{disease}) + p(\text{no disease})p(\text{positive}|\text{no disease})] \tag{12.2}$$

Needless to say, it is very important to know for a particular test how frequently false positives and false negatives occur, as can be seen by considering the results of an imaginary toxicological test.

We assume that the results of the test are either positive (polluted) or negative (unpolluted) and that 10% of the samples are actually polluted. Let us also assume that prior assessment of the test has shown that the 2% of the positive results and 2% of the negative results are false. We follow Gigerenzer's (2002) suggestion and construct a 'frequency tree' (Table 12.2). The diagram shows that the false negatives have a large impact on the result, despite being only 2% of the total negatives. Out of 1000 samples, there were 98 true positives and 18 false negatives (which show as positive), giving a total of 116 apparently positive results. The true figure is 100, so the result was 16% too large. Worse still, if we work out $p(\text{pollution}|\text{positive})$ from the Bayes Equation, we find that only 98/116 of the samples which tested positive were truly positive and need to ask if a decision could really be based on a test which was only 84% accurate. If this kind of test is to be used in risk assessment, the specificity is vitally important. A specificity of 98% sounds good, but the other 2% caused a lot of trouble. A specificity of 90%, which still sounds reasonable, would have implied that only half the samples which tested positive were truly positive. This aspect of risk testing almost certainly needs to be understood better than it is at present.

Table 12.2. Frequency tree for 1000 samples. P, polluted; N, non-polluted; t, true; f, false. (After Gigerenzer, 2002.)

	1000			
Truth	100 P (t)		900 N (t)	
Test	98 P (t)	2 N (f)	18 P (f)	882 N (t)

The Precautionary Principle

The precautionary principle sounds to be a prudent idea, but what does it really mean? The definition usually used by policy makers is that set out in the Rio declaration in 1992:

> Where there are threats of serious or irreversible damage, lack of full scientific certainty shall not be used as a reason for postponing cost-effective measures to prevent environmental degradation.

Greenpeace's version of the principle, cited by Tren and Bate (2001), is less woolly but more extreme:

> Do not admit a substance until you have proof that it will do no harm to the environment.

It has to be said with regard to the Rio version of the principle that science can never offer certainty, but only probabilities, so this principle is invoked frequently. This is fine if it means that any development progresses cautiously with careful risk assessment but, if not, the precautionary principle itself carries definite risks. Science can never prove the complete absence of harm, so the Greenpeace version means in effect that nothing can be admitted, and this too carries risks.

Risks from the precautionary principle

One risk is that the principle can be used in a bureaucratic way as a means of preventing progress. In effect, the uncertainty surrounding risk assessment is made a justification for stopping progress. Indeed, the whole process of risk assessment is virtually side-lined, as seems to be happening to an increasing extent. Table 12.3 lists just a few of the developments that would not have occurred had the precautionary principle been in force at the time. Several of these are crucial developments in medicine that are now taken completely for granted and without which life would be far more risky. Those on the top line of the table suggest that the precautionary principle would put an end to the evolution of the human race.

Another risk is that the principle will be invoked without any consider-

Table 12.3. A few of the innovations and developments that would have been ruled out by the application of the precautionary principle in earlier times.

Eating unknown fruit	Hunting animals	The use of fire
The bicycle	The supply of gas	The supply of electricity
The motor car	The steam train	The aeroplane
Radio	Radar	Nuclear power
Water chlorination	The smallpox vaccine	The polio vaccine
X-rays	CAT scans	MRI scans
Radiotherapy	Chemotherapy	Blood transfusions
Antibiotics	Steroids	Pasteurization
Anaesthetics	Most surgery	The contraceptive pill
DDT	Malaria control	Pesticides

Many of these examples came from a list on the sp!ked-online website.

ation of the 'reciprocal risk' that was discussed in an earlier section of the chapter. This can have dire consequences.

What happens to the precautionary principle if you subject it to the precautionary principle?

One of the oddities of the precautionary principle is that it arguably carries the seeds of its own destruction. Most risks, as we saw above, have an associated 'reciprocal risk'. There are risks attached to not doing something as well as to doing it. The 'lack of full scientific certainty' probably applies to both the risk and the reciprocal risk, which surely leads to the restatement of Greenpeace version of the precautionary principle as:

> Do not admit the use of the precautionary principle until you have proof that it will do no harm to the environment.

This proof of lack of harm is as scientifically impossible as it would be for some chemical. This is an essentially pedantic point, but it needs to be kept in mind as a balance to some of the more extreme uses of the precautionary principle.

Risk Assessment for Nitrate

Nitrate has not yet been mentioned in this chapter, because we have been discussing the ideas underlying risk and risk assessment. In reality, virtually no risk assessments were made on nitrate, in respect of the environment or health, before problems or legislation emerged. This is mainly because nitrate has been around a very long time and only becomes a problem because of human folly. It is also because the idea of risk assessment is relatively new. We can, however, examine how these ideas might have applied to the environmental risks from nitrate discussed in Chapters 7 and 8 and the health risks examined in Chapters 9 and 10.

Environmental risk

The time when an environmental risk assessment on nitrate should have been made in Europe was arguably in 1957 when the Treaty of Rome was signed and the Common Agricultural Policy established. But given the circumstances then prevailing, it is not even remotely surprising that nothing of the kind was considered. It was, after all, only 10 years since the last famine in Europe. However, the nitrate problem does provide a warning to politicians that a risk assessment should be made on any legislation they propose, and it is unfortunate that in the UK the time available for scrutiny of legislation is declining. Many pieces of legislation have been found subsequently to have hidden problems or risks.

The tiering procedure should have been useful in 1957. The 'back of an envelope' assessment might have shown only benefit from expanding food production. But more detailed consideration would surely have suggested that intervention buying would greatly increase production, that greatly increased production would necessitate greatly increased use of nitrogen fertilizer and that this would in the long run mean greater losses to the environment.

The main flaw in this argument is that 'the environment' as we currently understand it had not been invented in 1957. But this was not the case when the EC's 50 mg/l limit was imposed in the 1980s. The line of reasoning outlined in the previous paragraph would surely have indicated that an overhaul of the entire Common Agricultural Programme was needed. The limit was applied to the symptom but the cause was ignored.

Health risk

There is no evidence that the EC's nitrate limit for potable water was based on any kind of risk assessment. If one had been done, it would have made a distinction, I hope, between water from the mains supply and well-water. Applying the tiering procedure to the mains supply would have shown at the 'back of the envelope' stage that there was no evidence of any risk of any kind and that nitrate posed no threat to health at any age when the concentration was less than 200 mg/l. The well-water might have had to go one stage further, but it would have shown that, provided the well was kept free of the bacteria associated with human and animal excrement, nitrate was no problem at concentrations less than 200 mg/l.

Is an Agency of Risk Assessment Needed?

North (2000) suggested that there was a need for an Agency of Risk Assessment (ARA). He thought the ARA should bring together the committees that advise the government on risk and scientists working in ministries. It would have two main purposes: to facilitate disinterested enquiry into matters involving risk and to disseminate its conclusions. The ARA should report directly to Parliament and be designated the Government's official source of advice on risk and the associated legislation. Further details can be found on pages 72–75 of North's (2000) book.

The ARA could also have an educative function. Politicians and the public and maybe even some scientists and economists would all benefit from a better understanding of risk. There may also be a need to expand our understanding of risk by finding out what 'risk' means to people working in a range of occupations. What a trader on the stock market means by 'risk' may well differ from the ideas held by someone working in the food industry. We need to look no further than the word 'organic' to see how different meanings accrue to the same word in different situations. And perhaps the ARA could work on the rationalization of the legislation about nitrate.

13 Coming to Terms with Nitrate: Public Attitudes to Science

Almost all allusions to nitrate in the media refer to 'nitrates' in the plural. This is wrong for the reason pointed out in Chapter 2 – that we are always concerned with the nitrate ion in water, which is always a single entity. As Gertrude Stein might have said, a nitrate is a nitrate is a nitrate. Nitrates in the plural perhaps sound more threatening than the single nitrate. (A monstrous army of nitrates waiting to commit mayhem with our health and our environment?) But the persistent reference to nitrates in the plural seems to speak of a casual attitude to scientific fact in the media. It is almost as if feelings about nitrate matter more than facts about it. This may be part of a wider problem faced by science as a whole and one with which much of this chapter is concerned.

Nitrate in the Water Supply

The UK spends millions of pounds removing nitrate from the domestic water supply (Table 13.1) and other countries do much the same. This happens because someone no longer identifiable decreed that, for our safety, our intake of nitrate should be limited. This edict was pronounced, unsupported by any definite evidence and in spite of our having known since 1916 that our bodies produce nitrate within themselves and since 1975 that our kidneys make a determined effort to prevent loss of nitrate from the body. We have learnt more recently that nitrate is central to our bodies' defences against bacterial gastroenteritis. Our bodies are clearly more interested in maintaining their stock of nitrate than in being protected against it. We should have taken more notice of what they were telling us.

Having spent so much money removing nitrate from our water supply, notionally to make it fit for drinking and cooking, what do we do with the water? We bath ourselves in it so that the nitrate on our skin (Chapter 9) dissolves in it. Cleopatra was considered an extravagant hussy because she bathed

Table 13.1. (a) Costs in the UK of removing nitrate from water supplies, 1992–1997. (Based on Pretty *et al.*, 2000.) (b) What the FARM-Africa charity could have done with the money in Africa.

(a) **UK costs**	
Annual capital expenditure (average)	£18.7 million
Annual operating costs	£1.7 million
Total expenditure in 6 years	£122 million
(b) **FARM-Africa could have provided**:	
A goat to 1 million poor families or widows	£27 million
Goat-breeding stations[a] to 50,000 villages	£21 million
Post-drought recovery packages[b] to 25,000 families	£15 million
250 kg of Lyamunga 90 bean seeds[c] to 100,000 farmers	£10 million
Training for 100,000 community-based 'bare-foot vets'	£14 million
Improved water supply to thousands of communities[d]	£35 million
Total	£122 million

[a]A goat-breeding station comprises a buck and three does, all Toggenburgs.
[b]Hand tools, seed and fertilizer for a family with five children.
[c]Bred for Tanzania. Drought-resistant and yields up to four times better than local varieties.
[d]No unit cost available for 'a water supply' because cost depends on locality, etc.

in asses' milk, but are we much better? We use it to clean our cars so that nitrate, gleaned by the paintwork from oxides of nitrogen in the atmosphere, dissolves in it (Chapter 7). We water our gardens with it so that nitrate from the soil dissolves in it. Most gardeners apply too much water, giving this nitrate an excellent start on its way to the aquifer and thence to the nitrate-removal plant to complete the circle. And we refer to ourselves as *Homo sapiens*?

The Misunderstanding About Nitrate and Health

L'hirondel and L'hirondel (2001) concluded their thoughtful book on the toxicity or otherwise of nitrate with the comment that, 'The history of nitrate is that of a world-scale scientific error that has lasted more than 50 years. The time has now come to rectify this regrettable and costly misunderstanding.' Their comment was commendably restrained, but it does lead to certain questions. What are the consequences of this misunderstanding? Who or what was responsible for it? And what steps do we need to take to avoid similar misunderstandings in the future?

Consequences of the misunderstanding of nitrate

The most obvious consequence of the misunderstanding lies in the costs of removing nitrate from the water supply in the UK (Table 13.1a). If we accept the evidence in Chapter 9 that nitrate, rather than being a threat to health, is beneficial to it, these vast sums of money have been largely wasted. Table 13.1b details what the FARM-Africa charity, mentioned in Chapter 11, could have done with the money spent on removing nitrate from water each year. There is no evidence that any consumer of water, young or old, would have suffered if we had removed only enough nitrate from water supplies to bring the concentration down to 100 mg/l, the original World Health Organization 'maximum'. Indeed, those at risk of bacterial gastroenteritis might have benefited somewhat. The money saved could have done a great deal for the poor, in the UK or in Africa. The most cogent argument I have heard in defence of the 50 mg/l nitrate limit came, curiously enough, from the fertilizer consultant Chris Dawson. He commented at a meeting in London late in 2003 that the limit had probably encouraged farmers to exercise moderation in applying nitrogen.

Lomborg (2001) warned that those who exaggerate environmental problems may, by scaring us, make us squander our resources on non-existent problems and ignore real problems. Has this happened with nitrate? In the UK and probably in the European Union generally and in North America, nitrate has dominated the agro-ecological research agenda during the past 25 years. Were resources squandered or problems ignored as a result?

I have to say at once that my immediate colleagues and I were for many years kept in employment largely by the perceived nitrate problem, so I am not an unbiased commentator. But I do not feel we squandered resources. We know a great deal more about the behaviour of nitrate and other forms of nitrogen in the soil than we did 25 years ago and I regard this knowledge as a 'public good'. Nitrate is not a health problem but it is not blameless in the environment. It can cause the problems in coastal waters described in Chapter 8, and nitrous oxide emitted from nitrate in the soil is implicated in global warming and the destruction of stratospheric ozone (Chapter 7). We needed a clear understanding of how nitrate and other forms of nitrogen behaved in the soil to tackle these problems.

Research on nitrate was useful in other ways. The methodologies for measuring and modelling leaching advanced appreciably, and these developments were relevant with appropriate modification to the leaching of pesticides and other problem solutes and to the transmission of bacteria through the soil. The 'nitrate problem' provided a major stimulus to research on this topic.

Were any environmental problems ignored as a result of the attention paid to nitrate? The most obvious candidate is phosphate. We now know that phosphate is the nutrient that limits the development of algal blooms in freshwater systems (e.g. Ferguson *et al.*, 1996). In the UK at least, there was not a great deal of research on phosphate until the interest in nitrate had begun to subside. But the main reason for this was probably that it was not until the 1990s that it became apparent that the leaching of phosphate was a matter of concern (e.g. Heckrath *et al.*, 1995). Up until then it had generally been assumed that phosphate did not leach. It may just be a coincidence that the decline in interest in nitrate and the realization that research was needed on phosphate occurred at the same time. If so, the concentration of attention and resources on nitrate did not actually hold up research on phosphate leaching. Whatever happened, research on phosphate in the UK has never been at the scale of that on nitrate.

Responsibility for the misunderstanding of nitrate

Did anyone deliberately misunderstand nitrate? It seems very unlikely that anyone in an official position did so, although certain unknown legislators seem to have failed conspicuously to ask some simple and rather obvious questions about the provenance and quality of the well-water data associated with methaemoglobinaemia cases in the mid-west of the USA. Or about the absence of cases involving water from the mains supply. Nobody setting the limit seems to have questioned the basic tenet that nitrate was a hazard or given any thought to the odd facts that our bodies manufacture this notionally hazardous substance and our kidneys seem keen to retain it.

The first investigator of well-water methaemoglobinaemia, Comly (1945), was clear that the problem wells contained both nitrate and bacteria, but Avery (1999) suggested that awareness of the presence of the bacteria was lost somewhere in the chain of reports that used Comly's data and those of other early investigators to assess the problem. Once the bacteria were forgotten, nitrate was left to bear the blame alone. A process akin to the precautionary principle perhaps then occurred. Nitrate appeared to threaten the health of very young infants, and lack of full scientific certainty was not going impede measures to counter the threat. The fact that the lack of scientific certainty arose from the lack of scientific investigation was not noticed.

Avoiding similar misunderstandings in the future

How we avoid similar misunderstandings in the future has become a very complicated question. This is because of the changes that have occurred during the past 25 years in the relationships between scientists, politicians, pressure groups and the public. Before we can answer the question, we need to examine this relationship.

Science, Politics, Pressure Groups and the Public

In November 2003, the Royal Institution and the Scientific Alliance organized a well-attended joint meeting entitled 'Science meets Politics'. Two comments made at that meeting stand out in my mind. One was by the Liberal Democrat MP, Dr Evan Harris, who said, 'There is now a battle between science and anti-science'. This showed, he added, not only in the decreasing trust in science but also in the increasing trust in the anti-scientific approach. The low esteem in which scientists are held is a disaster, he said, and something needs to be done about the attack on the independence of scientists.

The other comment was from Professor John Lawton, the Chief Executive of the Natural Environment Research Council (NERC). Professor Lawton, who had previously been the Chairman of the Royal Society for the Protection of Birds (RSPB), observed sadly that as Chairman of the RSPB, which has a million members, he had access to a government minister whenever he liked, but as the Chief Executive of NERC and with a better grasp of a whole range of environmental problems, he had never once been invited to speak to a minister. He said that his experience indicated that ministers were more interested in speaking to lobby groups than to scientific bodies. Professor Lawton described this as 'deeply regrettable' and felt that as a nation the UK paid dearly for it. He added that this lack of access for senior scientists to policy makers is almost unique in the world. His comment reminds me that a friend and fellow soil scientist, Professor Johan Bouma, was for a time seconded to a group of Dutch scientists, whose remit it was to advise the Dutch government on scientific matters. The British government does not have a similar arrangement and, to judge by Professor Lawton's experience, seems unlikely to bother with one.

The battle between science and anti-science

Evan Harris startled me with his remark about the battle between science and anti-science but unfortunately he is quite right. There is a battle at several levels. At the purely physical level there were more than 400 demonstrations by 'animal rights' activists at Huntingdon Life Sciences in the 10 months up to September 2002. The attacks on staff included the beating of the Director of the Laboratory with baseball bats. Huntingdon is far from being the only laboratory to be attacked. Battle tactics in attacks on laboratories have included throwing rape alarms on to the roofs of the houses of staff, bomb threats to the schools of their children, planting burning crosses in gardens, pouring acid on cars, smashing homes and daubing on walls accusations that the scientists were rapists or paedophiles. Three Japanese pharmaceutical companies came close to closing their UK operations, and plans for a centre for research on primates in Cambridge were abandoned at the end of January 2004, apparently because of the anticipated cost of security (*The Observer*, 1 February 2004).

Medical science is not alone in being attacked. Experiments on genetically modified crops have also come under attack, but the experiments rather than the experimenters have been the victims. Experiments at Rothamsted were among those that suffered. Designing and running an experiment in such a way as to tease out the information required is very difficult and it must have been intensely frustrating for my colleagues to see all the work wasted when a gang of crop-wreckers, guided by the grid reference placed on the World-wide Web by the government, arrived and smashed everything up. The crops destroyed at Rothamsted were annual crops, and worse happened elsewhere. I remember going to a scientific meeting at which one talk was turned by events into a rather sad account of how some potentially very interesting genetically modified trees had been chopped down by a hit-squad from a pressure group.

The assailants

What sort of people beat up medical scientists, wreck crops or vandalize trees? The common theme running through all these activities is the anti-science feeling to which Evan Harris referred. It seems these people had a strong feeling that, although what they were doing was illegal and highly distressing to their victims, it was part of a cause and therefore right. The Bible would have called them zealots. Their activities also reflect the increasing and dangerous tendency to view politics as irrelevant and direct action as preferable, a trend accompanied by a widespread decline in membership of political parties. Frank Furedi, a sociologist at the University of Kent, noted recently that since 1980 membership of the major political parties has fallen by about a half in the UK and Italy and by two-thirds in France.

Sir Karl Popper (Box 13.1) is the philosopher who contributed most to scientific thinking. His ideas about subjecting a theory or model to the possibility of falsification (Popper, 1959) are fundamental to the understanding of the 'scientific process'. He was also described as the most formidable critic of Marxism. In his book *The Open Society and its Enemies* (Popper, 1945), he related psycho-sociological ideas deriving from Greece of the 6th century BC to Western society of the 20th century AD. Among these ideas was that of the 'strain of civilization'. This concept may offer us a clue about the assailants of science whom we are trying to understand.

The 'strain of civilization'

A closed society is a tribal or feudal one, based on class, caste or even slavery, and this was the kind of society that existed in Greece before the 6th century BC. In that century, two forces were beginning to change this society to a more open one, commerce and the invention of critical discussion. Popper described how the influence of these changes brought the first symptoms of a new uneasiness, adding, 'The strain of civilization was beginning to be felt'. He went on to write of our own time:

> This strain … is still felt in our day, especially in times of social change. It is the strain created by the effort which life in an open and partially abstract society continually demands from us – by the endeavour to be rational, to forego at least some of our emotional social needs, to look after ourselves and to accept responsibilities. We must, I believe, bear this strain as the price to be paid for every increase in knowledge, in reasonableness, in cooperation and in

Box 13.1. Sir Karl Popper, 1902–1994.

Sir Karl Popper was born of Jewish parents and educated in Vienna. In his early professional life he combined school-teaching with a keen interest in philosophy. He mixed with, but did not become a member of, the Vienna circle of philosophers, the exponents of logical positivism. His career was interrupted by Hitler's rise to power, and he emigrated to teach philosophy in New Zealand. In 1946, he was appointed Reader in Logic and Scientific Method at the London School of Economics, later becoming Professor. This appointment presumably reflected his seminal work on the methodology of science summed up in his book *Die Logik der Forschung*. This was published in German in 1934 as one of a series of books put out by the Vienna Circle and later in English as *The Logic of Scientific Discovery* in 1959.

The essence of his argument (Popper, 1959) was that the distinction between science and non-science lay in the way that scientific theories make testable predictions and are discarded if the predictions fail. The logical positivists believed that the distinction could be made by linguistic analysis. Not everyone agreed with Popper. His work on the scientific process was later criticized by T.S. Kuhn, and Wittgenstein famously threatened him with a poker in 1946 after Popper had presented a paper in Cambridge directed against the anti-scientific philosophizing of 'Wittgenstein and his school'.

Popper was unquestionably the scientist's philosopher, but he is probably remembered more in non-scientific circles for his relentless criticism of Communism, Fascism and all forms of modern dogmatism, and for his championing of the critical scrutiny of ideas. In *The Open Society and Its Enemies* (published in 1945) and *The Poverty of Historicism* (published in 1957), he demolished one of the central tenets of Marxism, the idea that the course of history has a discoverable pattern.

Popper died 10 years ago, but his understanding of the nature of science remains absolutely crucial in the current struggle between science and anti-science.

mutual help, and consequently in our chances of survival and in the size of the population. It is the price we have to pay for being human.

All reputable scientists accept the need to be rational, to accept responsibilities and to forgo some of their own emotional needs in pursuing the goals of science, even if they do not totally achieve it. Science is, after all, a discipline. But discipline is unfashionable. We live in the 'me' society that was taught by Mrs Thatcher to put self first. Frank Furedi, mentioned above, also sees it as a society in which how you feel has become more important than what you believe in. Scientists implicitly expect other parts of society to accept the discipline they themselves have accepted, to be rational and to forgo some of their own needs. But a proportion of society wants nothing of it. They reject the discipline but they expect their views, often emotional rather than rational and reflecting their own feelings and needs, to be given the same weight as those of scientists who do accept the discipline. And two items about which people get particularly emotional are animals and the environment. This has resulted in the messy confrontations we have seen.

Magee (1985), in his book on Popper's philosophy, commented on the reaction against the strain of civilization, which he saw as having produced two kinds of response:

- A desire to return to the womb-like security of the closed tribal society that existed before the development of critical discussion.
- A wish to advance to some kind of Utopia.

The reactionary and utopian ideals both reject existing society and claim that a more perfect one is to be found at some other point in time. Both therefore tend to be 'violent and yet romantic' (Magee's words). Those who attack scientists and wreck crops and trees are indeed violent and they seem to see their activities in a romantic light.

Non-violent opposition to science

There was once a time when the main opposition to science came from the Church. Most people have heard of the banning of Galileo's writings by the Catholic Church or of the celebrated confrontation in 1860 between Bishop Samuel Wilberforce (Soapy Sam) and the zoologist T.H. Huxley over Darwin's theory of evolution. Huxley is widely perceived as having won, but it is interesting to note in this chapter, which is concerned with Karl Popper's ideas, that Huxley regarded natural selection as unproven, albeit the most probable mechanism of evolution.

Creationism, the belief that the world was created in 7 days as described in the first chapter of Genesis, still persists, mainly in the USA, but there are theological as well as scientific reasons for discounting it. (If you were God, would you really try to convey present-day ideas about biology or geology to a group of rather primitive shepherds on a middle-eastern hillside? It would be far better to give them a story in the narrative tradition with which they were familiar, and such a story might well start with, 'In the beginning …'.) The Creation narratives contain important theological truths in poetic form, but they were never meant to be understood completely literally in the context of a culture very different from that in which they originated.

I have been a professional scientist for about 40 years and a Reader (lay preacher) for the past 26 years. At no point have I been aware of any conflict between the intellectual demands of Science and Christianity, but I have recently become aware that they share a common adversary. This adversary has been described to me as 'post-modernism', but I am reluctant to use the term because I have yet to find a clear definition of it.

The problem is essentially one of authority. It is the idea that one person's ideas about a subject are as good as another's, regardless of what they actually know about it. And that feelings, irrational or not, are as important as properly evaluated scientific evidence or well tested theological doctrine. The views of someone who *feels deeply* about the MMR vaccine, for example, seem often to be perceived as having equal weight with those of a medical researcher who *knows a lot* about it. And it is impossible not to feel sympathy with the mother of an autistic child who needs to believe that the autism was caused by *something* and blames the vaccine, even when you know that it almost certainly was not the vaccine. The trend that seems to be emerging in the UK is that those who have feelings about scientific research should be given some authority in its direction. This is discussed below in the section on 'stakeholders'.

The problem of authority has been compounded with the 'celeb' problem. People who become famous or even just 'famous for being famous' are perceived to have credibility in fields about which they know little. The film star Meryl Streep ('famous' category) was transmogrified into an environmental campaigner and became part of the 1986 campaign against the chemical Alar, used to control the ripening of apples. The campaign, initiated on a TV programme, caused a panic about apples that spread in the USA from the president downwards, cost apple growers in Washington State alone $135 million and put many orchards out of business (Fumento, 1993). Alar had been in use for 21 years without any problems of any kind, and was never in any properly conducted tests shown to cause cancer. But that did not prevent it from being stigmatized as a carcinogen and banned (Fumento, 1993).

'Leaning' on businesses

Among those involved in the campaign against Alar was the consumer activist Ralph Nader. Fumento (1993) records how Nader boasted on American national television how he had gotten supermarkets to ban Alar-treated apples:

> So I decided to go direct to the supermarkets. I called up the head of Safeway one day, in Oakland, California… [and] I said, 'We're going to start a campaign to get Alar out of apples but why don't you save us a lot of trouble and yourself by saying that you are not going to buy any apples or apple products with Alar from your growers.' A week later [he] puts out a press release saying no more Alar apple products are being bought … So then we called up Grand Union, Kroger, A&P and guess what, we said, 'Safeway's not selling Alar-treated apples any more.' So they got them out.

The stores stopped selling apples treated with Alar not because there was evidence that Alar was a hazard, or because of consumer demand, but because they were threatened by someone more powerful than themselves. Nader was enormously powerful because of his reputation as a campaigner, notably on car safety, where his campaigning led to legislation. This brought him the title of the 'scourge of corporate morality', but he chose to make himself the scourge of the orchards as well. His campaigns on car safety were probably necessary, but his campaign on Alar sounds, from what he said on television, to have been an exercise in power play which did great damage, not to large corporations but to family businesses.

Nader exerted enormous power but he was accountable to no one. He did not have to face election and, because no one had appointed him, no one could fire him. He was obviously a very effective campaigner, but his role in the Alar affair raises important questions about individuals and organizations who exert power in public affairs without clear accountability to the electorate.

Pressure groups

We saw in Chapter 12 that Greenpeace felt sufficiently confident to issue its own version of the precautionary principle which was more extreme than the Rio one. Environmental pressure groups such as Greenpeace and Friends of the Earth now enjoy considerable power and their influence and financial clout put them in the 'multinational' class. Greenpeace International had an income of €165 million in 2002. But to whom are they accountable? And by whom are they funded? Asking these questions does not imply that they are necessarily undesirable – we all benefit from a decent environment. But power without accountability does not sit easily in an electoral democracy.

Pressure groups like Greenpeace are funded from the subscriptions of large numbers of people. Greenpeace has 2.8 million members worldwide. But the executive of a pressure group can no more be accountable to all its subscribers than the leadership of a political party can be to all its members, less so in the case of an international pressure group. It is impractical for the executive (or the leadership) to consult the subscribers (or the members) on all policy issues, so those at the top have in practice to take decisions for the organization. These decisions may have an impact not only on their subscribers but also on people throughout the country who have no affiliation to the pressure group.

I accept that it is in the nature of politics that I am affected by the decisions of leaders of parties other than the one of which I am a member, because those leaders can be removed at the next election. What is not acceptable is that I am affected by decisions made by the executive of a pressure group that has not been elected, cannot be removed at the ballot box and whose members make up less than 1% of the electorate. If Professor Lawton, as chairman of the RSPB, had access to a government minister whenever he wanted it, we can be certain that leaders of Greenpeace and other large multinational pressure groups have access too. And they will almost certainly have more access than Professor Lawton has now in his capacity as Chief Executive of the NERC.

The 221,000 Greenpeace members in the UK make up only about 0.5% of the electorate but, according to the Greenpeace website, their views receive an amount of attention disproportionate to their numbers. (See 'Political Corner' on the website.) If Greenpeace and other unelected and unaccountable multinational pressure groups are exerting influence on the British and other governments without any mandate from the electorate, they represent at least as much of a threat to democracy as other large multinational corporations against whom they sometimes campaign. We need to examine how they get their funding.

Political parties and pressure groups both depend on the subscriptions of their members, but there is one important difference between them. Although political parties have agendas that evolve over time, they ask their members to support activities that are stimulated by a fairly regular pattern of elections. People support a party by giving both money and time because they want that party to be in power. Pressure groups do not have this leverage and, although they get income from regular subscriptions, they usually have to raise a fair proportion of their funding by campaigning on risk-related issues. And when the income stream from one issue dries up, they campaign on another.

There is an obvious temptation to present something as a risk when there is no risk, and Bjørn Lomborg, probably the most famous ex-member of Greenpeace (but not its favourite one), has perceived this. He has argued (Lomborg, 2001) that many indicators suggest that, environmentally speaking, things are getting better, rather than worse as the pressure groups claim. He has also pointed out that if we spend money on non-existing risks, that money will not be there when we face a real and serious risk. We can add, with the Alar story in mind, that the pursuit of a non-existing risk can do great damage to people's livelihoods. These issues are important because Greenpeace and other pressure groups are now so powerful.

Those who doubt the reality of the anti-science movement need look no further than the torrent of abuse Lomborg had to endure when his book was published, some of it in the scientific press. As a colleague of mine remarked, the book had some 'evil reviews'. Lomborg himself was arraigned before the Danish Committee on Scientific Dishonesty and effectively found 'guilty'. Fortunately for Denmark's scientific credibility, the Danish Government declared the panel to have failed to back up their criticism of Lomborg or to give him a fair hearing.

Are the pressure groups anti-science?

Those who attack scientists and their experiments physically obviously have an anti-science bias. So far as I know, the main pressure groups do not condone physical attacks on individual scientists by their members, but where do they feature in the anti-science agenda? Their most effective anti-science activity has arguably been the propagation of what Lomborg (2001) described as the 'litany of our ever deteriorating environment'. This 'litany' is misleading when examined scientifically, but it has been widely accepted by the public worldwide. People now *feel they know* that the environment is in poor shape because they have heard the litany so often, and they are sceptical if anyone suggests that things are not as bad as they think.

The problem with the pressure groups may be not so much that they are anti-science but that they present 'alternative science'. If you look at the list of publications on the Greenpeace website, you will find a lot of publica-

Box 13.2. Peer review.

A paper submitted to a scientific journal is sent for assessment by two or three 'reviewers' or 'referees' before it is published. These are other scientists working on the same topic. The reviewers send their comments, which may be a recommendation to reject the paper, to the editor who sends them to the author(s) so that they can argue with the reviewer or change the paper. The system often works very well, with the authors feeling that the process has improved the paper, but on other occasions they are less happy. The system is not infallible, but it is the best system we have for the purpose, and peer review is regarded as central to the maintenance of standards in the scientific world. A paper that has not undergone peer review has appreciably less scientific standing than a paper that has.

tions described as 'Greenpeace Research Laboratories Technical Notes'. There are also publications in journals whose names I do not recognize, but that may be a failing on my part. I understand Greenpeace's laboratories to be very good, but I do not really know what the standing of their publications is. The standing of a scientific paper depends a great deal on whether or not it has undergone 'peer review' (Box 13.2). If their publications have gone through this process, they have scientific status but, without the process, they do not. Technical Notes in general would not necessarily be expected to be peer-reviewed and those from Greenpeace may not have this standing. There is no question but that the pressure groups have succeeded in gaining the ear of the public. The 'litany' is widely believed, but whether it is backed up by sound science is far from clear.

The pressure groups campaign on issues involving the welfare of wildlife and the environment, but surely human welfare is more important? About half the people who ever lived died from malaria (Pollock, 2004). They no longer die from it in the Western world but many do in parts of sub-Saharan and southern Africa. There the Greenpeace campaign against DDT, which is part of is their heavily funded 'toxics' campaign, has benefited wildlife, in the form of the malarial mosquito, at the expense of the local African population (Tren and Bate, 2001; Pollock, 2004). This is 'eco-imperialism' at its worst, and Box 13.3 gives some background to what Pollock (2004) described as 'the story of a scandal that has killed millions'.

The pressure groups have been enormously successful in making 'the environment' a political issue in a short space of time. According to the *Shorter Oxford English Dictionary*, the idea of an organism in, or adapting to, its environment did not emerge until 1874. Entries in books of quotations suggest that it was about another 100 years before 'the environment' came to the forefront of public awareness. There are no references to 'Environment' in the *Penguin Book of Quotations* or in my copy of the *Penguin Book of Modern Quotations*, which was reprinted in 1977. But we know from the *Oxford Dictionary of Humorous Quotations* that the environment was on the political agenda by the time of the Falklands War in 1982. Mrs Thatcher,

Box 13.3. Pressure groups, malaria and DDT.

Those who support the pressure groups no doubt mean well, but well-intentioned people can be a hazard, particularly if they concentrate on one issue to the exclusion of all others. The environment is not as important an issue as human health, but the phobia about DDT in the environment, for which Rachael Carson's book *Silent Spring* bears a lot of responsibility, threatens its use to prevent deaths from malaria in parts of Africa. In 1999, I visited the Dareda Mission Hospital in central Tanzania and was impressed by the matron and nursing staff (all Tanzanian) who were charming and well organized, albeit clearly short of resources. But my main memory, which still haunts me, is of ward after ward of small children sitting very still in their beds, their eyes listless and hopeless. Like many others, they were suffering from malaria, and some would soon be dead. Malaria is still a major killer in parts of Africa and has a huge economic impact (Tren and Bate, 2001).

Malaria is transmitted by *Anopheles* mosquitos, particularly *A. funestus* in southern Africa. When they bite, they infect humans with one of the various strains of a parasite described as a *plasmodium*, which causes the fever and other symptoms of malaria. Malaria control takes several forms:

- Eliminating the vector – that is, the mosquito.
- Destruction of the stagnant water in which the mosquito larvae hatch.
- Control of the parasite and its effects in humans through drugs.

It is obviously more effective to prevent infection by dealing with the vector or the water in which its larvae hatch than to deal with the consequences of infection. The use of DDT to kill mosquitos and thereby prevent malaria saved 500 million lives between 1950 and 1970 alone (Ames and Gold, 1997), and it continues to save lives in Africa now.

Despite DDT's life-saving record, the pressure groups demand that it should be banned, with Greenpeace describing it as a 'dangerous life threatening chemical'. But DDT is not particularly toxic to humans. In fact, one of the distinguished eccentrics who enlivened Rothamsted in the past carried a jar of DDT and a teaspoon in his jacket pocket and would eat a teaspoon of DDT to illustrate how safe it was. Pollock (2004) relates a similar story about the entomologist Gordon Edwards. Unfortunately, however, DDT persists in the environment and gets into body fat and thence into food chains, causing problems such as the thinning of the shells of the eggs of birds of prey. It is therefore on the list of persistent organic pollutants (POPs) defined by the United Nations Environmental Programme.

The Intergovernmental Negotiating Committee established in 1998 to agree a treaty on POPs was strongly attended by the wealthy and the politically correct but not by impoverished malaria sufferers (Tren and Bate, 2001). Seven highly developed countries sent five times as many delegates as 17 sub-Saharan African countries. And the pressure groups sent roughly twice as many delegates as all the African countries (raising interesting questions about whom they represented). The pressure groups campaigned to get the POPs treaty to ban DDT for all uses and initially seemed likely to succeed. But they were unable to sustain their argument and an appendix to the POPs treaty now allows countries to use DDT where no safe, effective or affordable alternative is available locally (Tren and Bate, 2001). But the World Wildlife Fund for Nature (WWF) has called for a total phase-out and ban on DDT production and use by 2007 (Pollock, 2004).

A ban on DDT would have been a disaster for parts of Africa where resistance to insecticides such as Deltamethrin has made DDT the only means of stopping mosquitoes from transmitting malaria. Those advocating its use (e.g. Tren and Bate, 2001) are not proposing widespread spraying. They concentrate on areas within and around people's homes where mosquitoes are mainly to be found, in the thatch, for example, or near the tops of inside walls. Used in this way DDT will not accumulate extensively in the environment, get into food webs or cause the thinning of the shells of birds of prey, the problem that sparked the initial concern about its use.

The risk to the environment of using DDT in this way is minute compared with the reciprocal risk of children dying of malaria, but the WWF still wants a total phase-out and ban on DDT production and use by 2007. And Greenpeace continues to devote part of its 'toxics' campaign against DDT and particularly against the factory of Hindustan Insecticides Limited of Kerala, India, one of the few remaining producers of DDT. Greenpeace's UK supporters and those of WWF are, no doubt, comfortably off people living in malaria-free luxury. But do they realize that they are donating money for the preservation of the malarial mosquito in its natural habitat, the African hut, free to feed on its natural prey, the African child? In so doing, do they not become accessories to what the American Congress for Racial Equality recently described as 'eco-manslaughter' (Pollock, 2004)?

who had then been in power for 3 years, said of the war that, 'It is exciting to have a real crisis on your hands, when you have spent half of your political life dealing with humdrum issues like the environment'. She was the last Prime Minister to refer to the environment as a humdrum issue. Twenty-two years later, environment*alism* has become the issue, a life and death issue for those with an interest in Africa. Pollock (2004) cited the popular author Michael Crichton, who recently described environmentalism as 'a religion that has killed millions', and commented that, 'Lives – mostly African mothers and infants – were sacrificed at the altar of green political correctness'.

Communicating Science to Politicians and the Public

The nitrate debacle could be seen as a triumph of anti-science over science, but it is probably more aptly described as the triumph of bureaucracy over science. Either way, we need to make sure that nothing similar happens again, but how can scientists ensure that it does not? Mr Blair famously declared, when he became Prime Minister in 1997, that his priorities were 'Education, education, education'. The priorities for scientists in 2005 must surely be, in the same spirit, 'Communication, communication, communication'. Mr Blair's comment reflected a recognition that his government was likely to stand or fall on educational issues. In the cultural climate in which science now finds itself, scientists are now judged not just on how good their science is when reviewed by their scientific peers, but also on how well they communicate it to politicians and the general public.

Communication with the public

I attend scientific seminars at Rothamsted but have to admit that I often do not fully understand those involving DNA and the technology surrounding it. This highlights the challenge scientists face in communicating with those outside science. If I as a scientist cannot always understand my fellow scientists, what hope is there for the rest of the community? The standard of communication in scientific meetings has improved over the years, aided by Powerpoint and similar systems, and a few prominent scientists communicate very effectively on television. Those funding research also put a much greater emphasis on the dissemination of the results. But the challenge of communication remains formidable.

In 1966, Harold Wilson enthused about the 'white heat of the technological revolution' and a substantial part of the UK enthused with him. Today, members of the public are increasingly sceptical about science and as likely to listen to pressure groups as to scientists. The pressure groups communicate very effectively and the public has no reliable criteria for knowing whether they are right when they disagree with scientists. Indeed, the public probably does not even know that they are disagreeing with scientists because they have no idea what scientists think. It is essential that scientists get their viewpoints into public view. And they have to communicate as well as, or preferably better than, the pressure groups.

Communication with politicians

All of us can gain access to our local MP, but some MPs are more scientifically oriented than others. There is more we can do. The Labour MP Dr Brian Iddon was also at the 'Science meets Politics' meeting cited above and he identified some of the ways in which scientists can interact with politicians. These include the Science and Technology Committee and the Parliamentary and Scientific Committee. The latter has a variety of events and lectures that facilitate contact between scientists and politicians. Scientists clearly need to keep these in mind as a means of communication with the world of politics.

Communication with the media

It is vital that scientists should be prepared to communicate with or through the media, not just because the media communicates with the public, but also for a reason identified by Evan Harris. The relationship between politicians and the electorate in the UK is becoming increasingly distant and indirect, surely a factor in the declining interest in politics and the dwindling membership of political parties. The relationship, such as it is, operates increasingly by way of the media and therefore according to media rules. This, incidentally, raises serious questions about the ownership of UK newspapers or television channels by those who are not UK citizens. More immediately, it makes it essential that scientists should be ready to communicate with the media and are fully prepared when they do so, with a clear and uncomplicated idea of the message they want to get across.

Communicating through the media is not without risk. Editors are far more concerned with selling newspapers than with providing a platform for science, and some of them are relentlessly hostile to some branches of science. (Remember Frankenfoods?) A quote out of context or an article that is altered slightly can be a considerable embarrassment. My colleague David Powlson and I have vivid memories of being accused by an irate fellow scientist of 'being economical with the truth' after *New Scientist* changed the wording and the meaning of an article we had written and inserted an extra diagram, all without telling us. But it is a risk we have to take.

The best route into the media is probably through a science journalist. Most of them are interesting characters and seem to do a good job, but apparently there are not many of them. It emerged at the 'Science meets Politics' meeting mentioned earlier in the chapter that there are only about 12 qualified science journalists in the UK. There are far more than 12 newspapers and television channels in the country, which suggests that finding people who can report science properly is a major problem. The media might look at the problem from a different angle and retort that there is a shortage of scientists who can write to the standard they require!

Transparency in communication

There is a long tradition in science of open publication in scientific journals. Your colleagues and your competitors can see clearly what you did, what you found out, and how you interpreted the results. And they can test whether your results can be repeated. This transparency is an essential part of the scientific process, and one particular aspect of it, openness about uncertainties and risks to the public, is vital during the communication of certain types of science. If the transparency is not there, the public and the media will assume that something is being hidden from them, probably something far worse than the reality.

The lack of transparency of the latter type during the BSE crisis was among the problems that gave the reputation of science a severe mauling from which it has not fully recovered. Lord Phillips' report into the crisis implicated three major problems in the disaster:

- Decisions were made apparently for political expediency rather than public health.
- All the decisions were made in complete secrecy, with no transparency at all.
- There was a 'culture of sedation'. Reassurances of certainty were given where there was no certainty.

Sir John Krebs, Chairman of the Food Standards Agency (FSA), speaking at the 'Science meets Politics' meeting described these problems and explained how the FSA, which was set up in the aftermath of the BSE crisis, was responding to the points raised by Lord Phillips' report. He said that the FSA is at arm's length from ministers and does not need permission to publish its views. It has resolved to do all its work in a completely transparent way. And it aims to be completely honest about uncertainty. Other scientists in the UK need to make similar resolutions, not just to repair the damage done by the BSE crisis to the reputation of science, but as a matter of scientific principle.

The relation between mainstream science and the pressure groups

Mainstream scientists and pressure groups are bound to differ over some issues involving animals and the environment, sometimes fiercely. But there will be some areas on which they agree. Scientists want a decent environment too, and they get angry when the landscape is despoiled for the wrong reasons. They also believe in the humane treatment of animals, which is not the same as treating animals as if they were human – a distinction whose neglect has led to a lot of trouble. I know that one of my closest colleagues buys 'free range' chickens and eggs because 'the bird had a better life'. I do not eat much chicken but I buy 'barn' eggs for the same reason. But we both bought shares in Huntingdon Life Sciences as a gesture of solidarity with the staff when they were under attack from 'animal rights' terrorists.

In the same way, there must be scientists working in pressure groups who subscribe to many of the convictions of those in the mainstream, even if they do not agree with them all. It is important that we get together to try to sort out the areas on which we can agree and then discuss sensibly the areas on which we do not agree. This will not be easy but it will be important in the long run. The Scientific Alliance has moved in this direction by organizing meetings on topics such as genetic modification and the long-term supply of electricity with speakers on both sides of particular issues. These were interesting, but the standard of communication was variable.

Avoiding Further 'Misunderstandings'

Nitrate in water supplies is subject to limits at least partly because some wells on farms in the mid-west of the USA were not constructed properly and were too close to pit privies. The 'panel of experts' that labelled Alar as a carcinogen included a TV presenter, a film star and a consumer activist. Our methods for determining which substances are harmful to us seem to lack precision. If we are to avoid further misunderstandings like that over nitrate, we need to improve them considerably. But how?

Using pre-existing knowledge properly

We have known for centuries that nitrate and nitrite guarded us against some peril in stored meat, now identified as the botulinum toxin. We have known for 88 years that our bodies manufacture nitrate within themselves. These two facts alone should have alerted us to the probability that nitrate was useful in the human body, particularly when a further fact emerged 29 years ago, that our kidneys make strenuous efforts to retain nitrate within the body. All this information should have informed the decision to limit nitrate in water supplies, but there is no evidence that it did. It is essential that in any regulation of a chemical in the environment we do not neglect to bring together what we know about it already. This can be seen as the first tier (Chapter 12) of any risk assessment. In doing so we need to be aware of the problems that can arise from the compartmentalization of knowledge. All the facts about nitrate were common knowledge, but in different spheres of activity.

Proper use of toxicology

We obviously need to know how toxic particular chemicals are to us and to our fellow human beings. But human toxicology has to cope with the fact that the experiments that really need to be done involve manslaughter and cannot be done. We can use animals in place of humans. Rats and guinea pigs, for example, are widely used in toxicity testing, but there are problems. One is that the dose has to be appropriate. If you force enough of any substance into the unfortunate animal, you can kill it, but the information you gain will be useless unless the dose was scientifically relevant. The best you can do is to calculate the dose on a body weight basis, but this leads to another problem. There are substantial differences between small mammals in their reactions to chemicals. Guinea pigs are nearly 100 times more sensitive, per kg of body weight, to the commonest dioxin (TCCD) than rats. Rats, on the same basis, are about 25 times more sensitive than human beings (Müller, 1997). These statistics imply that no small mammal is likely to be an entirely satisfactory model for the human species. The closer the animal is to us genetically, the more useful it will probably be as a model, but the less acceptable the experiment will be to the public.

Another issue to be resolved is that of the appropriate toxicological model (Barnett, 2004). Tables 12.1 and 12.2 and the associated text showed the impact that false negatives and positives have on the credibility of such tests, and we need a model that minimizes the proportion of false results. Calabrese (2004) argued that the hormetic model is superior on this and other criteria to the threshold and 'linear at low dose' models, but further research is needed.

Given all these problems, the best expedient may be to retrospectively collate information about chemicals that did kill or harm people, and there is already a massive amount of this kind of data on file. The data do not, of course, give advance warning that a new chemical is risky, but a lot can be inferred from the behaviour of related chemical species. The nitrate debacle gives an awful warning that this kind of data needs to be organized and handled meticulously and careful records kept of the provenance of the data.

Some new chemicals will carry an identifiable risk and will need careful handling. Others will carry no identifiable risk but they cannot be *proved* to be totally without risk. Should the precautionary principle be invoked to ban them? Perhaps we should recall that if the precautionary principle had been used in 1950 to ban just such a chemical, 500 million people would have died. The chemical was, of course, DDT (Box 13.3).

Nitrate is one of the oldest chemicals in use by humans. Its main problem was neatly summarized by an anonymous medical researcher at a very useful meeting organized at Essex University by Wilson *et al.* (1999) to bring together scientists working on nitrate in the fields of medical and soil science research. I was following two medical scientists along a corridor, and I heard one say to the other, 'Pity they didn't do the toxicology before the legislation'. We must hope that in the future the toxicology always comes first.

Stakeholders

A stakeholder in an enterprise is usually someone who has a financial interest in it, but the term can be used to embrace non-financial interests as well. The concept of the 'stakeholder' emerged in science about a decade ago as a means of taking account of the individuals and institutions that have some kind of interest, either financial or societal, in a scientific enterprise.

When Sir John Bennet Lawes started research in 1843, he used his own money to make experiments on his own land. Both he and the public, especially the farming public, regarded this work as being for the public good. Had the stakeholder concept been extant, Sir John might have regarded the farming public as stakeholders with him and his co-workers in the scientific enterprise. He did not patent the results of the experiments, and I doubt if the idea even occurred to him. Reasonably enough, he did patent the chemical processes he developed, the proceeds of which supported the experiments, but the results of the experiments were free for anyone to use. Most importantly, nobody questioned his right to experiment or doubted that the results would be beneficial. The farming public was well aware that the work was for their benefit even if no one had told them they were stakeholders.

When I joined Rothamsted 38 years ago, the funding had changed to mainly government funding, but not very much else had changed. The government had become a stakeholder, but a passive one. The work was still regarded as for the public good, we didn't patent our results, and nobody questioned our right to experiment or doubted that the work was beneficial. We also had a very small administration, amounting to the Director and Secretary of the institute and six others.

Much changed in the aftermath of Mrs Thatcher's coming to power in 1979 when the government stakeholder began to flex its muscles. We suffered a massive increase in bureaucracy mainly involving financial auditing of our work. This wasted time for which taxpayers had paid, and our administration had to increase roughly fourfold to deal with the bureaucracy. Scientists, who were also stakeholders, began to ask questions like, 'Who's auditing the auditors?', but no one seemed interested in them or that kind of question, and the slide has continued. Many of the more senior scientists now spend almost as much time looking for funding as they do on research, and most younger scientists have had a series of temporary jobs, no security and no sense of being a stakeholder at all. The molecular biologist who hit the headlines by quitting biology to become a plumber was part of a trend. Many good, well-motivated younger scientists have had enough and are getting out of research, and older scientists are often relieved when their children do not follow them into the profession.

We need to remind the government occasionally that the funds they disburse come from taxpayers. Taxpayers are therefore stakeholders too and arguably have a right to demand a say in the direction of the research for which they are paying. The present British government has been keen on 'focus groups' through which the people can feed policy ideas to politicians and this process has been extended to the formation of groups of 'ordinary people' to comment on priorities in science. These groups can be seen as a useful extension of democracy, but there are two questions about them that need to be asked.

- Do the people in the groups actually know anything about science or about the processes involved in its working? To ask this question is, of course, to invite accusations of scientific elitism. But would you really want to be operated on by a surgeon who was under the direction of some randomly selected members of the public?
- To what extent have those in the group have been influenced by the 'litany' promulgated by the pressure groups? The litany is not usually supported by scientific data and, as we saw above, the structure of the pressure groups is essentially undemocratic. And is it appropriate for people who have been influenced by unelected pressure groups with an anti-science bias to comment on scientific priorities?

The interaction between scientists and non-scientists may help to avoid the kind of misunderstandings like that over nitrate, but there may be problems when it comes to deciding priorities or lines of action. The stakeholders may reach an agreement or at least a compromise, but if they do not, the problem of finding an agreed balance between the views of scientific and non-scientific stakeholders could prove to be a thorny one.

The concept of the stakeholder is probably here to stay and it does provide useful opportunities for discussions between those identified as stakeholders. It may prove particularly useful where there is a large and disparate group of stakeholders, who would not normally talk to another because they do not realize they have a common interest. Identifying them and bringing them together could be a crucial first step in solving a problem. The approach might have been useful, for example, in determining the limit for nitrate concentration in potable water.

14 Coming to Terms with Nitrate: Land Use

Medical evidence (Chapters 9 and 10) suggests that if the EU limit for nitrate had been set at 100 rather than 50 mg/l, there would have been no extra deaths from methaemoglobinaemia or stomach cancer and there might have been slightly fewer deaths from bacterial gastroenteritis. Land use, and the lives of farmers and water suppliers, would have been made much easier by a 100 mg/l limit. It is difficult to restrain nitrate concentrations in drainage from the soil to less than 50 mg/l, particularly in areas with lower than average rainfall, but fairly easy to restrain them to less than 100 mg/l. (Table 14.1 shows how this applies when the restraint is on the percentage of land in arable cropping.) Restricting these concentrations is made more difficult by the fact that they vary widely with time. But the precautionary principle, together with the fears fuelled by the media and the pressure groups of anything that can be interpreted as a risk, mean that a change to the limit is virtually impossible. The nitrate limit is almost certainly here to stay.

Nitrate became a problem initially because of changes in land use resulting from the political decision to establish the CAP (Common Agricultural Policy) (Chapter 10). The CAP is now facing reform and the proposals will lead to further changes in land use. We need to look at the reform proposals and their possible consequences, but we need to consider first some issues that are at their heart, sustainability and organic farming, and another practice, no-till farming, which may be due for a return to popularity in the UK.

Sustainability

Sustainability is perhaps *the* buzz-word as we embark on the 21st century, but whether there is a consensus as to what it means is less clear. Most people would accept the definition of sustainable development given by the

Table 14.1. (a) Percentage of UK land growing various crops and estimated annual loss of nitrate-nitrogen and nitrate concentration in drainage assuming 200 mm through drainage. (b) Percentage of land that can be used for arable cropping with various target nitrate concentrations. (From Addiscott, 1989.)

(a)

Crop	% Arable land	Estimated	
		Nitrate-N loss (kg/ha)	Nitrate concentration (mg/l)
Winter cereals	62	40	89
Spring cereals	11	55	122
Rape	7	60	133
Sugarbeet	5	70[a]	155
Potatoes	3	80	177
Vegetables	3	120[a]	266
Other crops	9	60	133
Weighted mean		**50**	**111**

(b)

Target nitrate concentration (mg/l)	Percentage of land that can grow	
	Mix of crops above	Winter cereals only
50	41	53
75	65	83
100	89	100

[a]Assumes return of crop debris to soil.

1987 Brundtland Commission: 'Development that meets the needs of the present without compromising the ability of future generations to meet their own needs'. But what does this mean in practice? Enthusiasts for organic agriculture feel that 'sustainable' means 'organic'. Others see its meaning in low-input agriculture. Being by training a chemist with a definite preference for physical chemistry, I see sustainability in thermodynamic terms (Addiscott, 1995). The Principle of Minimum Entropy Production provides a valuable analogue of ecosystem function and a clear pointer to what sustainability actually means – provided you understand entropy. Entropy is explained briefly in the following section and in more detail in Box 14.1 for anyone not familiar with the concept. Box 14.1 also discusses the relation between entropy, information and work and explains how continuous work permits the self-organization of systems, and that this principle forms the basis of the ordering of the biosphere (Morowitz, 1970).

Ordering and dissipative processes

Entropy is *inter alia* a measure of disorder. Dissipative processes increase entropy while ordering processes lessen it. We see these processes all

Box 14.1. Entropy and its relation with information and work.

Anyone seeking a neat box to define the concept of entropy is doomed to disappointment. The chaos theorist Çambel (1993), who was also interested in entropy, said that when he was asked which was the best definition of entropy he would reply: 'There is no one best definition of entropy. Use the one best suited for your purposes. Also, please don't overlook the possibility that describing complex systems may require evaluating several entropies, not just one.' He went on to list eight types of entropy that may be relevant in different circumstances. The fact that the soil–plant system is an open system means that the type of entropy best suited to our purposes is the entropy implicit in Onsager's (1931) theory, which was one of the forms of entropy used by Prigogine (1947). It is this form that is discussed in the main text. The best general definition of entropy is that it is a measure of disorder. The molecules of a solid have a more ordered configuration than that of the corresponding liquid, so melting a solid causes an increase in entropy (Glasstone, 1947).

Entropy, information and work

An ordered state contains information. There is therefore an equivalence between entropy and information such that entropy increases when order is lost. Information is also related to thermodynamic work with interchangeability in both directions, work being converted to information in some circumstances and information to work in others. (Maxwell's demon is an example of the latter. He sat at a trap-door between the two compartments of a box and opened it to high-energy molecules but not to low-energy ones. To do so, however, he had to know which molecules were which. Thus information was converted into the work done when the temperature of the favoured compartment was raised.) This equivalence has the important consequence that continuous work permits the self-organization of systems. Morowitz (1970) considered this principle to be the basis of the ordering of the biosphere.

Thermodynamic work is performed when energy in the form of heat is transferred from a source at a high temperature to a sink at a low temperature. Continuous work therefore requires effectively infinite isothermal reservoirs at high and low temperatures. These are provided for the biosphere by the sun and outer space, respectively. The performance of this continuous work involves a flow of heat energy from the sun to outer space during which entropy is produced, but the work done at the surface of the earth can lead to considerable increases in order, and therefore decreases in entropy at the local scale. These result in the ordering processes listed in Table 14.2. Some of the dissipative processes listed in the table are a consequence of the random thermal energy generated when heat in the form of radiation passes through the atmosphere or strikes the soil surface.

around us, even if we do not think of them as ordering and dissipative. Biology provides many examples. Photosynthesis and its associated processes build complex ordered structures containing substances of large molecular weight from small molecules such as carbon dioxide, water and ammonia. Dissipative processes such as respiration and senescence degrade these structures back into the small molecules. Table 14.2 lists some processes in agricultural systems that have ordering or dissipative effects; these processes were discussed in greater detail by Addiscott (1994). The

Table 14.2. Ordering and dissipative processes in agricultural systems, listed as biological or physical. The pairs are not necessarily exact opposites. (From Addiscott, 1994.)

Ordering processes – entropy decreases	Dissipative processes – entropy increases
Biological	
Photosynthesis	Respiration
Growth	Senescence
Formation of humus	Decomposition of humus
Physical	
Water flow (profile development)	Water flow (erosion, leaching)
Flocculation	Dispersion
Aggregation	Disaggregation
Development of structure	Breakdown of structure
Larger units	Smaller units
Fewer of them	More of them
More ordered	Less ordered

pairs listed are not necessarily exact opposites, and water can have either effect.

Morowitz (1970) saw these ordering and dissipative processes as the drivers of the great ecological cycles of the biosphere, and the sustainability of agricultural ecosystems depends on the maintenance of a balance between order and dissipation. The same is true of the sustainability of many other systems which the human race creates or on which it depends.

The Principle of Minimum Entropy Production

Thermodynamic systems fall into three categories. An *adiabatic* system can exchange neither matter nor energy with its surroundings, a *closed* system can exchange energy but not matter, and an *open* system can exchange both. An adiabatic system is a theoretical construct that is not readily amenable to experimental investigation, but a closed system can be realized quite readily. However, most of the biosphere comprises open systems that exchange both energy and matter with their surroundings and the soil is obviously an open system (Johnson and Watson-Stegner, 1987).

Closed and open systems have different types of thermodynamics. A closed system is described by *equilibrium* thermodynamics and tends towards an equilibrium characterized by minimum energy and *maximum entropy*. An open system can be described by *linear non-equilibrium* thermodynamics (Onsager, 1931; Katchalsky and Curran, 1967). *Non-linear non-equilibrium* thermodynamics, also known as far-from-equilibrium thermodynamics, exists too (Prigogine, 1947) but is more complicated and falls beyond the scope of this book. Katchalsky and Curran (1967) examined the theory underlying the linear system and were able to prove that, in an open system allowed to mature, entropy production will decrease with time and reach a minimum. This *Principle of Minimum Entropy Production* in open systems is important because soils and ecosystems are open systems.

Katchalsky and Curran (1967) also showed that, when a flow in a linear non-equilibrium thermodynamic system is perturbed, the flow acts to decrease the perturbation so that the system returns towards its original state. Their theory also implied that the perturbation causes an increase in entropy production but that the removal of the perturbation allows entropy production to decline towards the original minimum.

Steady states, perturbations and sustainability

Katchalsky and Curran (1967) saw the implications of the theory outlined above for processes in the biosphere. 'There are several remarkable analogies', they wrote, 'between an open system approaching a steady state and living organisms in their development towards maturity.' They went on to suggest that The Principle of Minimum Entropy Production could be the physical principle underlying the evolution of the phenomena of life. They also pointed out that living organisms have regulatory mechanisms that preserve the steady state by countering perturbations as in the theory for steady-state systems.

But living organisms experience two major perturbations, birth and death, that limit the use of steady-state concepts to individual organisms. Addiscott (1995) suggested that it would be more relevant to apply these ideas to communities of organisms or to the ecosystems of which they form a part. For a given set of constraints, an ecosystem will, during a period of time, mature to a particular steady state. The soil is initially one of the constraints determining the direction in which the ecosystem matures but it remains part of the ecosystem and is itself changed during the process of maturing.

The thermodynamic analogue suggests that, if the ecosystem is perturbed, the flows in the system will act to counter the perturbation and restore the steady state and that this will probably involve the soil. This leads to two questions (Addiscott, 1995):

- How long does an ecosystem take to restore itself to the steady state?
- Can there be a catastrophic perturbation as a result of which the system cannot redirect itself to any steady state?

These questions are central to the understanding of sustainability and can best be answered by examining an ecosystem in a steady state, of which the most obvious example is the system often described as climax vegetation but perhaps better described as steady-state vegetation. This is frequently a forest and one characteristic is that the soil is an integral part of the ecosystem and would be totally different without the rest of it.

The answer to the first question obviously depends on the extent to which the soil is disturbed, but the study by Nye and Greenland (1960) of traditional shifting cultivation in West Africa indicates the kind of answer likely to be found. This system involved clearing the steady-state vegetation from an area of land, cultivating it for a few years and then abandoning it and moving to another area when crop yields declined too much because of loss of organic matter and fertility from the soil. The time needed for the area to recuperate – that is, for it to revert to natural vegetation and for fertility to be restored – depended on the extent of the perturbation. If the soil was not tilled and crops were planted in holes made with 'dibbling sticks', the soil returned to its original state and fertility in about 10 years. But if the soil was tilled by turning it over with hoes, the recuperation took about 50 years. The soil fauna played a vital role in the recuperative process after both degrees of perturbation (P.H. Nye, Oxford, 1993, personal communication). The time needed for recuperation obviously influences the number of people the land can support and the sustainability of the system declined as the population increased. (See also Chapter 11.)

The second question can be answered more specifically. A catastrophic perturbation is probably one in which the capacity for self-organization is seriously damaged or destroyed. This leads us to the question of where this capacity resides.

Where does the capacity for self-organization reside?

Jenkinson associated the capacity for self-organization with the 'biological potential' (D.S. Jenkinson, Rothamsted, 1993, personal communication). Nye's comment suggests that this is associated with the soil fauna and Chapter 3 emphasized the crucial role played by whole soil population in recycling organic matter and nutrients within an ecosystem. Without this population the ecosystem would run out of nutrients within a few generations of plants and collapse. But the soil population is equally dependent on the plants for its survival, so the soil population and the plants of the ecosystem act together as the stewards of the capacity for self-organization. This capacity is a collective function of all that lives in and by the soil. We know that it does not function without the soil population, but Box 14.2 provides a reminder that its function is impaired without plants.

Sustainable change is gradual

Many areas have lost their steady-state vegetation but have not reacted catastrophically. The reason for this is that the change was gradual. The theory which Katchalsky and Curran saw as an analogue for processes in the biosphere holds only for perturbations that are not too rapid and which do not push the system too far from its steady state. In particular, they do not affect the system's capacity for self-organization, so that the system is able to recuperate and return towards the steady state when the perturbation is removed. Rothamsted has sites that illustrate very well the gradual

Box 14.2. Gradual perturbation of steady-state ecosystems and the return towards the steady state with the removal of the perturbation at sites at Rothamsted.

The theory presented in the main text shows that perturbing a steady-state system leads to an increase in entropy production but that if the perturbation was gradual enough the system returns towards the steady state with entropy production declining towards a minimum when the perturbation is removed. This can be seen to apply to steady-state ecosystems at Rothamsted.

The steady-state vegetation of the Rothamsted area is thought to have been deciduous woodland and there have been two main perturbations.

- Tree clearance, which began in Saxon times and had brought about a timber famine by the 17th century (Hoskins, 1977).
- Ploughing of the grassland which initially prevailed after tree clearance.

Addiscott (1994) suggested how the timber clearance would have caused an increase in entropy as some of the wood was burnt and much of the rest eventually decomposed by microbes, destroying its ordered structure and producing carbon dioxide. The increase in entropy from the ploughing of grassland is readily identifiable as having arisen from the breakdown of organic matter of high molecular weight to small molecules. When some old grassland was ploughed at Rothamsted, 4 t of nitrogen were lost in the first 20 years from organic matter in the topsoil and much of this could be found as nitrate in the chalk underlying the site. About 40 t of carbon must have been lost as carbon dioxide. The soil profile would also have become less ordered, but the entropy this produced would have been small compared with the entropy production as ordered organic matter became small molecules.

The return towards the steady state with the removal of the perturbation and the accompanying decline in entropy production were illustrated most effectively by two areas that had long been in arable cultivation but which were left uncultivated from the 1880s to the present day. In 1957, Thurston (1958) wrote of the more acid of the two sites:

> The area has reverted to woodland, consisting chiefly of elm, ash and oak. The largest tree is an oak 81 inches (2.06 m) in diameter at 4 feet (1.22 m) from the ground, growing near the middle of the area that was cultivated. Of the 46 species of angiosperms present in 1957, thirty-two had been recorded previously and 14, including eight woodland species, had come in since 1913. In the same period, 55 species, all characteristic of grassland, had disappeared. All the arable weeds had already gone by 1913.

The steady state had clearly reasserted itself to a very large extent. The woodland had also reappeared on the other area but, because of the calcareous soil, the species found were different, with hawthorn rather than oak being predominant (Jenkinson, 1971).

The decline in entropy production was clear. By the time of Jenkinson's (1971) report, the acid site had accumulated 180 t/ha of trees and the calcareous site 274 t/ha of trees, representing between them a massive ordering of small molecules, including, Jenkinson estimated, about 0.7 and 1.1 t/ha of nitrogen, respectively. A further ordering of small molecules occurred in the accumulation of an average of 530 kg/ha of carbon and 45 kg/ha of nitrogen in organic matter in the soil each year in the calcareous area and rather less in the acid area. And another, rather smaller decline in entropy production came from the greater ordering of the soil profile. These accumulations and the ordering of the soil profile were discussed in greater detail by Addiscott (1994).

perturbation of steady states and the recuperation to the steady state when the perturbation is removed (Box 14.2).

Systems that are perturbed on a regular basis for a very long time eventually mature into a steady state, and the perturbation becomes one of the constraints maintaining the steady state. Ploughing and fertilizer application are perturbations of this kind. Broadbalk field at Rothamsted has been ploughed for a very long time and the plots have received the same fertilizer applications for more than 150 years (Chapter 1). (Broadbalk was established in 1843 but the current cropping pattern was not established until 1852.) The plots can now be regarded as being in a steady state and this makes them, as Jenkinson (1991) pointed out, an extraordinarily useful resource for studying nitrogen flows in soils. Powlson *et al.* (1986) obtained some of the interesting and important data described in Chapter 5 from the Broadbalk plots.

Sustainable Agriculture

The preceding discussion suggests that the key criterion of sustainability in an agricultural system is that the system is as close as possible to a steady state characterized by minimum production of entropy. For this to happen, changes must be gradual and the system must maintain its capacity for self-organization. These are the essential conditions from a theoretical point of view. The first condition implies that nutrient flows in the system, particularly those of nitrogen, need to be constrained to levels consistent with the needs of the crop. A further condition of sustainability is that the maintenance of the steady state within the system must not involve excessive expenditure of energy, and production of entropy, outside the system.

One outcome of this definition is that continuous arable agriculture, or even monoculture, may be one of the more sustainable options available. Broadbalk has after all been growing wheat for more than 160 years. Continuous arable farming seems to offer steady-state conditions, but it must be accompanied by careful checks that the capacity for self-organization is maintained.

Can sustainability be assessed objectively?

Most definitions of sustainability, such as that of handing on the land to posterity in good condition, are made in terms of human needs or feelings. This is no bad thing, for we are sentient beings. But these are subjective definitions of sustainability and, if different sets of needs or feelings clash, we may need a more objective assessment. Here the thermodynamic approach can help.

An audit of small molecules?

We saw above that the sustainability of agricultural systems depends on maintaining a balance between dissipative and ordering processes – that is, between processes that produce entropy and those that lessen it. In the biosphere, small molecules, such as carbon dioxide, water and ammonia, are produced when dissipative processes degrade complex, ordered structures and taken up and re-ordered when these structures are created. Addiscott (1995) therefore suggested that an audit of small molecules provided a means of assessing sustainability with some degree of objectivity. Such an audit cannot be fully objective because it can never have all the information needed, but it does offer a means of thinking semi-quantitatively about the sustainability of agricultural ecosystems.

Addiscott (1995) showed how the audit of small molecules might work by applying it to the results of Bertilsson's (1992) 'desk study' on data for flows of energy and material in Swedish agriculture. Bertilsson was interested in high- and low-intensity farming systems and he asked two questions of fundamental importance to discussions on sustainability:

- Is it better to farm *more land* less intensively (that is, with smaller inputs of fertilizers and other chemicals and smaller yields) or to farm *more intensively*, using the spare land for 'nature' or for the production of energy crops?
- Should we consider losses of nitrate, for example, as kilograms per unit area or kilograms per unit of production?

The latter question, also raised by Addiscott *et al.* (1991), springs from the issue, raised in Chapter 10, that all productive activity tends to pollute and the most productive tends to pollute most. The losses per unit of production are more relevant to sustainability if the aim of farming is production. If the land is being managed other than for production, to maintain the appearance of the countryside, the loss per unit area is the more useful. Sustainability and purpose cannot be entirely separated.

The farming system which Bertilsson (1992) studied was no less than the whole of Sweden. In his somewhat simplified study, he divided the country into two systems:

- A milk and meat production system. This depended on a rotation of clover/grass ley, ley, oats, barley, potatoes, peas, barley, potatoes, and the stocking rate was 0.53 ha per cow.
- A crop production system.

Each system was subjected to the following nitrogen regimes:

- Normal-N. This supplied 85 and 100 kg/ha of fertilizer nitrogen to the milk production and crop production systems, respectively.
- Low-N. This supplied 25 and 60 kg/ha, respectively.
- Zero-N. Neither system received any fertilizer nitrogen.

Farmyard manure was assumed to be used effectively, and the use of peas, beans and forage crops was adjusted to meet the demands of the rest of the system. All three systems had to meet the same target for production, this target being set in terms of milk and beef production, grain for sale and potatoes. In this simplified study, the potato target included sugarbeet, vegetables and other crops. An important aspect of the study was that the area used to meet the target was not fixed. This meant that if one of the nitrogen systems met the target without using all the land, the 'spare' land could be used for 'nature', as Bertilsson put it – others might prefer the term 'biodiversity' – or for the growth of willows for energy production. Bertilsson (1992) gives further details, and his paper was summarized by Addiscott (1993).

Bertilsson computed for each nitrogen system the emissions of carbon dioxide, nitrogen oxides, ammonia, nitrate and phosphate, all of which are relevant to the audit of small molecules, and the energy demand, which is not. He estimated the emissions of each small molecule on a per hectare basis or for the whole of the target agricultural production, showing that carbon dioxide was the dominant small molecule. Addiscott (1995) summed these emissions of small molecules by expressing them collectively in terms of molarity, as moles per hectare or as total moles for the target production (Table 14.3).

When the collective small molecule emissions in the nitrogen systems were examined on a per hectare basis, as mol/ha, the normal-N system emerged as clearly the largest emitter of small molecules and therefore the largest producer of entropy and the least sustainable. The low-N system was intermediate, and the zero-N system the smallest emitter and the most sustainable. But considering the small molecule emissions involved in achieving the target production gave a very different picture (Table 14.3). The zero-N system remained the smallest emitter of small molecules, but only by a small margin, and the low-N system became the largest emitter and, by inference the least sustainable, with the normal-N system intermediate.

The use to which the 'spare' land was put emerged as an important determinant of net small molecule emissions. If it was devoted to 'nature' the pattern of emissions remained as above, but if it was used to grow willows for energy production the pattern changed greatly (Table 14.3). This was because the willows were assumed to replace a fossil fuel source. Burning willows releases only the carbon dioxide recently fixed in photosynthesis, but burning fossil fuel releases carbon dioxide long kept out of the atmosphere. This meant that there could be a net decrease in carbon dioxide emission if willows were grown. The zero-N system had no spare land because it had all been used in meeting the production target, but the low-N system had a little spare land and the normal-N had plenty. Growing willows on the spare land made the normal-N system an important *net* fixer of small molecules rather than an emitter, but the zero-N system remained an emitter. Growing willows on the spare land therefore made the normal-N system the most sustainable. Bertilsson (1992) reached the same conclusion when he considered energy demand or net production.

Table 14.3. Emissions of small molecules from the normal-, low- and zero-N systems of Bertilsson (1992). (a) Emissions per hectare, (b) total emissions for achieving the target production.

Species	System		
	Normal-N	Low-N	Zero-N
(a) **Emissions per hectare (10^3 mol/ha)**			
CO_2-C	11.99	9.41	7.16
NO_x-N	0.38	0.37	0.37
NH_3-N	1.14	0.86	0.93
NO_3-N	2.43	2.07	1.93
PO_4-P	0.01	0.01	0.01
Sum	15.94	12.71	10.39
(b) **Total emissions (10^3 mol)**			
CO_2-C	18.0	18.7	15.7
NO_x-N	0.6	0.7	0.8
NH_3-N	1.7	1.7	1.8
NO_3-N	3.3	4.1	4.3
PO_4-P	0.00(1)	0.00(2)	0.00(2)
Sum	23.9	25.2	22.6
CO_2-C (willows grown on spare land)	–120.6	–20.1	15.7
Sum	–114.6	–13.5	22.6

The audit of small molecules – how useful?

The use of The Principle of Minimum Entropy Production as an analogue of ecosystem function presumes that entropy production can be assessed, and the proposed audit offers one approach to assessing it. The approach is over-simplified because it implies that the entropy contributions of all small molecules are similar. In practice, the entropies of gases and ions are assessed by differing means. But Glasstone (1947) gives values of the virtual molar entropies of the gases CO_2, NO, NO_2, N_2O, NH_3 and H_2O that all fall in the range 44–52 entropy units (184–220 J/mol/K) and a standard entropy of NO_3^- (calculated from the entropies of salts) of 35 entropy units (146 J/mol/K). Treating small molecules collectively may not therefore be a huge over-simplification. Glasstone (1947) gave no value for phosphate, and if sulphate had been included, its entropy of four entropy units would have appeared well out of range. Fortunately, phosphate contributed very little to the collective emissions (Table 14.3), and sulphate probably little more.

The audit of small molecules is also open to the criticism that it concentrates solely on the physical aspects of sustainability and takes no account of social aspects such as rural employment, but it is difficult to envisage a parameter capable of bringing together losses of gases and ions and social factors. Another obvious weakness is that detailed data for the audit will almost never be available. Bertilsson's (1992) study is probably unique, but the conclusions from that study can be used as a framework within which other systems can be studied. Without data, the audit can usually provide only a qualitative assessment of sustainability, but it is useful as an aid to informed guesswork.

Sustainability and purpose

If we are concerned with sustainable use of the countryside we may need to ask some fairly fundamental questions such as: What is the purpose of the countryside? And what do we expect to find there? For what types of use are different parts of the countryside most suitable? Bertilsson's second question, discussed above, led to the point that the sustainability of a farming system cannot readily be separated from its purpose. Sustainability will not mean the same for all purposes, so what are the criteria of sustainability for these purposes?

The list of purposes might include the following.

- Farmland whose primary aim is the intensive production of crops grown for food, energy production or use as a feedstock for the chemical industry. The criterion of sustainability for this land will be the entropy production per unit of crop produced. In terms of the contribution of nitrate loss to entropy production, the criterion will be mg nitrate lost per kg of crop produced.
- Farmland managed primarily to maintain the traditional qualities of the farmed landscape. Because production is not the key aim, the criterion of sustainability is simply the entropy production per unit area. For the contribution of nitrate, the criterion is mg nitrate lost per hectare.
- Farmland managed to enhance wildlife, particularly birds. This might involve just leaving large field margins on which weed species valuable to birds can grow, or it might involve more ambitious features such as coppices. Sustainability would be defined in terms of the population and diversity of the wildlife.
- Woodland managed to produce trees for commercial use or energy production. The length of time involved in the production of a tree crop makes it difficult to define the criterion of sustainability. 'Sustainable practice' may be easier to define. This would probably include the avoidance of clear-felling, by removing only a proportion of the trees in an area at any time on a grid pattern so that so that the roots of the remaining trees could retrieve nitrate and other nutrients from the whole area.
- Woodland, forest, downland or moorland for recreational or sporting use. For 'nature' in Bertilsson's terminology. This must include the dwindling stock of steady-state or 'climax' woodland – too much has been lost already. With land of this kind, sustainability depends as much on the deposition of nitrogen from the atmosphere as on nutrient losses from the system (Chapter 7). The real sustainability issues lie at the sources of the deposited nitrogen, which may be transport or industry.

Spare land and 'set-aside'

The spare land issue raised by Bertilsson (1992) is an important one. The increased production which led to the butter mountains and wine lakes of the 1980s needed to be cut back, and it is more difficult to decrease production by lessening productivity per hectare than by producing on fewer hectares. This is the basis of the 'set-aside' scheme, which basically pays farmers to take areas of land out of production. The land set aside must be eligible under the Arable Areas Payment Scheme and the minimum percentage that could be set aside in 2003 was 10%. More can be included on a voluntary basis. The land set aside must remain in that state for at least the period 15 January to 31 August. Bare fallow is allowed, but not for two consecutive years, and the aim is usually that appropriate crop cover should be established by 15 January. Natural regeneration is an option. There are various categories of setaside and the regulations covering them, though far too voluminous to be presented here, seem to offer some flexibility. For example, the scheme allows the land set aside to be used for 'nature', as Bertilsson put it, or for crops grown for the production of energy.

The setaside scheme was basically a response to the problem of overproduction. How sustainable setaside is in terms of entropy production depends on the extent and number of perturbations it experiences. Establishing a green cover and leaving it in place obviously allows a steady state to develop in which entropy declines towards a minimum. In terms of the audit of small molecules, any nitrate formed by mineralization will readily be taken up by the roots of the crop, which are there all the time once the cover has established. Allowing natural regeneration is an even more desirable option in terms of the theory, as we saw from those areas at Rothamsted which were allowed to regenerate. Bare fallow is not sustainable, as the substantial long-term losses of nitrate from Drain Gauges illustrated.

The key issue is how frequently the system is perturbed. The steady state needs time to establish, and if the green cover or the naturally regeneration is ploughed up after (say) 3 years, the overall entropy production may be greater than that which would have occurred if the land had been left in continuous cultivation.

Organic Farming

The word 'organic' was initially used by early chemists to describe substances that had been produced by living organisms under the influence of some kind of 'vital force' but today's chemists use it to describe the chemistry of carbon (Box 2.2). The word has, however, been expropriated to refer to a philosophy of agriculture. Although the original use of the word 'organic' goes back well over 200 years and the new use has been around for only about 80 years, the word will for the rest of this section refer to organic farming, unless accompanied by the word 'chemists'. Conford (2001)

Box 14.3. The role of the far right in the early organic movement in Britain.

The organic movement seems to have had its roots in the reaction against the threat to the life of the countryside posed by industrialization and its perceived ally, socialism (Conford, 2001). Its supporters believed there was a natural order, both political and biological, and many of them were therefore politically on the far right. They included the poet Edmund Blunden and the historian Arthur Bryant, who both sympathized with the Nazis. Another was Henry Williamson, author of *Tarka the Otter*, who was a member of the British Union of Fascists. The Soil Association is the official organization of the organic movement, and one editor of its journal, Jorian Jenks, was an active Mosleyite for a while. Concerns about food purity and racial purity may have more in common than is immediately apparent.

provided a history of the origins of the organic movement in Britain and the USA between the 1920s and the 1960s, and parts of this history will come as a considerable surprise to those who see organic food as a mark of political correctness (Box 14.3). Stockdale *et al.* (2001) gave an excellent account of the agronomic and environmental implications of organic farming, while Trewavas (2004) has provided a vigorous critique.

If one had to choose a word to sum up the philosophy of organic farming, it might be 'holistic'. The description 'organic' derives from Steiner's concept set out in 1924, of the whole farm as an organism (Stockdale *et al.*, 2001). Rudolf Steiner, a Swiss, was the founder of the anthroposophical movement and a man of broad concerns. Two communities caring for mentally disabled people were founded in the UK under his influence, at Thornbury in Gloucestershire and at Botton in Yorkshire. Steiner was also something of a mystic and was once described as a 'scientist of the invisible'.

The Steiner communities do good work, and I make occasional donations to one of them, but there are two reasons why I have not been convinced by what I have read of Steiner's ideas about farming. One was his use of unconventional terms such as 'capillary dynamolysis' that were not defined. The other was that his mysticism seemed to spill over into practices, such as the planting of the earth with cows' horns to capture the earth's rays, that seemed both scientifically and theologically questionable. (Though this practice might be seen to accord with the 'vital force' concept of the early organic chemists?)

Notable British exponents of organic farming include Sir George Stapledon and Sir Albert Howard, both of whom were influenced by Steiner (Conford, 2001; Stockdale *et al.* 2001) and, perhaps most notable of all, Lady Eve Balfour, who began the Haughley Experiment in 1939. The poet T.S. Eliot facilitated the publication of organic farming books by Faber and Faber, of which he was a director from 1925, and three of his 'Four Quartets' were published first in the *New English Weekly*, an 'organic' journal.

Is organic food preferable to conventionally produced food?

Are there any indications that organic food is better for the consumer? Stockdale *et al.* (2001), who set out the case for organic farming, noted that there is anecdotal evidence that organic food tastes better and contains a better balance of vitamins, but no consistent scientific evidence that this is so. Various other authors, including Bourn and Prescott (2002) and Woese *et al.* (1995), found no significant differences in composition between organic and conventionally produced food. Stockdale *et al.* add that with food crops it is difficult to separate the effects of the environment and the farming system on taste and composition, and with milk they cite Lund (1991) who found that the breed of cow had much more effect on the composition than whether the milk was produced in an organic or conventional system.

Bread is a key food item. The supermarket I use has made an inadvertent customer-preference survey on organic and conventionally produced bread, which has shown a consistent pattern over several years. When the bread is running out, the bread that remains at the end is almost always organic bread. This suggests that most of the customers prefer conventionally produced bread, and the reason is likely to be that organic wheat grain usually contains less protein than conventionally produced grain and probably mills and bakes less well (Stockdale *et al.*, 2001). I discovered recently that my friend and colleague Margaret Glendining had made the same observation about the bread counter completely independently (M.J. Glendining, 2004, personal communication).

The supermarket mentioned above, which carries and sells a lot of organic food, is in one of the wealthiest towns in south-east England, and its car park often contains substantial numbers of BMWs, Mercedes and large four-wheel-drive vehicles. Its customers can afford the 'organic premium' which arises from the extra costs of production incurred in organic farming. Organic food tends to be a preoccupation of the wealthy middle classes, but farmers must cater too for less prosperous members of the population for whom the price of food, rather than its organic credentials, is the key issue. That being said, the organic movement has stimulated a useful debate about food quality.

Is organic farming more sustainable than conventional farming?

For the organic faithful, the question in the title is redundant. 'Organic' and 'sustainable' are synonymous. But what does the thermodynamic approach suggest? Organic farming is generally less productive than conventional farming (Stockdale *et al.*, 2001). This implies that it is also intrinsically less polluting, and Stockdale *et al.* provide several references to smaller losses of nitrate per hectare, but not per unit of production (see above). It is not easy to determine which farming system is the more sustainable because nitrate can be lost other than from the main crop and small molecules other than nitrate, notably carbon dioxide, are lost too, some of them outside the system. Appreciable amounts of nitrate can be lost from organic systems when clover or clover/grass leys grown to get nitrogen into the system are ploughed but, on the other hand, carbon dioxide is lost to the atmosphere outside the system when fertilizers and other chemicals are manufactured. Each pass made with the tractor during cultivation also releases carbon dioxide from the fuel used by the tractor, so the number of passes made in each system is relevant. Inverting the soil in primary cultivation is particularly energy-consuming and prone to release carbon dioxide, and organic systems that require two passes with the plough per year raise questions.

Losses of nitrate and carbon dioxide are not the only sustainability issues. We need to consider too the retention of the capacity for self-organization in the system, together with the maintenance of the steady state and perturbations from it. There can be no doubt that organic farming is philosophically attuned to the retention of the capacity for self-organization. But there does not seem to be much evidence of the loss of this capacity in sensibly run arable agriculture. Broadbalk is still in good heart after the 160 years of the experiment, to which must be added the long period of arable agriculture that preceded it.

Broadbalk was cited above as an example of a steady state that was perturbed by ploughing and fertilizer application but over a long period so that the perturbation had become one of the constraints of a new steady state. The same is probably true of other conventionally managed arable farmland when treated consistently. Do organic farming systems allow the evolution of steady states in the same way? In the short term, the growth of (say) a grass/clover ley after an arable phase perturbs any arable steady state that may have begun to be established, and the ley itself will be perturbed later. Both perturbations will lead to an increase in entropy, at least in the short term. This appears to be a problem for any rotation, conventional or organic, that includes short-term grassland.

Overall, there does not seem to be a clear-cut answer as to which system, organic or conventional, is inherently the more sustainable in terms of the thermodynamic approach. It might be possible to make some progress towards an answer if a great deal of effort was put into gathering data and making an extensive audit of losses of small molecules, but that would be a major project that would be difficult to implement because of the variations in practice in both systems. Organic farming was and is a philosophy and its sustainability maybe needs to be evaluated by a philosopher using philosophical criteria.

The demand for organic food has increased during the last 20 years, reflecting the increasing wealth of the community and its preoccupation with 'lifestyle'. How can this demand be met in a sustainable way? The categories of land discussed earlier in the context of 'sustainability and purpose' included farmland managed primarily to maintain the traditional qualities of the farmed landscape. For this land, in which production is not the key aim, the criterion of sustainability is simply the minimum entropy production per unit area, which includes minimum loss of nitrate per hectare. We saw earlier that organic farming tends to lose less nitrate per hectare than conventional farming, but not per unit of production, so it should logically be located on the land managed to maintain traditional qualities. The fact that organic farms are more likely to be mixed farms complements the traditional qualities.

No-till Agriculture

We saw above that the number of passes made with the tractor, and particularly the energy used and carbon dioxide released in inversion tillage, emerge as a sustainability issue. This raises the question, 'Why plough?' No-till agriculture is practised quite widely in the USA and enjoyed some attention in the UK in the 1960s and 1970s, but it has not taken hold in the UK. Ploughing is a major perturbation of the soil and results in one of the largest increases in entropy caused by any agricultural operation. Nye's comment, quoted above, that the recovery of soil fertility in slash-and-burn agriculture was slower when the soil had been cultivated with hoes than when the crop had been planted straight into the soil using 'dibbling sticks' suggests that no-till deserves attention.

Trewavas (2004) recently reviewed the advantages of no-till over tilled organic and conventional systems. His conclusions suggest that if sustainability is assessed by the minimum of entropy production and the audit of small molecules, no-till is unquestionably the more sustainable. This is clear from the following indicators in his review.

- Fuel use and the accompanying emission of carbon dioxide per unit of production on a no-till field are one-third of that on a tilled field.
- Releases of carbon dioxide and nitrate by mineralization are stimulated by tillage, and nitrate mineralized in autumn is vulnerable to leaching. These releases are smaller under no-till. It should be noted that research by Catt *et al.* (2000) suggests that this may not always be the case.
- Soil erosion on a no-till field is about 5% of that on a tilled field.
- The structure and natural drainage of the soil are better under no-till.
- Losses in drainage or surface runoff of nutrient ions and herbicides are decreased by no-till.
- Large earthworms, which are an important part of the system's capacity for self-organization, thrive better under no-till. They contribute greatly to the structure and natural drainage of the soil.

Interestingly, Trewavas's (2004) review shows that some of the benefits claimed for organic farming seem to be provided at least as well, and sometimes better, by no-till systems, particularly when the no-till is implemented in the context of integrated farm management (IFM) which is defined by the LEAFUK audit. LEAF (Linking Environment And Farming) demands high standards of landscape and hedgerow management, large field margins, good soil management and attention to animal welfare. The review showed the combination of no-till and IFM to have further benefits.

- Natural pest predators, such as carabids and staphyglinids, can hide from birds and from sprays under the crop litter left on the surface of no-till systems and are six times more numerous in no-till than in ploughed systems.
- Bird territories increase greatly, anywhere from three- to 100-fold, in no-till systems.
- Small mammals are more abundant in no-till than in tilled fields.

One key point to emerge from Trewavas's review was that in the USA the global warming potential of no-till systems is one-third of that of organic systems, which in turn is about half that of conventional ones. If the same can be shown to hold true for the corresponding systems in the UK it will be a powerful argument for no-till.

Bennett *et al.* (2004) have used life cycle assessment (LCA) to compare the impacts on human health and the environment of growing conventional and genetically modified (GM) sugarbeet crops where the GM crop was grown without tillage. LCA has been defined as 'an objective process to evaluate the environmental burdens associated with a product, process or activity by identifying energy and materials used and wastes released to the environment, and to evaluate opportunities to effect environmental improvements' (SETAC, 1991). See also ISO (1997). LCA has mainly been applied to industrial products, but it has begun to be used in agriculture (e.g. Audsley *et al.*, 1997).

The results of Bennett *et al.* (2004) broadly agreed with those of Trewavas (2004) in that the LCA showed that with GM sugarbeet grown in the no-till system there was a considerable decrease in energy requirements, global warming potential and what they described as 'nutrification' (presumably losses of nitrate and phosphate). It also showed a useful reduction in ozone depletion potential and acidification, and ecotoxicity, calculated in chromium equivalents, was greatly decreased. The LCA further suggested health benefits, in that summer smog, toxic particulates and carcinogenicity, calculated in NO_x, PM10 and PAH equivalents, respectively, were all smaller in the system using GM with no-till.

Why did no-till fail to catch on in the UK during the earlier period of interest? Dudley Christian, who worked on no-till at the time, identified several possibilities (D.G. Christian, Rothamsted, 2004, personal communication).

- Weeds, particularly blackgrass, were a serious problem.
- Slugs, as well as pest predators, took advantage of the crop litter on the surface and damaged the crop.
- It was difficult to drill seeds directly into soil covered with the litter. Straw litter, for example, was sometimes forced into the slot with the seed and either kept it from proper contact with the soil or caused anoxic conditions which impaired germination.
- Some soils became compacted or suffered from poor drainage, but soil loosening through very light cultivation helped.

Christian considered that no-till was most viable before the ban on straw-burning was implemented. He thought too that it was best adapted to cracking and swelling clay soils, a view that the Department for Environment, Food and Rural Affairs seems to share.

Some of the problems on which he commented may now have been ameliorated by technological developments in the last 20 years. Trewavas's review and the LCA study of Bennett *et al.* (2004) assume that the no-till envisaged would involve crops that had been genetically modified to be resistant to glyphosate, and this should largely eliminate the weed problem. The problem of drilling seeds through the surface litter may be soluble through improved technology, if this has not already been achieved. The application of IFM standards should lessen soil compaction. As for the slugs, they will probably still be a problem, but we can hope that the increased number of birds and small mammals will help deal with it. Blackbirds reportedly deal with slugs 'messily but very effectively'.

Reforming Farming

The CAP reform proposals

We saw in Chapter 10 that the CAP and particularly the policy of intervention buying were responsible for much of the nitrate problem. To what extent will reform of the CAP resolve the problem? The reform proposals are not yet complete but the general nature of the reforms is becoming clearer. This section is based on a memo (MEMO/03/128) issued from Brussels on 10 June 2003. The overall objective of the memo was to give the EU's farmers a clear idea of its aims, and the financial framework in which they fit, up to 2013. In general, European agriculture will become more competitive and will be oriented more towards the market, and the CAP will be simplified. These changes will make the enlargement of the EU easier and the CAP easier to defend at the World Trade Organization (WTO), while providing farmers stable incomes and allowing them flexibility in decisions about what they produce. 'Environmentally negative' incentives in the current policy will be removed and sustainable farming practices encouraged. These general goals can be reduced to the following specific proposals.

- The link between direct payments and production will be cut. This process, described as decoupling, appears to put a complete end to the intervention buying which caused so many of the problems with nitrate. The EC proposes a single decoupled payment to each farm that will integrate existing payments. This step is logical but, with the benefit of hindsight, about 20 years too late (see Chapter 10).
- Payments will be linked not to production, but to the achievement of environmental, food safety, animal welfare, health and occupational safety standards. There will, for example, be incentive payments for participation in schemes designed improve the quality of agricultural produce.

- The EU will increase its support for rural development using money derived from a modulation (that is, an adjustment or, probably more to the point, a decrease) in direct payments, from which small farmers will be exempted.
- A new farm advisory system will be introduced.
- There will be new rural development measures to improve the quality of production, food safety and animal welfare and to cover the costs of the farm advisory system.

The enormous changes that resulted from the formation of the CAP were followed by major changes in society and in economic activity that resulted from the emergence and eventual dominance of market dogma. The changes were, in thermodynamic parlance, irreversible. There is therefore no prospect that these measures will return European agriculture to a state of 'pre-CAP innocence'. But would that be desirable anyway, given the wages and working conditions prevailing in the agricultural sector before the CAP? The changes appear, perhaps unsurprisingly, to be something of a compromise. European agriculture is apparently to be more 'market-oriented', but many of the proposals do not obviously appear characteristic of the operation of a free market. But they are none the worse for that.

One of the aims of the proposals was to make the CAP easier to defend at the WTO. This is all to the good if the WTO is mindful of the 'dumping' of agricultural produce in the poor countries of Africa and other parts of the developing world (Chapter 11), but not so good if the WTO is out to protect the subsidized agriculture of other wealthy countries. Those responsible for the reform process must make sure that mechanisms protecting the interests of poor countries are built into the proposals.

Farming and Food: the Policy Commission

The report of the Policy Commission on the Future of Farming and Food, chaired by Sir Donald Currie, was published in 2002. It covered issues way beyond the scope of this book, and it would be futile to try to comment on the report as a whole. But the report began with the Commission's vision for the food and farming industry and some items in that vision are worth highlighting. Not surprisingly, at least two of these accord closely with the proposals for the reform of the CAP. Farmers will continue to receive money from the public purse but only for benefits the public wants. And they will be rewarded for looking after their land and providing an attractive countryside.

I was brought up during and after the Second World War and remember rationing and also delicacies such as National Dried Egg, which arrived in large tins. I was therefore relieved to see the comment, 'Some [farmers] will have diversified from food production, but land and expertise remain available if greater quantities of home-produced food are suddenly needed'. The comment that food production will be largely based on supply contracts sounds reasonable, but supply contracts from whom? A fair proportion of these will presumably be from supermarkets, with whom farmers have not always fared well in their dealings. How well the requirements of the supermarkets will tie in with the objective of providing an attractive countryside remains to be seen, but a recent press report (Don, 2004) on the visual impact in Herefordshire, one of the jewels of the English landscape, of polytunnels growing strawberries for supermarkets suggests that the two may not be easy bedfellows.

Ultimately, there is no such thing as 'EC money' or 'government money'; there is only 'taxpayers' money'. It is taxpayers, many of them living in urban areas, who fund payments to farmers. They have to be recognized as stakeholders in the 'new' countryside. The report recognizes that they need to have access to the countryside for recreational purposes such as walking. They will provide income as tourists, but they must respect the rights of those who live in the countryside. Most of these urban stakeholders will be looking not only for an attractive countryside but also for good, cheap food in the supermarket, and these requirements need to be to reconciled with care.

Changes in land use

If the CAP reform proposals go ahead, farmers will be paid not for producing food but for achieving standards. It is not immediately clear how this is will change land use. The 'setaside' scheme presumed that we were producing too much food and that the best way of cutting production was to cut the area producing food, which in turn would decrease the overall loss of nitrate to the rest of the environment. Setaside was a fairly blunt instrument and needs to be refined, but refining it will involve tackling questions of the following kind.

- What changes in land use are suggested by environmental problems shown to be associated with nitrate?
- Are there areas of land on which it is desirable to change or stop agricultural production?
- What other problems are associated with land use?
- What other pressures are there on land use?
- What desirable landscape features might be introduced?

Changes in land use suggested by environmental nitrate problems

Nitrate is frequently assumed to be responsible for algal blooms in rivers and lakes, but it is phosphate that is the limiting nutrient for these blooms (Chapter 7). Phosphate also seems to limit the growth of larger water plants. The problem for which nitrate can properly be blamed is algal blooms and excessive benthic macroalgae in coastal and estuarine waters. It is also the source of much, but not all, of the nitrous oxide that contributes to global warming and the destruction of stratospheric ozone. It is probably prudent to ensure that not too much nitrate accumulates in aquifers, even if there is no evidence of any health risk from nitrate in drinking water. All these problems will be ameliorated by simply having less nitrate at risk in the soil.

The change in land use most likely to lessen the amount of vulnerable nitrate in the soil is the wider adoption of no-till, preferably with IFM. No-till can lessen the concentrations in the soil of nitrate and other oxidized nitrogen compounds to one-fifth or even less of the concentrations in tilled soils (Trewavas, 2004). But this change will, of course, involve the use of genetically modified crops, which will be unacceptable to the politically correct and probably to some others. These people arguably face a question. Are they more worried about nitrate problems, specifically algal blooms in coastal waters, stratospheric ozone damage and the contribution from nitrous oxide to global warming, or about genetic modification?

Most of the nitrate that causes problems in coastal or estuarine waters arrives there in a river. What else can we do to stop it getting into the river? Buffer zones, also known as riparian zones or strips (Haycock *et al.*, 1997), along the edges of rivers help to restrict losses of nitrate from land to water, partly because the plants in the zone can catch the nitrate before it gets to the water and partly by keeping agricultural operations such as fertilizer spreading away from the water's edge. Applications of manure and slurry also need to be kept away from the water's edge, as do cattle. These zones are commonly grass strips about 5 m wide running along the river bank, but a buffer zone can be turned into a landscape feature by planting trees or allowing natural regeneration of the vegetation (Box 14.2). The feature can be more than 5 m wide, of course. The trees planted should not be too different from those in the natural vegetation, which will tend to mean deciduous trees in most parts of England. Willows are an obvious tree to plant at the water's edge, where they help to stabilize the bank as well as yielding a variety of useful products.

It sometimes seems to be assumed that a buffer zone, particularly one with trees, will absorb an indefinite quantity of nutrients passing through the zone. This is not correct because, once the trees are fully grown, the nutrients taken up will be replenished in the soil by leaf fall and the re-mineralization of the nutrients in the leaves. The zone will become saturated and cease to be a 'sink' for nutrients. One solution to this problem is to harvest the trees and remove nitrogen and other nutrients as wood, for which there always seems to be a demand. The harvesting needs to be done on a systematic basis, so that no area of the zone is left without trees.

Areas of land on which agricultural production should be stopped or changed

English Nature has bought some farmland which is to be returned to fenland (Trewavas, 2004). Are there other initiatives of this nature which could be taken to remove land from production? Any land that used to be marshland but which was drained might be considered for return towards its original state. The North Kent marshes and parts of Otmoor in Oxfordshire come to mind as examples. In general, any land which is marginal in the sense that it is hard to farm profitably because of the soil or the natural drainage may be best allowed to return to its natural vegetation.

There is also some land that could usefully be taken out of arable use and put under grass. Some former grassland that went under the plough in the last century may now be ripe for a return to its former state, particularly if it is heavy land that gives management problems when wet. Some areas of chalk downland which have only thin soil over the chalk were also formerly under grass but have been ploughed. This land does not necessarily give management problems, but some of it may best be returned to grass for aesthetic as well as practical agricultural reasons.

The use to which an area of land can be allocated depends among other things on the slope. Sloping land that is ploughed is often at risk of erosion and may be better put under grass. Land that is so steep that it needs to be ploughed up and down the slope is particularly at risk (Quinton and Catt, 2004) and probably should be under grass.

Other problems: flooding

The most obvious 'other problem' of land use is flooding. This has caused great distress on recent occasions in the UK and Europe and is a major problem in some other parts of the world. The weather obviously contributes to floods, but land use is a factor too. One cause is that the way the land surface is managed encourages rain water to move laterally rather than downwards into the soil. Land covered by tarmac is an obvious problem, but a glance at a map suggests that in all but the big conurbations most flood water originates from rural land.

The nature of the soils from which the flood water comes is part of the problem, but inappropriate soil management is often at the root of it. The soil may have become compacted by the passage of agricultural machinery, and therefore less permeable, so that the water tends to move laterally rather than downwards, particularly during heavy rain. The rapid overland movement may be far more than ditches, streams or rivers can carry and flooding results. It is often accompanied by erosion. Ploughing the soil, however carefully it is done, leaves compacted soil at the base of the plough layer, and this too can encourage water to move laterally and rapidly. No-till may offer a way of resolving these problems, particularly if the larger worms it encourages improve drainage, but the resurgence of interest in it has not yet lasted long enough to enable a conclusion to be drawn.

Heavy clay soils often have field drains of the kind described in Chapter 2 and Figs 2.2 and 5.6. These are in place with the express purpose of removing water from the soil to a ditch or stream and efficient recently installed drains do so very effectively. The transit time from the soil surface to the ditch or stream may be only a few hours, and this rapid movement may be too much for other parts of the system. These drains gradually decrease in efficiency after they are installed, and need to be renewed from time to time. It has been suggested that, to avoid flooding, drains in certain parts of the landscape should simply be allowed to deteriorate so that the soil becomes waterlogged but holds back the water.

Drainage ditches feed streams, and streams feed rivers. The management of streams and rivers, particularly for avoiding floods, is a fascinating topic but takes us into river engineering, which is beyond the scope of this book. Purseglove (1988) has provided an entertaining and readily accessible introduction to the topic.

Non-agricultural pressures on land

Because of a rail strike, I recently used the new M6 toll motorway, and for £2 it transformed the journey by car from Hertfordshire to Manchester from a fairly grim experience to an appreciably more pleasant one. But that 43-km stretch of new motorway must have swallowed up roughly 2 km^2 of rural Warwickshire and Staffordshire, and Warwickshire has also given up land to the main M6, the M40, the M42 and the M69. The demand for ever more roads is an additional pressure on land.

There is also a political demand for large numbers of new homes in the UK. Some of these will be built on 'brown field' sites within existing towns and cities, but some will have to come from agricultural land. These homes should ideally be built on land identified as best removed from agricultural production, obviously so long as it is not designated for wetland or prone to flooding. Developers have presumably learnt that flood-plains are not suitable places to build homes? It would also be perverse to place housing developments where farmers are being paid to preserve the traditional qualities of the landscape. And it would be unwise to build on land needed for what is described below as 'core agricultural production'.

The idea of building homes on agricultural land will not worry one expert from English Nature who argued recently that the best way to defend the countryside is to build houses on it. This is not necessarily the oxymoron it appears at first sight, providing the housing includes sufficient green spaces to provide wildlife corridors and assuming that the gardens offer good habitats for birds, butterflies and other desirable species. It would need to be done with great sensitivity but it is not impossible, provided affordable new homes are made available to young people who already live in the countryside.

Desirable landscape features

Views will differ as to what features are desirable in the landscape. When I was a boy, I lived for 16 years in that western corner of Hertfordshire which makes up the northern end of the Chiltern Hills, so beech woods come high on my list. But the soil has an influence on the species of trees that thrive and may well decide which type of deciduous woodland is most appropriate. This is certainly the case if natural regeneration is allowed (Box 14.2). Farmland that has been limed regularly because of its natural acidity will not return to the original vegetation (Jenkinson, 1971).

Woodland is not just pleasant to the view. Once established, it is not leaky with respect to nitrate (unless the trees include an appreciable proportion of leguminous species). It may therefore be useful in areas such as nitrate vulnerable zones simply as a means of restricting the area of land that contributes nitrate. Establishing woodland on 25% of an area should usually cut nitrate losses from it by 25%.

Implications of Bertilsson's approach

Bertilsson's calculations show that, if there is a target for production of food crops, land can be 'bought' for 'nature', energy crops or other purposes by intensifying the production of the target crops so that there is 'spare land'. Energy crops can, of course, be included in the target production in the first place. The UK does not have a centrally planned economy or any of the associated trappings such as 5-year plans, and there are no targets for production, although the emphasis on market orientation in the CAP reform proposals and the mention of supply contracts in the Policy Commission Report imply targets of some sort.

Agriculture seems to be required to release land for homes and roads and may be expected to stop or change production on other land. Applying Bertilsson's approach shows that, unless there is a large surplus of land, we must retain a capacity for what could be described as 'core agricultural production', particularly if we want to have some land that is managed to retain the traditional qualities of the landscape. This core production would be intensive and would use fertilizers and agrochemicals, but ideally under an IFM regime. This type of regime, as we saw above, goes some way towards meeting the demand for traditional landscape qualities. Whether it would use no-till will depend on the crop and the soil and on the outcome of the current argument about GM crops.

The importance of this core agricultural production depends on the extent of the supposed surplus of land in the UK and on the degree of self-sufficiency we expect to achieve. The expansion of the EU increases the number of potential food suppliers available to us. Hungary, for example, is expected to become an important supplier of grain, but the cost of this grain will have to be counted not just in pounds or furmints but also in emissions of carbon dioxide and nitrogen oxides. There is a virtually linear relation between 'food miles' on the road and carbon dioxide released, and this will become more than a global warming issue as oil becomes more expensive and in shorter supply. The widespread transport of food by air is even more undesirable, because nitrogen oxides are then discharged straight from the jet engine into the stratosphere where they can do most damage.

It is 60 years or more since British and allied sailors were being torpedoed by U-boats on the North Atlantic convoys, but has the need for self-sufficiency in food totally abated? Another world war seems unlikely, but the threat of war has been replaced by the threat of terrorism. Oil production is an obvious target for terrorists, and an incident that seriously affected the supply of oil would also influence the movement of food. This is another reason for retaining the capacity for core agricultural production – in case it becomes essential to be self-sufficient in food. The possible threat to our oil supply is also another argument for no-till.

The most positive move we can make to lessen the threat of terrorism is to address the issues of trade justice and those WTO agreements that harm rather than help the poor (Chapter 11). This, unfortunately, is not the whole answer – terrorism also feeds off ancient hatreds, misconstrued religious doctrine and sheer wickedness.

Coming to Terms with Nitrate

The state of dependence on synthetic nitrogen fertilizer identified in Chapter 1 is largely a consequence of the increase in world population. It is therefore irreversible barring an almost unimaginable tragedy. Whether we like or not, we have to come to terms with this state and with the attendant nitrate problems. We cannot pretend that we can stop the production of nitrogen fertilizer and return to a state of innocence. Indeed, it is essential that, as oil and gas supplies gradually become scarcer, we recognize that food for the world is ultimately more important than the transport facilities of the rich and prepare to set aside supplies of energy for fertilizer production (C.J. Dawson, York, 2004, personal communication).

Coming to terms with nitrate means we need a clear understanding of the soil processes that result in nitrate losses to the rest of the environment and I hope this book has provided it. We also need to know which of the supposed nitrate problems are real and which are not. There is clear evidence that nitrate not a health threat and actually has an important positive role in health as part of our bodies' defence system against bacterial gastroenteritis. Limits as low as 44 and 50 mg/l for nitrate in potable water in the USA and EU are unnecessary and may be counter-productive. Legislation should reflect this reality rather than the myth generated by the imprudent juxtaposition of wells and pit privies in the mid-western states of the USA half a century ago. I hope that Jean-Louis L'hirondel will accept my use of the word 'myth' as an expression of 'the sincerest form of flattery' for his book and of respect for the work of his late father.

It is equally clear that nitrate does cause environmental problems. It causes algal blooms and excessive growth of benthic macroalgae in coastal and estuarine waters, but is not the limiting nutrient for freshwater algal blooms. Nitrous oxide released from nitrate plays an identifiable role in the destruction of the ozone layer and contributes to global warming.

Approaches for dealing with these problems have been outlined. The measures aimed at decreasing nitrate in coastal waters should eventually succeed, but it will take time. Global warming may turn out to be governed by forces beyond our control.

Science and anti-science

This book has been mainly about nitrate problems but it has touched other important problems, notably the 'battle between science and anti-science' to which Evan Harris referred. This is not just a problem for scientists in the UK and other affected countries, it is a matter of life and death in parts of Africa and affects other parts of the developing world (pp. 215–217). Evan Harris went on to say that, 'The low esteem in which scientists are held is a disaster and something needs to be done about the attack on the independence of scientists'. We scientists cannot expect that 'something' be done for us without any effort on our part. We need to make ourselves heard to a greater extent so that the public is aware when science does not agree with the 'green agenda' and why. We have to communicate better with the public and particularly the media. It may help to join organizations such as the Scientific Alliance or Sense about Science.

It is not just scientists who need a period of self-examination. Obesity has reached crisis proportions in the UK, but is it the only form of flabbiness? Have we become flabby about risk too? We are safer and more secure than at any time in history but have allowed ourselves to be manipulated by pressure groups into a state of paranoid chemophobia about some risks that are trivial beyond words compared with those faced on our behalf by those struggling to maintain peace in Iraq. Have we become intellectually flabby? Has the strain of civilization (Popper, 1945) become too much for us? Popper perceived that to remain civilized, we need to forgo at least some of our own emotional needs, to look after ourselves and to accept responsibilities, and above all to endeavour to be rational. If we have become too mentally flabby to be rational, particularly about the environment, science is in trouble in this country and so possibly is our civilization.

The last words in this book have to be about the EU nitrate limit on potable water. There is no medical case for it – indeed it would be a potential medical problem had our bodies not fortunately acquired the capacity to manufacture nitrate for themselves. But it sits unassailable, protected by the precautionary principle and by the EU, making anyone who questions it sound Europhobic. It is a great frustration to those who, like me, are instinctively pro-European and would vote in a referendum to stay in the EU, but wish the EU to behave rationally. If Popper (1945) is correct, and those who do not accept the need to be rational threaten civilization, could the EU be undermined by its own creation?

References

Acheson, E.D. (1985) *Nitrate in Drinking Water*. CMO(85)14. HMSO, London.

Addiscott, T.M. (1969) A method for measuring the phosphate potential of a Tanzanian soil. *East African Agriculture and Forestry Journal* 35, 21–27.

Addiscott, T.M. (1977) A simple computer model for leaching in structured soils. *Journal of Soil Science* 28, 554–563.

Addiscott, T.M. (1983) Kinetics and temperature relationships of mineralization and nitrification in Rothamsted soils with differing histories. *Journal of Soil Science* 34, 343–353.

Addiscott, T.M. (1988a) Long-term leakage of nitrate from bare, unmanured soil. *Soil Use and Management* 4, 91–95.

Addiscott, T.M. (1988b) Farmers, fertilizers and the nitrate flood. *New Scientist* 8 October, pp. 50–54.

Addiscott, T.M. (1989) Effect of permitted nitrate concentration on use of land for arable cropping. In: House of Lords Select Committee on the European Communities (ed.) *Nitrate in Water*. HMSO, London.

Addiscott, T.M. (1993) How high should low-intensity be on the agenda? *The Agronomist* No. 2, pp. 4, 5, 16.

Addiscott, T.M. (1994) Simulation, prediction, fore-telling or prophesy? Some thoughts on pedogenetic modelling. In: Bryant, R.B. and Arnold, R.W. (eds) *Quantitative Modeling of Soil Forming Processes*. SSSA Special Publication No. 39. Soil Science Society of America, Madison, Wisconsin, pp. 1–15.

Addiscott, T.M. (1995) Entropy and sustainability. *European Journal of Soil Science* 46, 161–168.

Addiscott, T.M. (1996) Fertilizers and nitrate leaching. In: Hester, R.E. and Harrison, R.M. (eds) *Agricultural Chemicals and the Environment*. Issues in Environmental Science and Technology No. 5. The Royal Society of Chemistry, Cambridge, pp. 1–26.

Addiscott, T.M. and Bailey, N.J. (1990) Relating the parameters of a leaching model to the percentages of clay and other components. In: Roth, K., Flühler, H., Jury, W.A. and Parker, J.C. (eds) *Field-scale Solute and Water Flux in Soils*. Birkhaüser Verlag, Basel, pp. 209–221.

Addiscott, T. and Brookes, P. (2002) What governs nitrogen loss from forest soils? *Nature* 418, 604.

Addiscott, T.M. and Cox, D. (1976) Winter leaching of nitrate from autumn-applied calcium nitrate, ammonium sulphate, urea and sulphur-coated urea. *Journal of Agricultural Science, Cambridge* 86, 381–389.

Addiscott, T.M. and Mirza, N.A. (1998) New paradigms for modelling mass transfers in soils. *Soil and Tillage Research* 47, 105–109.

Addiscott, T.M. and Powlson, D.S. (1992) Partitioning losses of nitrogen fertilizer between leaching and denitrification. *Journal of Agricultural Science, Cambridge* 118, 101–107.

Addiscott, T.M. and Smith, P. (1997) Using models in risk assessment of pollutant losses from agricultural land to water. In: Zelikoff, J.T. (ed.) *Ecotoxicology: Responses, Biomarkers and Risk Assessment*. SOS Publications, Fair Haven, New Jersey, pp. 489–506.

Addiscott, T.M. and Thomas, V.H. (1979) Glycoluril as a slow-release nitrogen source for plants. *Chemistry and Industry* 6 January, pp. 29–30.

Addiscott, T.M. and Whitmore, A.P. (1987) Computer simulation of changes in soil mineral nitrogen and crop nitrogen during autumn, winter and spring. *Journal of Agricultural Science, Cambridge* 109, 141–157.

Addiscott, T.M. and Whitmore, A.P. (1991) Simulation of solute leaching in soils of differing permeabilities. *Soil Use and Management* 7, 94–102.

Addiscott, T.M., Thomas, V.H. and Janjua, M.A. (1983) Measurement and simulation of anion diffusion in natural soil aggregates and clods. *Journal of Soil Science* 34, 709–721.

Addiscott, T.M., Whitmore, A.P. and Powlson, D.S. (1991) *Farming, Fertilizers and the Nitrate Problem*. CAB International, Wallingford, UK.

Addiscott, T., Smith, J. and Bradbury, N. (1995) Critical evaluation of models and their parameters. *Journal of Environmental Quality* 24, 803–807.

Addiscott, T.M., Armstrong, A.C. and Leeds-Harrison, P.B. (1998) Modelling the interaction between leaching and intraped diffusion. In: Selim, H.M. and Ma, L. (eds) *Physical Nonequilibrium in Soils. Modeling and Application*. Ann Arbor Press, Chelsea, Michigan, pp. 223–241.

Al-Dabbagh, S., Forman, D., Bryson, D., Stratton, I. and Doll, R. (1986) Mortality of nitrate fertilizer workers. *British Journal of Industrial Medicine* 43, 507–515.

Allingham, K.D., Cartwright, R., Donaghy, D., Conway, J.S., Goulding, K.W.T. and Jarvis, S.C. (2002) Nitrate leaching losses and their control in a mixed farm system in the Cotswold Hills, England. *Soil Use and Management* 18, 421–427.

Allison, F.E. (1955) The enigma of soil nitrogen balance sheets. *Advances in Agronomy* 7, 213–250.

Al-Sa'doni, H. and Ferro, A. (2000) *S*-nitroso-thiols: a class of nitric oxide-donor drugs. *Clinical Science* 98, 507–520.

Ames, B.N. and Gold, L.S. (1990) Misconceptions on pollution and the causes of cancer. *Angewandte Chemie* 29, 1197–1208 (international version in English).

Ames, B.N. and Gold, L.S. (1997) Pollution, pesticides and cancer misconceptions. In: Bate, R. (ed.) *What Risk? Science, Politics and Public Health.* Butterworth Heinemann, Oxford, UK.

APHA (1949–1950) Committee on Water Supply. Nitrate in potable waters and methaemoglobinaemia. *American Public Health Association Yearbook* 40(5), 110–115.

Audsley, E., Alber, S., Clift, R., Cowell, S., Crettaz, P., Gaillard, G., Hausheer, J., Joliot, O., Kleijn, R., Mortensen, B., Pearse, D., Roger, E., Teulon, H., Weidema, B. and Van Zeijts, H. (1997) *Harmonization of Environmental Life Cycle Assessment for Agriculture*. Final Report Concerted Action AIR3-CT94-2028. European Commission, Brussels.

Avery, A.A. (1999) Infantile methemoglobinemia: reexamining the role of drinking water nitrates. *Environmental Health Perspectives* 107, No. 7.

Avery, B.W. and Bullock, P. (1969) Morphology and classification of Broadbalk soils. *Report of the Rothamsted Experimental Station for 1968, Part II*. Lawes Agricultural Trust, Harpenden, UK, pp. 63–81.

Barnett, V. (2004) *Environmental Statistics: Methods and Applications*. John Wiley & Sons, Chichester, UK, 293 pp.

Bawden, F.C. (1969) Broadbalk: foreword. *Report of the Rothamsted Experimental Station for 1968, Part II.* Rothamsted Experimental Station, Harpenden, UK, pp. 7–11.

Bekunda, M.A., Bationo, A. and Ssali, H. (1997) Soil fertility management in Africa: a review of selected research trials. In: Buresh, R.J., Sanchez, P.A. and Calhoun, F. (eds) *Replenishing Soil Fertility in Africa.* SSSA Special Publication No. 51. Soil Science Society of America, Madison, Wisconsin, pp. 63–79.

Bell, S.G. and Codd, G.A. (1996) Detection, analysis and risk assessment of cyanobacterial toxins. In: Hester, R.E. and Harrison, R.M. (eds) *Agricultural Chemicals and the Environment.* Issues in Environmental Science and Technology No. 5. The Royal Society of Chemistry, Cambridge, UK, pp. 109–122.

Benjamin, N. (2000) Nitrates in the human diet – good or bad? *Annales de Zootechnologie* 49, 207–216.

Benjamin, N., O'Driscoll, F., Dougall, H., Duncan, C., Smith, L., Golden, M. and McKenzie, H. (1994) Stomach NO synthesis. *Nature* 368, 502.

Benjamin, N., Pattullo, S., Weller, R., Smith, L. and Ormerod, A. (1997) Wound licking and nitric oxide. *Lancet* 349, 1776.

Bennett, R., Phipps, R., Strange, A. and Grey, P. (2004) Environmental and human health impacts of growing genetically modified herbicide-tolerant sugar beet: a life cycle assessment. *Plant Biotechnology Journal* 2, 273–278.

Beresford, S.A. (1985) Is nitrate in drinking water associated with gastric cancer in the urban UK? *International Journal of Epidemiology* 14, 57–63.

Bernstein, P.L. (1996) *Against the Gods. The Remarkable Story of Risk.* John Wiley & Sons, New York.

Bertilsson, G. (1992) Environmental consequences of differing farming systems using good agricultural practices. In: *Proceedings of an International Conference of the Fertilizer Society,* Cambridge, 16–17 December 1992, pp. 1–27.

Beven, K. (1981) Micro-, meso- and macro-porosity and channelling phenomena in soils. *Soil Science Society of America Journal* 45, 1245.

Birch, H.F. (1958) The effect of soil drying on humus decomposition and nitrogen availability. *Plant and Soil* 10, 9–31.

Bosch, H.M., Rosenfield, A.B., Huston, R., Shipman, H.R. and Woodward, F.L. (1950) Methemoglobinemia and Minnesota well supplies. *Journal, American Water Works Association* 42, 161–170.

Bouman, A.F. and Booij, H. (1998) Global use and trade of feedstuffs and consequences for the nitrogen cycle. *Nutrient Cycling in Agroecosystems* 52, 261–267.

Bourn, D. and Prescott, J. (2002) A comparison of the nutritional value, sensory qualities and food safety of organically and conventionally produced foods. *Critical Reviews in Food Science and Nutrition* 42, 1–34.

Boyd, P.W., Watson, A.J., Law, C.S., Abraham, E.R., Trull, T., Murdoch, R., Bakker, D.C.E., Bowie, A.R., Buessler, K.O., Chang, H., Charette, M., Croot, P., Downing, K., Frew, R., Gall, M., Hadfield, M., Hall, J., Harvey, M., Jameson, G., LaRoche, J., Midlecoat, M., Ling, R., Macdonado, M.T., McKay, R.M., Nodder, S., Pickmere, S., Pridmore, R., Rintoul, S., Safi, K., Sutton, P., Strzepek, R., Tanneberer, D., Turner, S., Waite, A. and Zeldis, J. (2000) A mesoscale phytoplankton bloom in the polar southern ocean stimulated by iron fertilization. *Nature* 407, 695–702.

Bremner, J.M. (1997) Sources of nitrous oxide in soils. *Nutrient Cycling in Agroecosystems* 49, 7–16.

Brookes, P.C., Powlson, D.S., Jenkinson, D.S. and Tate, K.R. (1982) The secret life of soil. *New Scientist* 96, 564.

Brookes, P.C., Landman, A., Pruden, G. and Jenkinson, D.S. (1985) Chloroform fumigation and the release of soil nitrogen: a rapid direct extraction method to measure microbial biomass nitrogen in soil. *Soil Biology and Biochemistry* 17, 837–842.

Brown, L., Syed, B., Jarvis, S.C., Sneath, R.W., Phillips, V.R., Goulding, K.W.T. and Li, C. (2002) Development and application of a mechanistic model to evaluate emissions of nitrous oxide. *Atmospheric Environment* 36, 917–928.

Buesseler, K.O. (2001) Ocean biogeochemistry and the global carbon cycle: an introduction to the US joint global ocean flux study. *Oceanography* 14, 1–117.

Bumb, B.L. (1995) World nitrogen supply and demand: an overview. In: Bacon, P.E. (ed.) *Nitrogen Fertilization in the Environment*. Marcel Dekker, New York, pp. 1–35.

Buresh, R.J., Sanchez, P.A. and Calhoun, F. (eds) (1997a) *Replenishing Soil Fertility in Africa*. SSSA Special Publication No. 51. Soil Science Society of America, Madison, Wisconsin, 251 pp.

Buresh, R.J., Smithson, P.C. and Hellums, D.T. (1997b) Building soil phosphorus capital in Africa. In: Buresh, R.J., Sanchez, P.A. and Calhoun, F. (eds) *Replenishing Soil Fertility in Africa*. SSSA Special Publication No. 51. Soil Science Society of America, Madison, Wisconsin, pp. 111–149.

Burns, I.G. (1974) A model for predicting the redistribution of salts applied to shallow soils after excess rainfall or evaporation. *Journal of Soil Science* 25, 165–178.

Burns, I.G. (1975) An equation to predict the leaching of surface-applied nitrate. *Journal of Agricultural Science, Cambridge* 85, 443–454.

Busch, D. and Meyer, M. (1982) A case of infantile methaemoglobinaemia in South Dakota. *Journal of Environmental Health* 44, 310.

Calabrese, E.J. (2004) Hormesis: from marginalization to mainstream. A case for hormesis as the default dose–response model in risk assessment. *Toxicology and Applied Pharmacology* 197, 125–136.

Çambel, A.B. (1993) *Applied Chaos Theory. A Paradigm for Complexity*. Academic Press, London.

Cannell, R.Q., Goss, M.J., Harris, G.L., Jarvis, M.G., Douglas, J.T., Howse, K.R. and le Grice, S. (1984) A study of mole drainage with simplified cultivation for autumn-sown crops on a clay soil. 1. Background, experiment and site details, drainage systems, measurements of drainflow and summary of results, 1978–80. *Journal of Agricultural Science, Cambridge* 102, 539–559.

Catt, J.A., Howse, K.R., Christian, D.G., Lane, P.W., Harris, G.L. and Goss, M.J. (1998a) Strategies to decrease nitrate leaching in the Brimstone Farm Experiment, Oxfordshire, UK, 1988–93: the effects of winter cover crops and unfertilized grass leys. *Plant and Soil* 203, 57–69.

Catt, J.A., Howse, K.R., Christian, D.G., Lane, P.W., Harris, G.L. and Goss, M.J. (1998b) Strategies to decrease nitrate leaching in the Brimstone Farm Experiment, Oxfordshire, UK, 1988–93: the effect of straw incorporation. *Journal of Agricultural Science, Cambridge* 131, 309–319.

Catt, J.A., Howse, K.R., Christian, D.G., Lane, P.W., Harris, G.L. and Goss, M.J. (2000) Assessment of tillage strategies to decrease nitrate leaching in the Brimstone Farm Experiment, Oxfordshire, UK. *Soil and Tillage Research* 53, 185–200.

Cerco, C. (2000) Chesapeake Bay eutrophication model. In: Hobbie, J.E. (ed.) *Estuarine Science – a Synthetic Approach to Research and Practice*. Island Press, Washington, DC, pp. 363–404.

Chambers, B.J., Smith, K.A. and Pain, B.F. (2000) Strategies to encourage better use of nitrogen in organic manures. *Soil Use and Management* 16, 157–161.

Chaney, K. (1990) Effect of nitrogen fertilizer rate on soil nitrogen content after harvesting winter wheat. *Journal of Agricultural Science, Cambridge* 114, 171–176.

Chisholm, S.W. and Morel, F.M.M. (eds) (1991) What controls phytoplankton production in nutrient-rich areas of the open sea? *Limnology and Ocean* 36, 1507–1966.

Choquette, K. (1980) Nitrates: groundwater. *Journal of the Iowa Medical Society* 70, 309–311.

Christian, D., Goss, M.J., Howse, K.R., Powlson, D.S. and Pepper, T.J. (1990) *Leaching of Nitrate Through Soil*. IACR Report for 1989. Lawes Agricultural Trust, Harpenden, UK, pp. 67–68.

Clarkson, D.T. and Warner, A.J. (1979) Relationships between root temperature and transport of ammonium and nitrate ions by Italian and perennial ryegrass (*Lolium multiflorum* and *Lolium perenne*). *Plant Physiology* 64, 557–561.

Cloern, J.E. (1999) The relative importance of light and nutrient limitation of phytoplankton growth: a simple index of coastal ecosystem sensitivity to nutrient enrichment. *Aquatic Ecology* 33, 3–16.

Coale, K.H., Johnson, K.S., Fitzwater, S.I., Gordon, R.M., Tanner, S., Chavez, F.P., Ferioli, L., Sakamoto, C., Rogers, P., Millero, F., Steinberg, P., Nightingale, P., Cooper, D., Cochlan, W.P., Landry, M.R., Constantinou, J., Rollwagen, G., Trasvina, A. and Kudela, R. (1996) A massive phytoplankton bloom induced by an ecosystem-scale iron fertilization experiment in the equatorial Pacific Ocean. *Nature* 383, 495–501.

Comly, H.H. (1945) Cyanosis in infants caused by nitrates in well water. *Journal of the American Medical Association* 129, 112–116.

Committee on Risk Assessment of Hazardous Air Pollutants (1994) *Science and Judgement in Risk Assessment*. National Academy Press, Washington, DC.

Conen, F., Dobbie, K.E. and Smith, K.A. (2000) Predicting N_2O emissions from agricultural land through related soil parameters. *Global Change Biology* 6, 417–426.

Conford, P. (2001) *The Origins of the Organic Movement*. Floris Books, Edinburgh.

Cooke, G.W. (1986) The intercontinental transport of plant nutrients. In: *Proceedings of the 13th Conference of The International Potash Institute*. International Potash, Reims, France, pp. 267–287.

Cornblath, M. and Hartmann, A.F. (1948) Methaemoglobinaemia in young infants. *Journal of Paediatrics* 33, 421–425.

Corre, M.D., Van Kessel, C. and Pennock, D.J. (1996) Landscape and seasonal patterns of nitrous oxide emissions in a semiarid region. *Soil Science Society of America Journal* 60, 1806–1815.

Cotgreave, P. and Forseth, I. (2002) *Introductory Ecology*. Blackwell Science, Oxford, UK.

Cox, D. and Addiscott, T.M. (1976) Sulphur-coated urea as a fertilizer for potatoes. *Journal of the Science of Food and Agriculture* 27, 1015–1020.

Craig, J.K., Crowder, L.B., Gray, C.D., McDaniel, C.J., Henwood, T.A. and Hanifen, J.G. (2001) Ecological effects of hypoxia on fish, sea turtles and marine mammals in the north-western Gulf of Mexico. In: Rabalais, N.N. and Turner, E.R. (eds) *Coastal Hypoxia: Consequences for Living Resources and Ecosystems*. American Geophysical Union, Washington, DC, pp. 269–292.

Croll, B.T. and Hayes, C.R. (1988) Nitrate and water supplies in the United Kingdom. *Environmental Pollution* 50, 163–187.

Culotta, E. and Koshland, D.E. Jr (1992) NO news is good news. *Science* 258, 1862–1865.

Currie, J.A. (1961) Gaseous diffusion in the aeration of aggregated soils. *Soil Science* 92, 40–45.

Davidson, E.A. (1991) Fluxes of nitrous oxide and nitric oxide from terrestrial ecosystems. In: Rogers, J.E. and Whitman, W.B. (eds) *Microbial Production and Consumption of Greenhouse Gases: Methane, Nitrous Oxide and Halomethanes*. American Society of Microbiology, Washington, DC, pp. 219–235.

Darwin, C. (1881) *The Formation of Vegetable Mould Through the Action of Worms, with Observations of their Habits*. Murray, London.

Dawson, F.H., Newman, J.R., Gravelle, M.J., Rouen, K.J. and Henville, P. (1999) Assessment of the trophic status of rivers using macrophytes. Evaluation of the mean trophic rank. *R and D Technical Report E39*. Environment Agency, Swindon, UK, 177 pp.

D'Elia, C.F., Sanders, J.G. and Boynton, W.R. (1986) Nutrient enrichment studies in a coastal plain estuary: phytoplankton growth in large-scale continuous cultures. *Canadian Journal of Fisheries and Aquatic Science* 43, 397–406.

De Groote, M.A., Testerman, T., Xu, Y., Stauffer, G. and Fang, F.C. (1996) Homocysteine antagonism of nitric oxide-related cystostasis in *Salmonella typhimurium*. *Science* 272, 414–417.

De Nobili, M., Contin, M., Mondini, C. and Brookes, P.C. (2001) Soil microbial biomass is triggered into activity by trace amounts of substrate. *Soil Biology and Biochemistry* 33, 1163–1170.

De Smedt, F., Wauters, F. and Sevilla, J. (1986) Studies of tracer movement through unsaturated sand. *Geoderma* 38, 223–236.

Dobbie, K. and Smith, K.A. (2003) Nitrous oxide emission factors for agricultural soils in Great Britain: the impact of water-filled pore space and other controlling variables. *Global Change Biology* 9, 208–213.

Dobbie, K., McTaggart, I.P. and Smith, K.A. (1999) Nitrous oxide emissions from intensive agricultural systems: variations between crops and seasons, key driving variables, and mean emission factors. *Journal of Geophysical Research* 104, 26,891–26,899.

Doering, P.H., Oviatt, C., Nowicki, B.L., Klos, E.G. and Reed, L.W. (1995) Phosphorus and nitrogen limitation of primary production in a simulated estuarine gradient. *Marine Ecology Progress Series* 124, 271–287.

Don, M. (2004) Strawberry fields for ever? *The Observer*, 16 May, p. 31.

Dougall, H.T., Smith, L., Duncan, C. and Benjamin, N. (1995) The effect of amoxycillin on salivary nitrite concentrations: an important mechanism of adverse reactions? *British Journal of Clinical Pharmacology* 39, 460–462.

Dowdell, R.J., Webster, C.P., Hill, D. and Mercer, E.R. (1984) A lysimeter study of the fate of nitrogen fertilizer in spring barley crops grown on shallow soil overlying chalk: crop uptake and leaching losses. *Journal of Soil Science* 35, 169–181.

Dudal, R. and Deckers, J. (1993) Soil organic matter in relation to soil productivity. In: Mulongoy, K. and Merckx, R. (eds) *Soil Organic Matter Dynamics and Sustainability of Tropical Agriculture*. John Wiley & Sons, Chichester, UK, pp. 377–380.

Duncan, C., Dougall, H., Johnston, P., Green, S., Brogan, R., Leifert, C., Smith, L., Golden, M. and Benjamin, N. (1995) Chemical generation of nitric oxide in the mouth from the enterosalivary circulation of dietary nitrate. *Nature Medicine* 1, 546–551.

Duthie, D.W. (1953) Crop responses to fertilizers and manures in East Africa. *East African Agriculture and Forestry Journal* 19, 19–57.

Duwig, C., Becquer, T., Charlet, L. and Clothier, B.E. (2003) Estimation of nitrate retention in a Ferrosol by a transient-flow method. *European Journal of Soil Science* 54, 505–515.

Dyke, G.V. (1991) *John Bennet Lawes: the Record of his Genius*. Research Studies Press, Taunton, UK.

Dyke, G.V. (1993) *John Lawes of Rothamsted: Pioneer of Science, Farming and Industry*. Hoos Press, Harpenden, UK.

Dykhuizen, R.S., Frazer, R., Duncan, C., Smith, C.C., Golden, M., Benjamin, N. and Leifert, C. (1996) Antimicrobial effect of acidified nitrite on gut pathogens: importance of dietary nitrate in host defense. *Antimicrobial Agents & Chemotherapy* 40, 1422–1425.

Dykhuizen, R.S., Fraser, A., McKenzie, H., Golden, M., Leifert, C. and Benjamin, N. (1998) *Helicobacter pylori* is killed by nitrite under acidic conditions [see comments]. *Gut* 42, 334–337.

Edwards, C.A. and Lofty, J.R. (1972) *Biology of Earthworms*. Chapman & Hall, London.

Elmgren, R. (2001) Understanding human impact on the Baltic: changing views in recent decades. *Ambio* 30, 222–231.

Elmgren, R. and Larsson, U. (2001) Nitrogen and the Baltic Sea: managing nitrogen in relation to phosphorus. In: *Optimizing Nitrogen Management in Food and Energy Production and Environmental Protection. Proceedings of the Second International Nitrogen Conference on Science and Policy*. Ecological Society of America, Washingtpn, DC, pp. 371–377.

Ewing, M.C. and Mayon-White, R.M. (1951) Cyanosis in infancy from nitrates in drinking water. *Lancet* 260, 931–934.

Falkowski, P.G., Barber, R.T. and Smetucek, V. (1998) Biogeochemical controls and feedbacks on ocean primary production. *Science* 281, 200–205.

Fang, F.C. (1997) Perspectives series: host/pathogen interactions. Mechanisms of nitric oxide-related antimicrobial activity. *Journal of Clinical Investigation* 99, 2818–2825.

FARM-Africa (2001) *Farmer Participatory Research in Southern Ethiopia. The Experiences of the Farmers' Research Project.* FARM-Africa, London.

Farrell, R.E., Sandercock, P.J., Pennock, D.J. and Van Kessel, C. (1996) Landscape-scale variations in leached nitrate: relationship to denitrification and natural nitrogen-15 abundance. *Soil Science Society of America Journal* 60, 1410–1415.

Ferguson, A.J.D., Pearson, M.J. and Reynolds, C.S. (1996) Eutrophication of natural waters and toxic algal blooms. In: Hester, R.E. and Harrison, R.M. (eds) *Agricultural Chemicals and the Environment.* Issues in Environmental Science and Technology No. 5. The Royal Society of Chemistry, Cambridge, UK, pp. 27–41.

Fertilizer Manufacturers Association (2002) *The Fertilizer Review 2002.* Fertilizer Manufacturers Association, Peterborough, UK, 7 pp.

Flemer, D., Mackiernan, G., Nehlsen, W. and Tippie, V. (1983) *Chesapeake Bay: a Profile of Environmental Change.* US Environmental Protection Agency, Chesapeake Bay Program Office, Annapolis, Maryland, 200 pp.

Foloronso, O.A. and Rolston, D.E. (1985) Spatial and spectral relationships between field-measured denitrification gas fluxes and soil properties. *Soil Science Society of America Journal* 49, 1087–1093.

Forman, D., Al-Dabbagh, A. and Doll, R. (1985) Nitrate, nitrite and gastric cancer in Great Britain. *Nature* 313, 620–625.

Fowler, D., Sutton, M.A., Skiba, U. and Hargreaves, K.J. (1996) Agricultural nitrogen and emissions to the atmosphere. In: Hester, R.E. and Harrison, R.M. (eds) *Agricultural Chemicals and the Environment.* Issues in Environmental Science and Technology No. 5. The Royal Society of Chemistry, Cambridge, UK, pp. 57–84.

Fowler, D., Flechard, C., Skiba, U., Coyle, M. and Cape, J.N. (1998) The atmospheric budget of oxidized nitrogen and its role in ozone formation and deposition. *New Phytologist* 139, 11–23.

Froelich, P.N. (1988) Kinetic control of dissolved phosphate in natural rivers and estuaries: a primer on the phosphate buffer mechanism. *Limnology and Oceanography* 33, 649–668.

Frost, B.W. (1996) Phytoplankton bloom on iron rations. *Nature* 383, 475–476.

Fumento, M. (1993) *Science under Siege.* William Morrow, New York.

Garner, H.V (1957) *Manures and Fertilizers.* MAFF Bulletin No. 36. HMSO, London.

Garner, H.V. and Dyke, G.V. (1969) The Broadbalk: yields. *Report of the Rothamsted Experimental Station for 1968, Part II.* Rothamsted Experimental Station, Harpenden, UK, pp. 26–49.

Gigerenzer, G. (2002) *Reckoning with Risk. Learning to Live with Uncertainty.* Penguin Books, London.

Giller, K.E., Cadisch, G., Ehaliotis, C., Adams, E., Sakala, W.D. and Mafongoya, P.L. (1997) Building soil nitrogen capital in Africa. In: Buresh, R.J., Sanchez, P.A. and Calhoun, F. (eds) *Replenishing Soil Fertility in Africa.* SSSA Special Publication No. 51. Soil Science Society of America, Madison, Wisconsin, pp. 151–192.

Glasstone, S. (1947) *Textbook of Physical Chemistry.* Macmillan, London.

Gleick, J. (1987) *Chaos.* Penguin, New York.

Glendining, M.J., Poulton, P.R. and Powlson, D.S. (1992) The relationship between inorganic N in the soil and the rate of fertiliser N applied on the Broadbalk wheat experiment. Nitrate in farming systems. *Aspects of Applied Biology* 30, 95–102.

Goolsby, D.A., Battaglin, W.A., Lawrence, G.B., Artz, R.S., Aulenbach, B.T., Hooper, K.P., Keeney, D.R. and Stensland, G.J. (1999) Flux and sources of nutrients in the Mississippi–Atchafalaya River Basin. *Topic 3, Report of the Integrated Assessment of Hypoxia in the Gulf of Mexico*. NOAA Coastal Ocean Decision Analysis Series No. 17. NOAA, Silver Springs, Maryland.

Goss, M.J., Howse, K.R., Lane, P.W., Christian, D.G. and Harris, G.L. (1993) Losses of nitrate-nitrogen to water draining from under autumn-sown crops established by direct drilling or mouldboard ploughing. *Journal of Soil Science* 44, 35–48.

Goss, M.J., Beauchamp, E.G. and Miller, M.H. (1995) Can a farming systems approach help minimize nitrogen losses to the environment? *Journal of Contaminant Hydrology* 3, 285–297.

Goulding, K.W.T. (1990) Nitrogen deposition to land from the atmosphere. *Soil Use and Management* 6, 61–63.

Goulding, K.W.T., Bailey, N.J., Bradbury, N.J., Hargreaves, P., Howe, M., Murphy, D.V., Poulton, P.R. and Willison, T.W. (1998) Nitrogen deposition and its contribution to nitrogen cycling and associated soil processes. *New Phytologist* 139, 49–58.

Goulding, K.W.T., Poulton, P.R., Webster, C.P. and Howe, M.T. (2000) Nitrogen leaching from the Broadbalk Wheat Experiment, Rothamsted, UK, as influenced by fertilizer and manure inputs and the weather. *Soil Use and Management* 16, 244–250.

Graneli, E., Wallstrom, K., Larsson, U., Graneli, W. and Elmgren, R. (1990) Nutrient limitation of primary production in the Baltic Sea area. *Ambio* 19, 142–151.

Granli, T. and Bøckman, O.C. (1994) Nitrous oxide from agriculture. *Norwegian Journal of Agricultural Sciences* Supplement No. 12, 1–128.

Green, L.C., Ruiz de Luzuriaga, K., Wagner, D.A., Rand, W., Istfan, N., Young, V.R. and Tannenbaum, S.R. (1981) Nitrate biosynthesis in man. *Proceedings of the National Academy of Sciences USA* 78, 7764–7768.

Greenland, D. (2000) Effects on soils and plant nutrition. In: Tinker, P.B. (ed.) *Shades of Green. A Review of UK Farming Systems*. Royal Agricultural Society of England, Stoneleigh, UK, pp. 6–20.

Greenwood, D.J. and Goodman, D. (1965) Oxygen diffusion and aerobic respiration in columns of fine soil crumbs. *Journal of the Science of Food and Agriculture* 16, 152–160.

Grennfelt, P. and Thornelof, E. (eds) (1992) *Critical Loads for Nitrogen. Nord 1992,* 41. Nordic Council of Ministers, Copenhagen.

Groffman, P. and Tiedje, J.M. (1989) Denitrification in northern temperate forest soils: spatial and temporal patterns at the seasonal and landscape scales. *Soil Biology and Biochemistry* 21, 613–620.

Hammond, D.E., Fuller, C., Harmon, D., Hartman, B., Korosec, M., Miller, G., Rea, R., Warren, S., Berelson, W. and Hager, S.W. (1985) Benthic fluxes in San Francisco Bay. In: Cloern, J.E. and Nichols, F.H. (eds) *Temporal Dynamics of an Estuary: San Francisco Bay*. Dr W. Junk Publishers, Boston, Massachusetts.

Harlin, M.H. (1995) Changes in major plant groups following nutrient enrichment. In: McComb, A.J. (ed.) *Eutrophic Shallow Estuaries and Lagoons*. CRC Press, Boca Raton, Florida, pp. 173–187.

Harris, G.L., Goss, M.J., Dowdell, R.J., Howse, K.R. and Morgan, P. (1984) A study of mole drainage with simplified cultivation for autumn-sown crops on a clay soil. 2. Soil water regimes, water balances and nutrient loss in drain water. *Journal of Agricultural Science, Cambridge* 102, 561–581.

Haycock, N.E., Burt, T.P., Goulding, K.W.T. and Pinay, G. (eds) (1997) *Buffer Zones: Their Processes and Potential in Water Protection*. Quest Environmental, Harpenden, UK.

Heckrath, G.J., Brookes, P.C., Poulton, P.R. and Goulding, K.W.T. (1995) Phosphorus leaching from soils containing different phosphorus concentrations in the Broadbalk Experiment.

Journal of Environmental Quality 24, 904–910.

Hecky, R.E. and Kilham, P. (1988) Nutrient limitation of phytoplankton in freshwater and marine environments: a review of the recent evidence on the effects of enrichment. *Limnology and Oceanography* 33, 796–822.

Hegesh, E. and Shiloah, J. (1982) Blood nitrates and infantile methemoglobinemia. *Clinica Chimica Acta* 125, 107–115.

Hesketh, N., Brookes, P.C. and Addiscott, T.M. (1998) Chlordane transport in a sandy soil: effects of suspended soil material and pig slurry. *European Journal of Soil Science* 49, 709–716.

Heuvelink, G.B.M. (1998) *Error Propagation in Environmental Modeling with GIS.* Taylor and Francis, London.

Hoosbeek, M.R. and Bryant, R.B. (1992) Towards the quantitative modelling of pedogenesis – a review. *Geoderma* 55, 183–210.

Hoskins, W.G. (1977) *The Making of the English Landscape.* Hodder and Stoughton, London.

Houghton, J. (1992) *Global Warming. The Complete Briefing.* Lion Publishing, Oxford.

Howarth, R.W., Anderson, D.M., Church, T.M., Greening, H., Hopkinson, C.S., Huber, W.C., Marcus, N., Naiman, R.J., Segerson, K., Sharpley, A. and Wiseman, W.J. (2000) *Clean Coastal Waters: Understanding and Reducing the Effects of Nutrient Pollution.* National Academy Press, Washington, DC.

Hye-Knudsen, P. (1984) Nitrate in drinking water and methaemoglobinaemia. *Ukeskrift for Laeger* 141, 51–53.

IPCC (1997) *Revised 1996 Guidelines for National Greenhouse Gas Inventories,* Vol. 2. Meteorological Office, Bracknell, UK.

ISO (1997) *International Standard 14040. Environmental Management – Life Cycle Assessment – Principles and Framework.* International Organization for Standardization, Geneva.

Jarvis, M.G. (1973) *Soils of the Wantage and Abingdon District.* Soil Survey of England and Wales, Harpenden, UK.

Jarvis, S.C. (1993) Nitrogen cycling and losses from dairy farms. *Soil Use and Management* 9, 99–105.

Jarvis, S.C. (1999) Accounting for nutrients in grassland: challenges and needs. In: Corrall A.J. (ed.) *Accounting for Nutrients.* British Grassland Symposium Occasional Symposium No. 33. British Grassland Association, UK, pp. 3–12.

Jarvis, S.C. and Ledgard, S. (2002) Ammonia emissions from intensive dairying: a comparison of contrasting systems in the United Kingdom and New Zealand. *Agriculture, Ecosystems and the Environment* 92, 83–92.

Jarvis, S.C., Barraclough, D., Williams, J. and Rook, A.J. (1991) Patterns of denitrification loss from grazed grasslands: effects of fertilizer inputs at different sites. *Plant and Soil* 131, 77–88.

Jenkinson, D.S. (1971) The accumulation of organic matter in soil left uncultivated. *Rothamsted Experimental Station Report for 1970, Part 2.* Rothamsted Experimental Station, Harpenden, UK, pp. 113–137.

Jenkinson, D.S. (1977) The soil biomass. *New Zealand Soil News* 25, 213–218.

Jenkinson, D.S. (1990) An introduction to the global nitrogen cycle. *Soil Use and Management* 6, 56–61.

Jenkinson, D.S. (1991) Rothamsted long-term experiments: are they still of use? *Agronomy Journal* 83, 1–10.

Jenkinson, D.S. (2001) The impact of humans on the nitrogen cycle, with focus on temperate arable agriculture. *Plant and Soil* 228, 3–15.

Jenkinson, D.S. and Powlson, D.S. (1976a) The effects of biocidal treatments on metabolism in soil. 1. Fumigation with chloroform. *Soil Biology and Biochemistry* 8, 167–177.

Jenkinson, D.S. and Powlson, D.S. (1976b) The effects of biocidal treatments on metabolism in soil. V. A method for measuring soil biomass. *Soil Biology and Biochemistry* 8, 209–213.

Jenkinson, D.S., Fox, R.H. and Rayner, J.H. (1985) Interactions between fertilizer nitrogen and soil nitrogen – the so-called 'priming' effect. *Journal of Soil Science* 36, 425–444.

Johnson, C.J., Bonrud, P.A., Dosch, T.A., Kilness, A.W., Serger, K.A., Busch, D.C. and Meyer, M.R. (1987) Fatal outcome of methaemoglobinaemia in an infant. *Journal of the American Medical Association* 257, 2796–2797.

Johnson, D.L. and Watson-Stegner, D. (1987) Evolution model of pedogenesis. *Soil Science* 143, 349–366.

Johnston, A.E. and Garner, H.V. (1969) Broadbalk: historical introduction. *Report of the Rothamsted Experimental Station for 1968, Part II.* Rothamsted Experimental Station, Harpenden, UK, pp. 12–25.

Jones, K.C., Johnston, A.E. and McGrath, S.P. (1994) Historical monitoring of organic contaminants in soils. In: Leigh, R.A. and Johnston, A.E. (eds) *Long-term Experiments in Agricultural and Ecological Sciences.* CAB International, Wallingford, UK, pp. 147–163.

Jury, W.A. (1982) Simulation of solute transport using a transfer function model. *Water Resources Research* 18, 363–368.

Justic, D., Rabalais, N.N. and Turner, R.E. (1995) Stoichiometric nutrient balance and origin of coastal eutrophication. *Marine Pollution Bulletin* 30, 31–46.

Kahn, T., Bosch, J., Levitt, M.F. and Goldstein, M.H. (1975) Effect of sodium nitrate loading on electrolyte transport by the renal tubule. *American Journal of Physiology* 229, 746–753.

Katchalsky, A. and Curran, P.F. (1967) *Non-equilibrium Thermodynamics in Biophysics.* Harvard University Press, Cambridge, Massachusetts.

Kelliher, F.M., Reisinger, A.R., Martin, R.J., Harvey, M.J., Price, S.J. and Sherlock, R.R. (2002) Measuring nitrous oxide transmission rate from grazed pasture using Fourier-transform infrared spectroscopy in the nocturnal boundary layer. *Agricultural and Forest Meteorology* 111, 29–38.

Koning, N. (2002) *Should Africa Protect its Farmers to Revitalise its Economy?* Gatekeeper Series No. 105. International Institute for Environment and Development, London.

Kooijman, S.A.L.M. and Bedaux, J.J.M. (1996) *The Analysis of Aquatic Toxicity Data.* VU University Press, Amsterdam.

Korom, S.F. (1992) Natural denitrification in the saturated zone: a review. *Water Resources Research* 28, 1657–1668.

Kuo, C., Lindberg, C. and Thomson, D.J. (1990) Coherence established between atmospheric carbon dioxide and global temperature. *Nature* 343, 709–715.

Lægreid, M., Bøckman, O.C. and Kaarstad, O. (1999) *Agriculture, Fertilizers and the Environment.* CAB International, Wallingford, UK.

Lark, R.M. and Webster, R. (1999) Analysis and elucidation of soil variation using wavelets. *European Journal of Soil Science* 50, 185–206.

Lark, R.M., Milne, A.E., Addiscott, T.M., Goulding, K.W.T., Webster, C.P. and O'Flaherty. S. (2004a) Analysing spatially intermittent variation of nitrous oxide emissions from soil with wavelets and the implications for sampling. *European Journal of Soil Science* 55, 601–610.

Lark, R.M., Milne, A.E., Addiscott, T.M., Goulding, K.W.T., Webster, C.P. and O'Flaherty, S. (2004b) Scale- and location-dependent correlation of nitrous oxide emissions with soil properties: an analysis using wavelets. *European Journal of Soil Science* 55, 611–627.

Lawes, J.B., Gilbert, J.H. and Warington, R. (1881) On the amount and composition of the rainfall at Rothamsted. *Journal of the Royal Agricultural Society of England, 2nd Series* 17, 241–279.

Lawes, J.B., Gilbert, J.H. and Warington, R. (1882) On the amount and composition of drainage water collected at Rothamsted. The quantity of nitrogen lost by drainage. *Journal of the Royal Agricultural Society of England, 2nd Series* 18, 43–71.

Laws, J.A., Pain, B.F., Jarvis, S.C. and Schofield, D. (2000) Comparison of grassland management systems for beef cattle using self-contained farmlets: effects of contrasting nitrogen inputs and management strategies on nitrogen budgets, and herbage and animal production. *Agriculture, Ecosystems and Environment* 80, 243–254.

L'hirondel, J. and L'hirondel, J.-L. (2001) *Nitrate and Man. Toxic, Harmless or Beneficial?* CAB International, Wallingford, UK.

Loague, K. and Green, R.E. (1991) Statistical and graphical methods for evaluating solute transport models: overview and application. *Journal of Contaminant Hydrology* 7, 51–73.

Lomborg, B. (2001) *The Skeptical Environmentalist: Measuring the Real State of the World.* Cambridge University Press, Cambridge, UK.

Lovelock, J. (1995a) *Gaia. A New Look at Life on Earth.* Oxford Paperbacks, Oxford, UK.

Lovelock, J. (1995b) *The Ages of Gaia. A Biography of Our Living Earth.* Oxford Paperbacks, Oxford, UK.

Lowrison, G.C. (1989) *Fertilizer Technology.* Ellis Horwood Series in Applied Science and Industrial Technology, Ellis Horwood Ltd, Chichester, UK.

Lund, P. (1991) Characterization of alternatively produced milk. *Milchwissenschaft* 46, 166–169.

Lundberg, J.O.N., Weitzberg, E., Lundberg, J.M. and Alving, K. (1994) Intragastric nitric oxide production in humans: measurements in expelled air. *Gut* 35, 1543–1546.

Lunt, H.A. and Jacobson, G.M. (1944) The chemical composition of earthworm casts. *Soil Science* 58, 367.

MacDonald, A.J., Powlson, D.S., Poulton, P.R. and Jenkinson, D.S. (1989) Unused fertilizer nitrogen and its contribution to nitrate leaching. *Journal of the Science of Food and Agriculture* 46, 407–419.

McKinney, P.A., Paslow, R. and Bodansky, H.J. (1999) Nitrate exposure and childhood diabetes. In: Wilson, W.S., Ball, A.S. and Hinton, R.H. (eds) *Managing the Risks of Nitrates to Humans and the Environment.* The Royal Society of Chemistry, Cambridge, UK, pp. 327–339.

McKnight, G.M., Smith, L.M., Drummond, R.S., Duncan, C.W., Golden, M. and Benjamin, N. (1997) Chemical synthesis of nitric oxide in the stomach from dietary nitrate in humans. *Gut* 40, 211–214.

MAFF (2000) *Fertiliser Recommendations for Agricultural and Horticultural Crops (RB209),* 7th edn. The Stationery Office, London.

Magee, B. (1985) *Popper.* Fontana, London.

Marks, H.F. (1989) In: Britton, D.K. (ed.) *A Hundred Years of British Food and Farming. A Statistical Survey.* Taylor and Francis, London.

Martin, J.H. and Fitzwalter, S.E. (1988) Iron deficiency limits phytoplankton growth in the north-east Pacific subarctic. *Nature* 331, 341–343.

Martin, J.H., Coate, K.H., Johnson, K.S., Fitzwalter, S.E., Gordon, R.M., Tanner, S.J., Hunter, C.N., Elrod, V.A., Nowicki, J.H., Coley, T.L., Barber, R.T., Lindley, S., Watson, A.J., Van Scoy, K., Law, C.S., Liddlcoat, M.I., Ling, R., Stanton, T., Stockel, J., Collins, C., Anderson, A., Bidigare, R., Ondrusek, M., Latasa, M., Millero, F.J., Lee, K., Yoa, W., Zhang, J.Z., Friederich, G., Sacamoto, C., Chavez, F., Buck, K., Kolber, Z., Greene, R., Falkowske, P., Chisholm, S.W., Hoge, F., Swift, R., Yungel, J., Turner, J., Nightingale, P., Hatton, A., Liss, P. and Tindale, N.W. (1994) Testing the iron hypothesis in ecosystems of the equatorial Pacific Ocean. *Nature* 371, 123–129.

Medawar, P.B. (1969) *The Art of the Soluble.* Pelican Books, London.

Mendez, S.L.S., Allaker, R.P., Hardie, J.M. and Benjamin, N. (1999) Antimicrobial effect of acidified nitrite on cariogenic bacteria. *Oral Microbiolial Immunolology* 14, 391–392.

Mendum, T.A., Sockett, R.E. and Hirsch, P.R. (1999) Use of molecular and isotopic techniques to monitor the response of autotrophic ammonia-oxidizing populations of the β subdivision of the class proteobacteria in arable soils to nitrogen fertilizer. *Applied and Environmental Microbiology* 65, 4155–4162.

Mills, E. (1989) *Biological Oceanography – An Early History, 1870–1960.* Cornell University Press, Ithaca, New York, 378 pp.

Milne, A.E., Lark, R.M., Addiscott, T.M., Goulding, K.W.T., Webster, C.P. and O'Flaherty, S. (2004) Scale- and location-dependent correlation of nitrous oxide emissions soil properties: an analysis using wavelets. *European Journal of Soil Science* 55, 611–627.

Mitchell, H.H., Shonle, H.A. and Grindley, H.S. (1916) The origin of the nitrates in the urine. *Journal of Biological Chemistry* 24, 461–490.

Moelwyn-Hughes, E.A. (1957) *Physical Chemistry.* Pergamon Press, London.

Morowitz, H.J. (1970) *Entropy for Biologists.* Academic Press, New York.

Moss, B. (1996) A land awash with nutrients – the problem of eutrophication. *Chemistry and Industry* 3 June, pp. 407–411.

Müller, H.E. (1997) The risks of dioxin to human health. In: Bate, R. (ed.) *What Risk?* Butterworth-Heinemann, Oxford, UK, pp. 201–217.

Murphy, D.V., Fillery, I.R.P. and Sparling, G.P. (1998) Seasonal fluctuations in gross N mineralization, ammonium consumption, and microbial biomass in a Western Australian soil under different land uses. *Australian Journal of Agricultural Research* 49, 523–535.

Murphy, D.V., Bhogal, A., Shepherd, M., Goulding, K.W.T., Jarvis, S.C., Barraclough, D. and Gaunt, J.L. (1999) Comparison of ^{15}N-labelling methods to measure gross nitrogen mineralization. *Soil Biology and Biochemistry* 31, 2015–2024.

Newell, R.I.E. (1988) Ecological changes in Chesapeake Bay: are they the result of over-harvesting the American oyster, *Crassostrea virginica*? In: *Understanding the Estuary: Advances in Chesapeake Bay Research. Proceedings of a Conference.* Chesapeake Research Consortium Publication 129, CBP/TRS 24/88. Chesapeake Research Consortium, Baltimore, Maryland.

Nichols, F.H. (1979) Natural and anthropogenic influences on benthic community structure in San Francisco Bay. In: Conomos, T.S. (ed.) *San Francisco Bay, The Urbanized Estuary.* Pacific Division, AAAS, California Academy of Science, San Francisco, California, pp. 409–426.

Nixon, S.W. (1995) Coastal marine eutrophication: a definition, social causes and future concerns. *Ophelia, International Journal of Marine Biology* 41, 199–219.

North, R.D. (2000) *Risk: the Human Choice.* ESEF, Barton (Cambridge).

Nyamangara, J. (2001) Nitrogen leaching and recovery studies in a sandy soil amended with cattle manure and inorganic fertilizer N under high-rainfall conditions. DPhil thesis, University of Zimbabwe, Harare.

Nye, P.H. (1955) Some soil-forming processes in the humid tropics. IV. The action of soil fauna. *Journal of Soil Science* 6, 78.

Nye, P.H. and Greenland, D.J. (1960) *The Soil under Shifting Cultivation.* Technical Communication 51. Commonwealth Agricultural Bureau, Farnham Royal, UK.

OECD (1995) *Report of the OECD Workshop on Environmental Hazard/Risk Assessment, London, UK, 24–27 May 1994.* OECD Monograph, London.

Officer, C.B., Smayada, S.J. and Mann, R. (1982) Benthic filter feeding: a natural eutrophication control. *Marine Ecology Progress Series* 9, 203–210.

Onsager, L. (1931) Reciprocal relations in irreversible processes, I and II. *Physical Reviews* 37, 405–426; 38, 2265–2279.

Ormerod, P. (1994) *The Death of Economics.* Faber and Faber, London.

Orth, R.J. and Moore, K.A. (1983) Chesapeake Bay: an unprecedented decline in submerged aquatic vegetation. *Science* 22, 51–52.

Oviatt, C., Doering, P., Nowicki, L., Reed, L., Cole, J. and Frithsden, J. (1995) An ecosystem level experiment on nutrient limitation in temperate coastal marine environment. *Marine Ecology Progress Series* 116, 171–179.

Parkin, T.B. (1987) Soil microsites as a source of denitrification variability. *Soil Science Society of America Journal* 51, 1194–1199.

Parkinson, R.J., Griffiths, P. and Heathwaite, A.L. (2000) Transport of nitrogen in soil water following the application of animal manures to sloping grassland. *Hydrological Sciences – Journal des Sciences Hydrologiques* 45, 61–73.

Pearce, F. (2002) Botched botany. *New Scientist* 26 January, p. 11.

Penny, A., Addiscott, T.M. and Widdowson, F.V. (1984) Assessing the need of maincrop potatoes for late nitrogen by using isobutylidene diurea, by injecting nitrification inhibitors with aqueous nitrogen fertilizers, and by dividing dressings of 'Nitro-chalk'. *Journal of Agricultural Science, Cambridge* 103, 577–585.

Perakis, S.S. and Hedin, L.O. (2002) Nitrogen from unpolluted South American forests mainly via dissolved organic compounds. *Nature* 415, 416–419.

Persson, J. and Nasholm, T. (2001) Amino-acid uptake: a widespread ability among boreal forest plants. *Ecology Letters* 4, 434–438.

Pollock, J. (2004) DDT: the story of a scandal that has killed millions. *The Times*, 1 May, p. 30.

Popper, K.R. (1959) *The Logic of Scientific Discovery*. Hutchinson and Co., London.

Popper, K.R. (1962) *The Open Society and Its Enemies*. Routledge and Kegan Paul, London.

Powlson, D.S. (1994) Quantification of nutrient cycles. In: Leigh, R.A. and Johnston, A.E. (eds) *Long-term Experiments in Agricultural and Ecological Sciences*. CAB International, Wallingford, UK, pp. 95–115.

Powlson, D.S., Pruden, G., Johnston, A.E. and Jenkinson, D.S. (1986) The nitrogen cycle of the Broadbalk wheat experiment: recoveries and losses of ^{15}N-labelled fertilizer applied in spring and impact of nitrogen from the atmosphere. *Journal of Agricultural Science, Cambridge* 107, 591–609.

Powlson, D.S., Hart, P.B.S., Poulton, P.R., Johnston, A.E. and Jenkinson, D.S. (1992) Influence of soil type, crop management and weather on the recovery of ^{15}N-labelled fertilizer applied to winter wheat in spring. *Journal of Agricultural Science, Cambridge* 118, 83–100.

Pretty, J., Brett, C., Gee, D., Hine, R., Mason, C.F., Morison, J.I.L., Raven, H., Rayment, N. and van der Bijl, G. (2000) An assessment of the total external costs of UK agriculture. *Agricultural Systems* 65, 113–136.

Prigogine, I. (1947) *Étude Thermodynamique des Processus Irreversibles*. Desoer, Liège.

Prins, W.H., Dilz, K. and Neeteson, J.J. (1988) Current recommendations for nitrogen fertilization within the EEC in relation to nitrate leaching. *Proceedings of the Fertilizer Society* 276, 27 pp.

Purseglove, J. (1988) *Taming the Flood.* Oxford University Press, Oxford, UK.

Quinton, J.N. and Catt, J.A. (2004) The effects of minimum tillage and contour cultivation on surface runoff, soil loss and crop yields in the long-term Woburn Erosion Reference Experiment on sandy soil at Woburn, England. *Soil Use and Management* 20, 343–349.

Rabalais, N.N. and Turner, R.E. (2001) Hypoxia in the northern Gulf of Mexico: description, causes and change. In: Rabalais, N.N. and Turner, R.E. (eds) *Coastal Hypoxia: Consequences for Living Resources and Ecosystems.* American Geophysical Union, Washington, DC, pp. 1–36.

Rabalais, N.N., Smith, L.E., Harper, D.E. and Justic, D. (2001) Effects of seasonal hypoxia on continental shelf benthos. In: Rabalais, N.N. and Turner, R.E. (eds) *Coastal Hypoxia: Consequences for Living Resources and Ecosystems.* American Geophysical Union, Washington, DC, pp. 211–240.

Rasband, S.N. (1990) *Chaotic Dynamics of Nonlinear Systems*. Wiley, New York.

Recous, S., Fresneau, C., Faurie, G. and Mary, B. (1988) The fate of ^{15}N-labelled urea and ammonium nitrate applied to a winter wheat crop: II. Plant uptake and N efficiency. *Plant and Soil* 112, 215–224.

Rochette, P., Chantigny, M.H., Angers, D.A., Bertrand, N. and Cote, D. (2001) Ammonia volatilization and soil nitrogen dynamics following fall application of pig slurry on canola crop residues. *Canadian Journal of Soil Science* 81, 515–523.

Rolston, D.E., Rao, P.S.C. and Davidson, J.M. (1984) Simulation of denitrification losses of nitrate fertilizer applied to uncropped, cropped and manure-amended soils. *Soil Science* 137, 270–279.

Rosenburg, R., Agrenius, S., Hellman, B., Nilsson, H.C. and Norling, D.K. (2002) Recovery of marine benthic habitats and fauna in a Swedish fjord following improved oxygen conditions. *Marine Ecology Progress Series* 234, 43–53.

Ross, C.A. and Jarvis, S.C. (2001) Measurement of emission and deposition patterns of ammonia from urine in grass swards. *Atmospheric Environment* 35, 867–875.

Ryden, J.C., Ball, P.R. and Garwood, E.A. (1984) Nitrate leaching from grassland. *Nature (London)* 311, 50.

Sanchez, P.A., Shepherd, K.D., Soule, M.J., Place, F.M., Buresh, R.J., Izac, A.N., Mokwunye, A.U., Kwesiga, F.R., Ndiritu, C.G. and Woolmer, P.W. (1997) Soil fertility replenishment in Africa: an investment in natural resource *capital.* In: Buresh, R.J., Sanchez, P.A. and Calhoun, F. (eds) *Replenishing Soil Fertility in Africa.* SSSA Special Publication No. 51. Soil Science Society of America, Madison, Wisconsin, pp. 1–46.

Sasaki, T. and Matano, K. (1979) Formation of nitrite from nitrate at the dorsum linguae. *Journal of the Food Hygiene Society of Japan* 20, 363–369.

Scaife, M.A. (1968) Maize fertilizer experiments in Western Tanzania. *Journal of Agricultural Science* 70, 209–222.

Schindler, D.W. (1974) Eutrophication and recovery in experimental lakes: implications for lake management. *Science* 184, 897–899.

Scholefield, D., Tyson, K.C., Garwood, E.A., Armstrong, A.C., Hawkins, J. and Stone, A.C. (1993) Nitrate leaching from grazed grassland lysimeters; effects of fertilizer input, field drainage, age of sward and pattern of weather. *Journal of Soil Science* 44, 601–613.

Schroth, G., Rodrigues, M.R.L. and D'Angelo, S.A. (2000) Spatial patterns of nitrogen mineralization, fertilizer distribution and roots explain nitrate leaching from mature Amazonian oil palm plantation. *Soil Use and Management* 16, 222–229.

Sen Gupta, B.K., Turner, R.E. and Rabalais, N.N. (1996) Seasonal oxygen depletion in continental shelf waters of Louisiana: historical record of benthic foraminifers. *Geology* 24, 227–230.

SETAC (1991) *A Technical Framework for Life Cycle Assessments.* Society of Environmental Toxicology and Chemistry, Washington, DC.

Sharpley, A.N., Chapra, S.C., Wedepohl, R., Sims, J.T., Daniel, T.C. and Reddy, K.R. (1994) Managing agricultural phosphorus for protection of surface waters. *Journal of Environmental Quality* 23, 437–451.

Smaling, E.M.A., Nandwa, S.M. and Janssen, B.H. (1997) Soil fertility in Africa is at stake. In: Buresh, R.J., Sanchez, P.A. and Calhoun, F. (eds) *Replenishing Soil Fertility in Africa.* SSSA Special Publication No. 51. Soil Science Society of America, Madison, Wisconsin, pp. 47–61.

Smetacek, V., von Bodingen, B., Knoppers, B., Pollehne, F. and Zeitzschel, B. (1982) The plankton tower. IV. Interactions between water column and sediment in enclosures experiments in Kiel Bight. In: Grice, G.D. and Reeve, M.R. (eds) *Marine Mesocosms – Biological and Chemical Research in Experimental Ecosystems.* Springer-Verlag, New York, pp. 205–216.

Smil, V. (2001) *Enriching the Earth: Fritz Haber, Carl Bosch and the Transformation of World Food Production.* The MIT Press, Cambridge, Massachusetts.

Smith, J.U., Bradbury, N.J. and Addiscott, T.M. (1996a) SUNDIAL: a PC-based system for simulating nitrogen dynamics in arable land. *Agronomy Journal* 88, 38–43.

Smith, J.U., Smith, P. and Addiscott, T.M. (1996b) Quantitative methods to evaluate and compare soil organic matter (SOM) models. In: Powlson, D.S., Smith, P. and Smith, J.U. (eds) *Evaluation of Soil Organic Matter Models.* NATO ASI Series Vol. 138. Springer Verlag, Heidelberg, pp. 181–199.

Smith, J.U., Glendining, M.J. and Smith, P. (1997) The use of computer simulation models to optimise the use of nitrogen in whole farm systems. Optimising cereal inputs: its scientific basis. *Aspects of Applied Biology* 50, 147–154.

Smith, K.A. (1980) A model for the extent of anaerobic zones in aggregated soils, and its potential application to estimates of denitrification. *Journal of Soil Science* 31, 263–277.

Smith, K.A. and Dobbie, K.E. (2001) The impact of sampling frequency and sampling times on chamber-based measurements of N_2O emissions from fertilized soils. *Global Change Biology* 7, 933–945.

Smith, K.A., Elmes, A.E., Howard, R.S. and Franklin, M.F. (1984) The uptake of soil and fertilizer nitrogen by barley growing under Scottish climatic conditions. *Plant and Soil* 76, 49–57.

Smith, K.A., Thomson, P.E., Clayton, H., McTaggart, I.P. and Conen, F. (1998) Effects of temperature, water content and nitrogen fertilization on emissions of nitrous oxide by soils. *Atmospheric Environment* 32, 3301–3309.

Smith, J.W. (2001) Distribution of catch in the Gulf menhaden, *Breevortia patronus*, purse seine fishery in the northern Gulf of Mexico from logbook information: are there relationships to the hypoxic zone? In: Rabalais, N.N. and Turner, R.E. (eds) *Coastal Hypoxia: Consequences for Living Resources and Ecosystems.* American Geophysical Union, Washington, DC, pp. 311–320.

Smith, V.H. (1998) Cultural eutrophication of inland, estuarine and coastal waters. In: Pace, M.L. and Groffman, P.M. (eds) *Successes, Limitations and Frontiers in Ecosystem Science.* Springer-Verlag, New York, pp. 7–49.

Smithson, J.B., Edje, O.T. and Giller, K. (1993) Diagnosis and correction of soil nutrient problems of common bean (*Phaseolus vulgaris*) in the Usambara mountains of Tanzania. *Journal of Agricultural Science, Cambridge* 120, 233–240.

Stanford, G. and Smith, S.J. (1972) Nitrogen mineralization potentials of soils. *Soil Science Society of America Proceedings* 36, 462–472.

Stein, A., Staritsky, J., Bouma, J., van Eijnsbergen, A.C. and Bregt, A.K. (1992) Simulation of moisture deficits and areal interpolation by universal cokriging. *Water Resources Research* 27, 1963–1973.

Stewart, I. (1995) *Nature's Numbers.* Weidenfeld and Nicholson, London.

Stiglitz, J.E. (2002) *Globalization and Its Discontents.* Penguin Books, London.

Stockdale, E.A., Lampkin, N.H., Hovi, M., Keatinge, R., Lennartsson, E.K.M., Macdonald, D.W., Padel, S., Tattersall, F.H., Wolfe, M.S. and Watson, C.A. (2001) Agronomic and environmental implications of organic farming systems. *Advances in Agronomy* 70, 261–327.

Stout, W.L., Weaver, S.R., Gburek, W.J., Folmar, G.J. and Schnabel, R.R. (2000) Water quality implications of dairy slurry applied to cut pastures in the northeast USA. *Soil Use and Management* 16, 189–193.

Sylvester-Bradley, R., Addiscott, T.M., Vaidyanathan, L.V., Murray, A.W.A. and Whitmore, A.P. (1987) Nitrogen advice for cereals: present realities and future possibilities. *Proceedings of the Fertilizer Society* 363, 36 pp.

Tabatabai, M.A. and Al-Khafaji, A.A. (1980) Comparison of nitrogen and sulphur mineralization in soils. *Science Society of America Journal* 44, 1000–1006.

Tannenbaum, S.R. (1987) Endogenous formation of *N*-nitroso compounds: a current perspective. In: Bartsch, H., O'Neill, I.K. and Schulte-Hermann, R. (eds) *Relevance of N-Nitroso Compounds to Human Cancer: Exposure and Mechanisms.* IARC Scientific Publication No. 84. IARC, Lyon, pp. 292–298.

ten Berge, H.F.M., van der Meer, H.M., Carlier, L., Hofman, T.B. and Neeteson J.J. (2002) Limits to nitrogen use on grassland. *Environmental Pollution* 118, 225–238.

Thurston, J. (1958) Geescroft wilderness. *Rothamsted Experimental Station Report for 1957.* Rothamsted Experimental Station, Harpenden, p. 94.

Tren, R. and Bate, R. (2001) *Malaria and the DDT Story.* IEA, London.

Trewavas, A. (2004) A critical assessment of organic farming-and-food assertions with particular respect to the UK and the potential environmental benefits of no-till agriculture. *Crop Protection* 23, 757–781.

Trewavas, A.J. and Stewart, D. (2003) Paradoxical effect of chemicals in our diet on health. *Current Opinions in Plant Biology* 6, 185–191.

Tyler, K.B., Broadbent, F.E. and Hill, G.N. (1959) Low temperature effects on nitrification in four California soils. *Soil Science* 87, 123–129.

United Kingdom Review Group on Impacts of Atmospheric Nitrogen (1994) *Impacts of Nitrogen Deposition in Terrestrial Ecosystems.* DOE, London.

van Breemen, N. (2002) Natural organic tendency. *Nature* 415, 381–382.

van Burg, P.F.J., Prins, W.H., den Boer, D.J. and Sluiman, W.J. (1981) Nitrogen and intensification of livestock farming in EEC countries. *Proceedings of the Fertilizer Society* 199, 78 pp.

Vance, E.D., Brookes, P.C. and Jenkinson, D.S. (1987) An extraction method for measuring soil microbial C. *Soil Biology and Biochemistry* 19, 703–707.

van Genuchten, R. and Wierenga, P. (1976) Mass transfer studies in porous sorbing media. I. Analytical solutions. *Soil Science Society of America Journal* 40, 473–480.

van Loon, A.J., Botterweck, A.A., Goldbohm, R.A., Brants, H.A., van Klaveren, J.D. and van den Brandt, P.A. (1998) Intake of nitrate and nitrite and the risk of gastric cancer: a prospective cohort study. *British Journal of Cancer* 78, 129–135.

van Maanen, J.M.S., Albering, H.J., van Breda, S.G.J., Amberger, A.W., Wolffenbutel, B.H.R., Kleinjans, J.C.S. and Reeser, H.M. (1999) Nitrate in drinking water and risk of childhood diabetes in The Netherlands. *Diabetes Care* 22, 1750.

Velthof, G.L., van Groeningen, J.W., Gebauer, G., Pietzrak, P., Pinto, M., Corre, W. and Oenema, O. (2000) Temporal stability of patterns of nitrous oxide emission from sloping grassland. *Journal of Environmental Quality* 29, 1397–1407.

Velthof, G.L., Kuikman, P.J. and Oenema, O. (2003) Nitrous oxide emission from animal manures applied to soil under controlled conditions. *Biology and Fertility of Soils* 37, 221–230.

Vitousek, P.M., Aber, J., Howarth, R.W., Likens, G.E., Matson, P.A., Schindler, D.W., Schlesinger, W.H. and Tilman, G.D. (1997) *Human Alteration of the Global Nitrogen Cycle: Causes and Consequences.* Issues in Ecology 1. Ecological Society of America, Washington, DC.

Wagenet, R.J. (1983) Principles of salt movement in soils. In: Nelson, D.W., Elrick, D.W. and Tanji, K.K. (eds) *Chemical Mobility and Reactivity in Soil Systems.* Special Publication 11. American Society of Agronomy, Madison, Wisconsin.

Wagenet, R.J. (1998) Scale issues in agroecological research chains. *Nutrient Cycling in Agroecosystems* 50, 23–34.

Wagenet, R.J. and Addiscott, T.M. (1987) Estimating the variability of unsaturated hydraulic conductivity using simple equations. *Soil Science Society of America Journal* 51, 42–47.

Waldrop, M.M. (1993) *Complexity: the Emerging Science at the Edge of Order and Chaos.* Viking, London.

Walton, G. (1951) Survey of literature relating to infant methemoglobinemia due to nitrate-contaminated water. *American Journal of Public Health* 41, 986–996.

Weast, R.C. (ed.) (1964) *Handbook of Chemistry and Physics*, 45th edn. Chemical Rubber Co., Cleveland, Ohio, pp. B148–B225.

Webster, C.P. and Goulding, K.W.T. (1989) Influence of soil carbon content on denitrification from fallow land during autumn. *Journal of the Science of Food and Agriculture* 49, 131–142.

Weller, R., Pattullo, S., Smith, L., Golden, M., Ormerod, A. and Benjamin, N. (1996) Nitric oxide is generated on the skin surface by reduction of sweat nitrate. *Journal of Investigative Dermatology* 107, 327–331.

Weller, R., Ormerod, A.D., Hobson, R.P. and Benjamin, N.J. (1998) A randomized trial of acidified nitrite cream in the treatment of tinea pedis. *Journal of the American Academy of Dermatology* 38, 559–563.

Wetzel, R.G. (2001) *Limnology: Lake and River Ecosystems*, 3rd edn. Academic Press, New York.

White, R.E. (1997) *Principles and Practice of Soil Science. The Soil as a Natural Resource*, 3rd edn. Blackwell Science, Oxford, UK.

Whitmore, A.P. (1991) A method for assessing the goodness of computer simulation of soil processes. *Journal of Soil Science* 42, 289–299.

Whitmore, A.P., Addiscott, T.M., Webster, R. and Thomas, V.H. (1983) Spatial variation of soil nitrogen and related factors. *Journal of the Science of Food and Agriculture* 34, 268–269.

Whitmore, A.P., Bradbury, N.J. and Johnston, P.A. (1992) Potential contribution of ploughed grassland to nitrate leaching. *Agriculture, Ecosystems and Environment* 39, 221–233.

Widdowson, F.W., Penny, A., Darby, R.J., Bird, E. and Hewitt, M.V. (1987) Amounts of NO_3-N and NH_4-N in soil, from autumn to spring, under winter wheat and their relation to soil type, sowing date, previous crop and N uptake at Rothamsted, Woburn and Saxmundham, 1979–1985. *Journal of Agricultural Science, Cambridge* 108, 73–95.

Williams, L.E. and Miller, A.J. (2001) Transporters responsible for the uptake and partitioning of nitrogenous solutes. *Annual Review of Plant Physiology and Plant Molecular Biology* 52, 659–688.

Wilson, W.S., Ball, A.S. and Hinton, R.H. (eds) (1999) *Managing the Risks of Nitrates to Humans and the Environment*. The Royal Society of Chemistry, Cambridge, UK.

Woese, K., Lange, D., Boess, C. and Bögl, K.W. (1995) Producte des öcological Landbaus. Eine Zusammenfassung von Unterschungen zur Qualität dieser lebensmittel. Teil II. *Bundesgesundheitblatt* 38, 265–273.

Wong, M.F.T., Wild, A. and Juo, A.S.R. (1987) Retarded leaching of nitrate measured in monolith lysimeters in south-east Nigeria. *Journal of Soil Science* 38, 511–518.

Wu, J.J. and Babcock, B.A. (1999) Metamodelling potential nitrate water pollution in the central United States. *Journal of Environmental Quality* 28, 1916–1928.

Wu, J., Sunda, W., Boyle, E.A. and Karl, D.M. (2000) Phosphate depletion in the western North Atlantic Ocean. *Science* 289, 759–762.

Yamulki, S., Jarvis, S.C. and Owen, P. (1998) Nitrous oxide emissions from excreta applied in a simulated grazing pattern. *Soil Biology and Biochemistry* 30, 491–500.

Young, C.P., Oakes, D.B. and Wilkinson, W.B. (1976) Prediction of future nitrate concentrations in groundwater. *Groundwater* 14, 426–438.

Zimmerman, R.J. and Nance, J.M. (2001) Effects of hypoxia on the shrimp fishery of Louisiana and Texas. In: Rabalais, N.N. and Turner, R.E. (eds) *Coastal Hypoxia: Consequences for Living Resources and Ecosystems.* American Geophysical Union, Washington, DC, pp. 293–310.

Index